THE BLUE FUNNEL LEGEND

The Blue Funnel Legend

A History of the Ocean Steam Ship Company, 1865–1973

Malcolm Falkus

Professor of Economic History
University of New England, New South Wales, Australia

MACMILLAN

First published 1990

Published by
MACMILLAN ACADEMIC AND PROFESSIONAL LTD
Houndmills, Basingstoke, Hampshire RG21 2XS
and London
Companies and representatives
throughout the world

Typeset by Footnote Graphics,
Warminster, Wilts

Printed and bound in Great Britain by
Butler & Tanner Ltd, Frome and London

British Library Cataloguing in Publication Data
Falkus, Malcolm, *1940–*
A history of the Ocean Steam Ship Company, 1865–1973
1. Great Britain, Shipping services. Ocean Steam Ship
Company, history
I. Title
387.5'06541
ISBN 0–333–52283–4

Contents

List of Tables   vi
List of Plates   viii
Foreword by Nicholas Barber   xi
Preface   xiii
Map of Trade Routes 1866 and 1965   xviii

1 Introduction: Ocean to Ocean   1

2 A Unique Style of Management   9

3 Sailings and Services   26

4 Representatives Abroad   54

5 Eastward Ho! The Beginnings of Ocean Steam Ship   83

6 Combined Efforts: Holts and the Conference System   117

7 The New Century: Profits and Perils   136

8 After the War: An Uncertain World   171

9 The 1930s: Collapse and Revival   202

10 The Company at War   235

11 Picking up the Pieces   250

12 Into the 1960s: Calm before the Storm   282

13 Organising a Group   333

14 Blue Funnel Contained: New Ships and New Enterprises   356

Epilogue   375

Notes   378

Sources and Bibliography   381

Appendix I   Ocean Managers and Directors, 1865–1973   383
Appendix II   List of 'Co-owners', 1886   384
Appendix III   List of 'Correspondents', February 1902   385
Appendix IV   The *Phemius* Epic, 1932   389
Appendix V   Letter to Midshipmen on Appointment, 1953   396

Index (General, Places, Persons, Ships)   399

List of Tables

3.1	Far-East–Europe Trade, Conference allocations	33
3.2	Homeward Straits rates (per ton) 26 February 1907	36
3.3	Pilgrim traffic from the Straits, 1905–14	38
3.4	Passenger fares, Australian service, 1914, *Nestor* and *Ulysses*	48
4.1	Principal Blue Funnel agents, 1923	54
4.2	Straits Steamship, fleet and tonnage, 1914–32	81
5.1	British exports to China and Hong Kong, 1854–70	88
5.2	Earnings on early voyages, 1866–7	100
5.3	Blue Funnel fleet and financial results, 1881–1900	110
5.4	New ships built, 1892–1900	113
6.1	Far East trades, homeward freight rates, 1874 and 1879	123
6.2	Far Eastern Conference: west-coast loadings	130
6.3	1911 Lancashire and Yorkshire Agreement, pool shares, 1911 and 1936	130
6.4	Blue Funnel Far Eastern homeward pool allocation, 1938	131
7.1	Voyages and earnings, 1901–13	140
7.2	Net voyage earnings, 1909–13	141
7.3	Blue Funnel fleet, 1913–18	158
7.4	Holt dividend payments, 1913–18	160
7.5	Net steamer earnings, 1914–17	160
7.6	Ocean Steam Ship war losses, 1914–18	168
8.1	Blue Funnel: net steamer earnings, 1919–39	178
8.2	Net voyage earnings, 1920	184
8.3	Steamers and motorships in the Blue Funnel fleet, May 1934	191
8.4	Post-war steamers and motorships, 1919–34	191
8.5	Development of diesel engines	193
8.6	Net voyage earnings, 1924	194
8.7	China's foreign trade, 1920–8	197
9.1	Book value of fleet, 1932–9	206
9.2	Selected voyage costs, inter-war years	212
9.3	Alfred Holt and Co., dividends paid, 1924–39	212
9.4	Net earnings, 'China' trade, 1927–39	213
9.5	Net earnings per voyage, 1929–36	215
9.6	Java voyages, 1929–31	215
9.7	Index of Australian Conference freight rates, greasy wool, 1922–37	218
9.8	West Australian trade, 1922–39, net earnings	219

9.9	China: British tonnage entered and cleared, 1909–34	219
9.10	British share of UK–Far-East trade, 1937	220
9.11	Average first-class single fares, 1913–38	222
9.12	Straits Steamship profits, 1932–9	225
10.1	Holt war losses, 1939–45	239
10.2	Ocean accounts, 1939–45	243
11.1	Age and size of Blue Funnel fleet, 1934–45	260
11.2	Cost of Blue Funnel ships, 1930–47	262
11.3	Far East weekly service, 1939 and 1950	264
11.4	New ships delivered, 1947–59	264
11.5	Blue Funnel fleet, 1947–50 (gross tonnage, end of year)	266
11.6	Blue Funnel fleet, size and age, 1934–61	267
11.7	Cost of certain vessels, 1947–60	272
12.1	Ocean managers, main areas of responsibility, June 1963	285
12.2	UK–Far East trades, 1939–53	295
12.3	Consolidated gross trading profit, Ocean Group, 1947–61	298
12.4	Merchant shipping tonnage (non-tanker) 1939–69	303
12.5	Depreciation and amount transferred to reserves, 1957–64	305
12.6	Index of various cost increases, 1938–68	306
12.7	Composition of voyage costs, 1947–66	307
12.8	Index of Blue Funnel freight rates, 1938–68	307
12.9	Return on capital employed, 1963–7	308
12.10	Changes in manning levels, Blue Funnel ships, 1960–7	311
12.11	Blue Funnel losses since 1871	313
12.12	Ocean Steam Ship Co., net trading and investment income, 1954–64	319
12.13	Ocean Steam Ship Co., variation in stock prices, 1956–66	320
12.14	New building programme, 1947–60	321
12.15	Priam and Glenalmond vessels: promised and actual deliveries, Vickers-Armstrong	324
12.16	Time taken between keel-laying and delivery, Priam class	330
13.1	Ocean Group turnover, 1970	345
13.2	Group operating profits, 1969–71	347
14.1	Wm Cory & Son operations, 1970–1	372

List of Plates

1 *Agamemnon* I, 1865   *facing foreword*
2 Philip Henry Holt (1931–1914), Manager, 1866–97   2
3 Alfred Holt (1829–1911), Manager, 1866–1904   3
4 Sir Richard Durning Holt, Manager, 1895–1941   17
5 Sailing notice for the West Australian trade, 1892   45
6 *Nestor*, built 1913   49
7 J. S. Swire (1825–98)   61
8 At Hazelwood (the Butterfield & Swire Shanghai Manager's
   mansion) first week in April 1920   68
9 New Ocean Buildings, Singapore, 1920s   80
10 *Nestor*, I, 1868–94   102
11 *Idomeneus*, I, 1899   112
12 *Telamon*, I, 1885   115
13 The launch of *Menelaus*   115
14 Lawrence Durning Holt, Manager, 1908–53   146
15 Sir Charles Sydney Jones, Manager, 1901–30   152
16 William Clibbett Stapledon, Manager, 1901–30   153
17 *Bellerophon*, c.1910 (built 1906), one of the famous 'goal-posters' 154
18 Nurses on board the *Charon*, First World War   169
19 Illustration by James Mann   172
20 Hon. Leonard Harrison Cripps, Manager, 1920–44   179
21 Sir John Richard Hobhouse, Manager, 1920–57   180
22 Roland Hobhouse Thornton, Manager, 1929–53   181
23 Launch of *Diomed*, June 1917   182
24 Yokohama after the earthquake, 1923   188
25 *Sarpedon* at Liverpool landing stage, 1923   189
26 Cargo for the *Centaur* arriving by camel   211
27 Advertisement for the Java trade from the *Java Gazette*, 1933   214
28 *Maron* embarking troops, Kowloon Bay, 1937   221
29 Typical advertising poster in the inter-war years   223
30 British troops being evacuated from Dunkirk, 1940   237
31 HM Australian hospital ship, *Centaur*, sunk by Japanese
   submarine, 14 May 1943   247
32 India Buildings after blitz, May 1941   248
33 A Victory ship, *Memnon* (renamed *Glaucus*, 1957)   250
34 *Menestheus* at Kure, Japan   253
35 Charles Douglas Storrs, Manager, 1944–60   259
36 Passenger lounge, Peleus class, 1949   263
37 Discharging cargo from the *Gorgon* at Singapore in the 1950s   265
38 Blue Funnel ships at Singapore, May 1953   265

39	Seamen's Mess, 'A' class	266
40	*Sarpedon* at Holt's Wharf, 1962	267
41	William Hugh Dickie, Manager, 1944–54	268
42	Launch of *Calchas*, 27 August 1946	270
43	Holt's Wharf, Kowloon, August 1945	276
44	Sir John Norris Nicholson, Manager and Director, 1944–76; Chairman, 1957–71	288
45	India Buildings after rebuilding, 1950s	293
46	*Gunung Djati*, dining room, 1959	297
47	*Gunung Djati* in 1960	298
48	Vittoria Dock Berth, 1967	309
49	The *Pyrrhus* fire, 1964	315
50	After the hurricane: *Phemius* at anchor at Kingston, Jamaica	316
51	*Centaur* III, 1964	326
52	List of sailings, Bangkok Exhibition, 1966	331
53	Sir John Lindsay Alexander, Manager and Director, 1955–86; Chairman, 1971–80	336
54	George Palmer Holt, Manager and Director, 1949–71	337
55	Sir Ronald Oliver Carless Swayne, Manager and Director, 1955–82	338
56	Ocean Buildings, Singapore, c.1970	346

1 *Agamemnon* I, 1865

Foreword

Founded 125 years ago, the Blue Funnel Line grew into the world's finest cargo liner company. That it achieved this pinnacle of consistent success was thanks to its people, both afloat and ashore. Their story and the story of the Far East trades which Blue Funnel primarily served provide a record of commercial success and human achievement which should not be lost.

Following the founding of Ocean's joint venture company Overseas Containers Limited in 1965, Blue Funnel's trades were progressively passed over to OCL. In 1986 we decided to withdraw from our OCL investment. This was the moment to commission a historian to tell the Blue Funnel story before individuals' memories grew too faint. In Malcolm Falkus we found just the author we had hoped for – a scholar who could write, a historian with a strong feeling for the Far East and its mercantile history. The book is not a history of Ocean as such but of the Blue Funnel Line and it deliberately stops in 1973, once the core of Blue Funnel's Far East trades had been passed to OCL. Since 1973 Ocean has changed enormously, especially in the last five years. It is no longer a shipping company and no longer managed from Liverpool. Yet fundamentally Ocean remains close to its Blue Funnel inheritance. Its customers are still to be found in the freight and marine markets. Its activities still span the globe. Above all it has remained a thriving services business with its people still dedicated to providing quality services to the highest professional standards, still pursuing the goal of excellence set by the Company's founder, Alfred Holt. Let me quote from a recent letter to the company newspaper from Sir John Nicholson, the remarkable man who was Chairman of Ocean when I joined the Company in the 1960s:

> We should remind ourselves that Alfred Holt was an adventurous innovator of wide vision who inspired the search for perfection which has underpinned all our technical and commercial achievements through over 100 years and which now lies behind the exciting developments of recent months.
>
> I fancy that he would have preferred Ocean's chosen way ahead to the fate of other cargo liner companies.

This book tells a story which is fascinating to read. At the same time it provides a fitting tribute to all those from Ocean's Blue Funnel past who for over a century, in peace and war, made the Company one of Britain's great enterprises.

NICHOLAS BARBER
April 1990   *Chief Executive, Ocean Group plc*

Preface

'We have always eschewed a conventional popular history.'
(Sir John Nicholson)

When in 1947 a future Chairman, Lindsay (later Sir Lindsay) Alexander, was invited to join Alfred Holt and Co. he had heard neither of Holts nor of the Blue Funnel Line. My situation, when asked to write the company's history, was marginally better. Largely through Francis Hyde's scholarly book *Blue Funnel* published in 1957, covering the years before the First World War, I knew that company and line had a distinguished place in British shipping history and that Blue Funnel was considered among the aristrocrats of British cargo lines. Beyond that, though, I knew scarcely more than Sir Lindsay. Certainly I had no idea of the colourful history which accompanied Blue Funnel's progress through two world wars, through the great slump of the 1930s, and through the age of containerisation. Nor was I aware of the unique management system, of the cluster of singular and talented individuals who comprised the firm's senior management in Liverpool, or of the many notable contributions to Liverpool which both collectively and individually was an indelible legacy of the Holt Managers.

I hope this book gives some flavour of the character of a remarkable British company as it faced in turn progress and opportunity and then wars, depression, and the challenge of the container revolution. A shipping enterprise is a kaleidoscope, touching on a bewildering variety of subjects and objects: the ships themselves, the seafarers, the cargoes carried to and from often exotic and evocative places, the ports, the agents, the head office and other home staff – the list could go on and on. In this book my focus has been above all upon how the company was managed, how it grew and changed, how it sustained its great reputation, how it succeeded as a commercial concern, and how it faced a changing economic environment. The view, as it were, is from Liverpool rather than from on board a ship. I should add, perhaps, that not being a maritime historian and being unversed in subjects such as marine engineering, I have not attempted to give a detailed history of ship design or marine technology.

I have had the fullest possible help from Ocean Transport and Trading (now Ocean Group), successors of Ocean Steam Ship, who have allowed me complete access to the company's archives. The company has in no way sought to influence what I have written, and I remain entirely responsible for the content and judgements contained here.

A few points need just a little discussion. First, in view of the book by

Professor Hyde already mentioned, perhaps it should be explained why this is not simply a 'Volume 2', taking the story beyond 1914. The answer is twofold. Since so many aspects of Blue Funnel's history are rooted in the pre-1914 era it would be difficult to write a history without constant reference to this earlier period. The twentieth century cannot be understood without the nineteenth. Also, Professor Hyde's book is different in emphasis and style from the present one, and some themes he stressed have received only muted treatment here, while others have been given greater emphasis.

Another point concerns the plan of the book. I have chosen to mix themes and chronology which, while it leads to some overlap, allows me to bring out what seems to me to be the triple-pillared structure upon which the entire enterprise rested: management, agencies, and the conference system. Thus some of these topics are dealt with, as is the overall structure of Blue Funnel's routes and services, before the detailed early history of the company is surveyed in Chapter 5. Again, conferences are discussed before taking up the chronological threads once more in Chapter 7. My reason for organising the material in this way is simply that a straightforward chronological account would make it difficult to do justice to some of the larger themes.

Two points of presentation. First, throughout the book I have kept prices and costs in their original values and currencies, for to give modern equivalents would probably spawn as much confusion as it solved. However, in order to give some perspective on changing values of money in Britain over the years covered in this book, a few points may briefly be made. From the 1860s until the Second World War a striking feature of price changes was their relatively long-term stability (apart from cyclical upturns and downturns, as in the depression of the early 1930s, and during the First World War and its immediate aftermath). Thus, average price levels in 1913 were little different from those reigning in the 1860s. By the end of the post-war boom in 1920, prices were some three times higher than in 1913, but after a sharp collapse, prices in the inter-war years averaged around 50 per cent more than their 1913 levels. By 1945 prices had reached about double their 1913 level. Then there was a slow rise until the mid-1960s, followed by much more rapid inflation, especially after 1970, with prices trebling between 1966 and 1976. Very roughly, £1 in 1865 was still worth £1 in 1913, but only about 66 new pence in 1939, 32 pence in 1945, 22 pence in 1966, 7 pence in 1976, and less than 3 pence in 1989.

The second point of presentation is that detailed footnote references to material drawn from archival sources have been omitted. This is not simply to make the text more readable, but because the bulk of archival material comes from the Ocean archives deposited at the Liverpool Maritime Museum, and there are early plans to reclassify the Ocean collection there. It should be stressed, therefore, that unless otherwise stated, all the quotations from individual Managers in the text come from written material available in the archives and not from verbal comment in interviews with me.

Finally, and unfortunately, we must grasp the nettle of names. What do we call the company and what do we call the ships? The commonly used company name Alfred Holt and Co. has a rather insubstantial claim to existence, 'the company that never was' as Roland Thornton termed it. Until around 1903 the Company traded under the name of, simply, Alfred Holt. Then informally the name began to be used as a way in which Managers other than Alfred Holt himself might sign Company letters (Alfred Holt retired in 1904). Although the name was registered in 1917, no company of that name was ever incorporated, and the name was quietly dropped in 1967. The Ocean Steam Ship Company is another title which can appropriately be used, with one slight proviso. For various reasons the company assigned the Blue Funnel ships to three 'companies' (there never was an official 'Blue Funnel Line' until after the First World War, and the name was not registered until the 1960s), Ocean Steam Ship, China Mutual, and the Dutch-registered NSMO. China Mutual and NSMO ranked as 'investments' in the Ocean Steam Ship accounts, although all the ships were in effect part of one Blue Funnel fleet. However, sometimes there may be ambiguity. For example to say 'Ocean Steam Ship's vessels made 11 Java voyages in 1921' is technically wrong, since some of these ships belonged to the China Mutual and NSMO fleets. But here I have used 'Holts', and 'Ocean', and 'Blue Funnel' freely, believing that any slight looseness of expression is more than compensated by convenience.

The ships' names are another problem. Not content with using sometimes rather tongue-twisting names from among Greek legendary heroes, *Philoctetes*, *Talthybius* and the like, the same names were given repeatedly. Strictly, we should distinguish such ships by a chronological number, for example '*Pyrrhus III* caught fire in 1964', but this seems cumbersome and once again in this book I have chosen simplicity at the expense of a slight loss of precision.

It is with genuine pleasure that I acknowledge the help I have received from many individuals in the preparation of this book. Nicholas Barber has been a source of encouragement and support throughout, and gave many helpful comments on an earlier draft of the book. Mr Ken Wright's help and advice has also been invaluable. Others, past and present staff of Ocean and their agents, have helped in various ways: discussing their own recollections of Blue Funnel history, lending me books and documents, checking drafts of the manuscript and answering particular queries. Here I must thank particularly Sir Lindsay Alexander, Mr Harry Chrimes, Mr George Holt, Sir John Nicholson and Sir Ronald Swayne.

I am also most grateful for the help and advice I have received from Mr Joop Dietrich, Mr Christopher Gawler, Mr Theo Gleichman, Mr John Greenwood, Mr Julian Holt, Mr Philippe Hughes, Mr Charles Medcalf, Mr A. N. Stimson, Mr Harold Smyth, Sir Adrian Swire, Sir John Swire, Mr John Utley, and Mr Hugh Wylie. I would like to acknowledge, too, the help I received from Sally Furlong in facilitating my research in Liverpool, and guiding me to various sources in India Buildings. The Archivist, Mr David Ryan, and staff at the Liverpool Maritime Museum have assisted greatly throughout. I am grateful, too, to John Swire & Sons for giving me access to their archives held at the London School of Oriental and African Studies, and for allowing me to use material still in the firm's London headquarters. Charlotte Havilland, archivist at John Swire & Sons, has been most helpful and has located several rare illustrations which are included in this book. I acknowledge, too the generosity of the Philip Holt Trust who have helped to finance the research and publication of the book. I am indebted to Tim Farmiloe and others at Macmillan who have made the process of publication a real pleasure. Finally I would like to say a heartfelt thank you to Fay Hardingham and Jeanettee Tan at the University of New England who so cheerfully and expertly translated my handwriting into a presentable typescript, and also to Gerda van Houtert for additional secretarial help.

MALCOM FALKUS

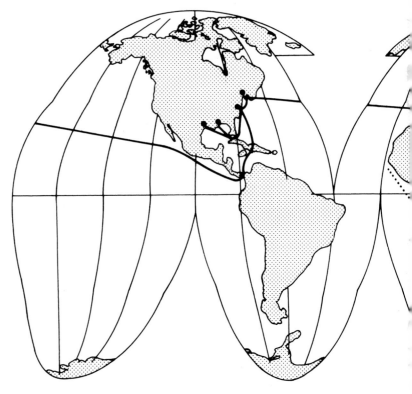

1986 trade routes (.....)

Agamemnon Voyage 1

Sailed from Liverpool for China
19th April 1866
Arrived London from China 26th
October 1866
Length of voyage 190 days
Distance steamed 28,400 miles

Days	From Liverpool
39	Mauritius
54	Penang
59	Singapore
68	Hong Kong

Days	From Shanghai
2	Foochow
15	Hong Kong
26	Singapore
30	Penang
46	Mauritius
86	London

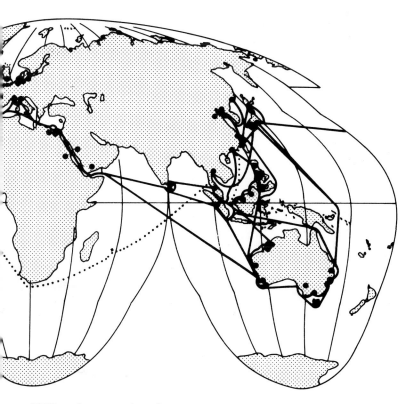

1986 trade routes (.....)

Patroclus Voyage 45

Sailed from Birkenhead for Japan via Rotterdam October 1964 Arrived Dublin from Japan 15th January 1965 Length of voyage 90 days Distance steamed 24,600 miles		44	Kobe
		48	Nagoya
		49	Yokohama

Days	From Birkenhead	Days	From Yokohama
2	Rotterdam	1	Shimizu
14	Port Said	2	Kobe
15	Suez	9	Hong Kong
19	Aden	15	Singapore
28	Singapore	18	Port Swettenham
34	Manila	27	Aden
38	Hong Kong	31	Suez
		32	Port Said
		39	Dublin

1
Introduction: Ocean to Ocean

Blue Funnel ... is one of the most personal and most individual of all the large shipping lines.

(S. G. Sturmey, 1962)

On 19 April 1866, the 2300 ton steamer *Agamemnon*, set out from Liverpool on a historic voyage. *Agamemnon* was a square-rigged barque, powered by high pressure compound engines, and she carried cargo for Penang, Hong Kong, and Shanghai. Within a few months two sister ships, *Ajax* and *Achilles*, had set out on similar journeys. The trio were owned by the Ocean Steam Ship Company, a new enterprise set up in January 1865 by the brothers Alfred and Philip Holt to operate the steamers. The Holt brothers were principal shareholders of the Company and 'managers' of the three steamers, while Alfred Holt had designed the ships and their engines.

Agamemnon's voyage opened a new chapter in shipping history. Her journey marked the first attempt to run a line of steamers directly from the United Kingdom to China; it was the first time a steamship had ever attempted a non-stop journey so great (Liverpool to Mauritius via the Cape); and it marked the start of the Blue Funnel Line, long known on Merseyside as 'the China Company', and destined to become one of the great names in British merchant shipping.

This book is about the history of the Blue Funnel Line and its management. The ships, with their sleek silhouettes and azure-blue funnels topped in black, were once part and parcel of Britain's flourishing maritime enterprise in the east, familiar sights in the great ports of south-east Asia and the Far East. The Ocean Steam Ship Company, which ran the line, was a close-knit, highly individual organisation, with a secure place in Britain's maritime history, and a management which for long was able to make the twentieth century come to terms with the nineteenth. The Blue Funnel Line was, for something like a hundred years, one of Britain's and the world's leading lines of cargo vessels.

As we pass into the 1990s it is all to easy to forget how recently ocean-going liners carried most long-distance passengers and the bulk of goods traded internationally. There was a time, not long ago, when Britain's merchant marine was, almost literally, the nation's life-blood.

1

2 Philip Henry Holt (1831–1914), Manager, 1866–97 (*photograph courtesy of Mrs Margaret Anderson*)

British-built ships, British registered, carried not only merchandise to and from this country, but participated in much of the trade carried on between other countries as well. This pattern had grown up as part of Britain's economic expansion in the nineteenth century. We are all familiar with the idea that Britain was the home of the industrial revolution, the pioneer of the steam-engine, railways, cotton-spinning machinery, and a thousand and one other developments which so profoundly altered the social and economic landscape. But it should not be overlooked, too, that Britain was also the world's first truly commercial nation, the first to forsake self-sufficiency so completely that by the end of the nineteenth century less than one in ten of the population worked in agriculture. The nation was totally dependent on its merchant

3 Alfred Holt (1829–1911) Manager, 1866–1904 (*photograph courtesy of Mrs Margaret Anderson*)

ships to bring the food and raw materials on which industrial prosperity rested.

The age of conventional liners passed quite suddenly, although the relative decline of Britain's merchant marine among shipping nations was a much lengthier process. Mass air travel arrived only in the 1960s, spelling the end of the great passenger liners, and by the end of the 1970s the age of container-ships likewise heralded the end of conventional cargo liners as principal carriers across the oceans. The Blue Funnel cargo ships, along with the rest, have now largely left the seas to the new carriers and to those tramps and liners which could still find a niche in the new environment.

Before this, Blue Funnel cargo ships were long a familiar sight in merchant shipping lanes throughout the world, east of Suez above all. The Liverpool and Birkenhead docks and the great harbours in Singapore and Hong Kong were forested with familiar blue funnels. 'There was a saying among the Britons in China', wrote an expatriate Englishman in Hong Kong in 1919, 'that whenever one of the Far Eastern ports might be visited, there would always be found a ship of the Blue Funnel Line discharging or taking in cargo'. He added, 'of all the British companies in the Orient, the Blue Funnel Line has the first reputation for efficiency'.[1]

The line belonged to a Liverpool company known as the Ocean Steam Ship Co., also, recalling its founder, simply as Alfred Holt & Co. From a small beginning in 1865, the Liverpool line had developed by 1914 into one of the country's largest shipping concerns. It maintained regular sailings between Britain and most leading international ports in Asia and Australia, with a network of feeder and joint services reaching to the remotest corners; and it served via the Cape the Pacific ports of Canada and the USA, and during the First World War spread its services to New York and other east-coast American ports. In 1914 Blue Funnel vessels were easily the largest single users of Liverpool berths among the numerous cargo lines, they passed more frequently through the Suez Canal than the ships of any other line, and they were by far the major carriers of Britain's cotton and woollen goods to Far Eastern markets.

The Ocean Steam Ship Company was notable for many things besides its size. It built up and long maintained a unique working arrangement with its agents abroad, above all with Butterfield and Swire in Hong Kong and with Mansfields in Singapore. Indeed, as will become clear, the history of Blue Funnel cannot be separated from the history of Swires and Mansfields, nor can it be separated from the history of the conference system which was pioneered by John Swire.

The reputation established by the Blue Funnel ships and maintained for more than a century was an enviable one. Holts were renowned for the high quality of their ships and crews, meticulous attention to safety and standards of crews' accommodation, the loyalty and excellence of Masters, and the efficient handling of cargoes. Beyond this, though, the Company was renowned for its highly individual management, a group of managers who controlled what was, until 1965, a firm still retaining much of its Victorian family origins. Ocean Steam Ship became a public limited liability company in 1902, but its shares were not publicly quoted until 1965. For this reason the Company was sometimes wrongly referred to as 'private', and, indeed, the epithet is not inappropriate. Holts never sought publicity and the shareholders' meetings were for long very much family gatherings. Throughout the inter-war years and beyond, both in organisation and ethos, Holts remained a family business, and the *Financial Times* (May 1964) was right in spirit if not in letter in speaking of 'not so much a business as a way of trade, a private company which is also a public institution'.

Three examples can probably provide a better flavour of what we might call the Blue Funnel 'character', than a mass of detail. First, uniquely, Holts never built their ships to the standard classifications drawn up by Lloyds or other classification societies. Instead they maintained their own classification, superior to Lloyds A1. During the First World War at the time of government requisitioning of ships for war purposes, it became necessary to return official forms, and the Ocean managers simply wrote 'Holts Class' for their vessels. This phrase, 'Holts Class', remained as a standing reminder both of the Company's own perception of its superiority, and of the acknowledgment of its standards by others. Second, again uniquely among shipowners, Holts fought the government's legal right to requisition the services of shipowners during the First World War. The matter was pursued by Richard (later Sir Richard) Holt, a redoubtable man who dominated Blue Funnel for more than forty years, and who remained staunchly true to the Gladstonian Liberalism and individualism of his youth. Third we may note the Managers' meticulous inspection of ships and the practice of detailed interviews with Masters and other senior officers. All officers were individually interviewed on arrival and before departure, the Masters by a senior Manager. Prior to sailing deep-sea every ship was subject to a detailed inspection by a Manager accompanied by all superintendents responsible for her outfitting. Insistence on high standards of skill and conduct extended from Masters to the newest recruits. The letter of appointment to young midshipmen drawn up by Lawrence Holt before the war and still sent in 1953 is a remarkable document. 'The Managers expect you to have a high ideal of pure and upright manhood', the midshipmen were told, being warned also of the Chinese crews 'that natives do not understand skylarking and resent deeply being struck or otherwise ill-treated. Be careful, therefore, that your conduct towards them, whether on board ship or ashore, is never familiar and always dignified.' This wholly characteristic epistle is reprinted in full in an appendix.

We may add that the classical Greek names nearly always given to Blue Funnel ships gave, perhaps, an unconscious stamp of superior quality and tradition to the line. Alfred and Philip Holt chose such names since they considered Homer's *Odyssey* a record of the greatest adventures ever undertaken, and so gave the names of Homer's legendary heroes to the ships participating in their own epic adventure. And they were particular in their choice of names, 'human, male, and on the whole good'.

This story of the Blue Funnel line is thus the story of a leading and singular shipping company. Part of its singularity, if not leadership, certainly stemmed from its role as the pioneer of steam-shipping between Britain and the Far East. This achievement has been the foundation of the company's own great sense of tradition and has given Blue Funnel an indelible place in mercantile history. The small line of three steamers founded by Alfred Holt in 1865 was established specifically to serve a route then still the sole preserve of sailing ships. The fact that the line went to China was in itself unimportant. Steamships belonging to various companies and individuals had long plied between the Red and China seas, fuelling at the numerous coaling stations established in Britain's conveniently-located colonial ports. What was significant was the long 8500 mile haul made non-stop between Liverpool and Mauritius, so bringing the prospect of direct steam services between Britain and the Orient without the cumbersome and costly land transport of goods and passengers from Alexandria to Suez. Few thought that a steamship could undertake such a journey, and that if it could, there would be no room for cargo since so much coal would be needed aboard. Alfred Holt showed otherwise.

Yet this particular piece of pioneering was not in itself of much long-term significance. Within a few years the new Suez Canal, opened in 1869, had made the Cape route to Asia largely obsolete for steamships. China was suddenly more than 3000 miles closer to Europe, and steamships flocked to the new route. Nor could we say that the early start in the China trade had given the Ocean Steam Ship company an impenetrable lead. Already by the late 1870s there were signs that Blue Funnel vessels were becoming obsolete in relation to their more dynamic competitors. A decade later the company faced a crisis.

In retrospect the most significant aspects of Ocean's history were two: the ability of the company to surmount the crisis caused by competition and obsolescence in the late 1880s, when so many family concerns might have foundered; and, even more remarkable, to sustain and strengthen the firm as a 'private' family company in the inter-war years during a period of intense depression and innumerable failures within the mercantile community. At a time when the economic air was filled with schemes for rationalisation and amalgamation, Blue Funnel remained steadfast in its adherence to Victorian individualism. Richard Holt, it may be said, confronted the worst the twentieth century had to offer with the best of the nineteenth century, and emerged triumphant.

As we examine Blue Funnel's success in surmounting the crisis of the 1880s, the inter-war depression, and the post-war challenges, it will become clear that much rested on the shoulders of the small group of Managers in Liverpool. There are those who see history as the product of impersonal forces, and who would accordingly minimise the roles of individuals. Let it be said unequivocally that the Ocean story cannot be understood except in terms of the achievements, aspirations, and personalities of the Company's Managers and other senior executives. When we come to examine the structure and style of management we will see that it gave full scope to the talents of some outstanding figures. Yet individualism was combined always with a strong corporate tradition, and long after the Second World War the ghosts of the founders still haunted the corridors of the company's headquarters in India Buildings.

Indeed, even in the early 1950s, some one hundred years after Alfred Holt's early essays with marine engineering which were to lead to the founding of the Blue Funnel Line, there was still much to recall the past. The Ocean Steam Ship Company still traded under its original name, still was managed by 'Alfred Holt and Company', still was concerned almost exclusively with operating conventional cargo liners from the company's Liverpool headquarters in India Buildings, though management of the Glen Line, acquired in 1935, was undertaken in London. The basis of the firm's business lay, as it had done from the beginning, in trade with the Far East. The company was still not publicly quoted, and had a restricted and carefully vetted list of shareholders. And this company, vast as was its business and extensive its world-wide connections, with a fleet of 61 cargo liners, was still to a striking extent a family enterprise. Although matters were soon to change, at the opening of 1953 the Managers (as they had been termed from the outset) were a close-knit group of only seven, of whom the eldest, Lawrence Holt, a nephew of the founders, had joined the firm in 1904; another, Sir John Hobhouse, a cousin of Lawrence, had joined in 1912; another, Roland Hobhouse Thornton, a cousin of John Hobhouse, had joined in 1919; while a fourth family member, George Palmer Holt, a nephew of Lawrence, had entered the firm in 1932.

Twenty years later, at the opening of 1973, it would be little exaggeration to say that the firm, as it had existed, had now vanished. There was a different name, Ocean Transport and Trading Ltd, adopted in January 1973; a different organisation, now a publicly quoted company replete with Chairman and Board of Directors; a 'Group' rather than a firm, with numerous and diverse subsidiary companies and interests, and the company organised into a management group structure reflecting these changes. The acquisition of Wm. Cory & Son in 1972 marked a huge move towards land rather than marine transport interests, while the growth of containerisation and bulk shipping meant an even further shift away from traditional cargo liners and added London to Liverpool as a twin focus of group activity. By 1973, the Blue Funnel fleet was fast declining, an inexorable movement which meant that by the end of the decade the fleet had all but disappeared. In 1980 only two conventional Blue Funnel cargo ships remained. The Holt name had been dropped for official purposes already in 1965. And the old family connections had all but gone. Within the management the last Holt, George Palmer, retired in 1971; only Richard Hobhouse, son of Sir John, remained to recall the great family tradition, and his retirement in 1976 brought the long connection finally to a close.

The new Ocean which emerged in the 1970s was an exciting creation, soon called upon to face circumstances arguably as testing as had ever affected its parent. The recession and virtual collapse of conventional liner shipping in 1982, consequent threatened survival and bitterly fought takeover battles, subsequent revival, restructuring, and renewed growth – all have been highlights in a brief and colourful history. But this cannot be the history with which we are primarily concerned, if only because the multiplicity of subsidiaries and interests after about 1973 makes any overall picture well-nigh impossible within the confines of the present volume. More fundamentally, though, the history of the new Ocean is a separate one, the strands of the past snapping at various places between 1960 and 1973, while many of the strands then remaining were soon to disapear.

This story, then, is from the old Ocean to the new Ocean, from the steamship enterprise of Alfred Holt in the 1860s to the growing land-based and specialised shipping concern which emerged in the 1970s. Beyond this is another story.

2

A Unique Style of Management

True family 'spirit', with the Flag at the top of the mast, is ever the road to righteous victory.

(Lawrence Holt, 1958)

I consider the management should have an establishment of eight persons, normally six Managers and two Assistant Managers, though the distribution may vary from time to time. This staffing allows sufficient spare time for outside jobs round Liverpool and for visiting foreign parts by ship. Two of the Managers should always be designated to make "Ship Management" the first charge on their time and energy ... the other four Managers should be responsible for general commercial operation of the ships and other duties.

These words were written, remarkably, in 1955 by Sir John Hobhouse, a Manager since 1920, in a memorandum submitted to his colleagues during the first faint stirrings of concern about the future management organisation of Ocean. Remarkably, because the words could almost equally have been penned in 1920 in the days of Richard Holt, or even twenty years before that in the era of Alfred and Philip Holt. Only the phrase 'by ship' (underlined in the original) reminds us that by the 1950s air travel was possible, and that the traditional lengthy journeys by members of the senior management by Blue Funnel vessels to their stamping-grounds in the Far East were still to be preserved.

The enduring structure of management was the single factor more than any other which gave the Ocean Steam Ship Company its unique character, a management team which was small, centralised, elitist, non-technical, and remarkably successful. In 1968 an article in *The Observer* could note with awe, 'Ocean Steam's style of management has hardly changed since the company was founded in Liverpool in the latter part of the nineteenth century by Alfred and Philip Holt.[1] Even as late as 1976, the *Financial Times* could be more specific, though with precision at the expense of accuracy: 'Ocean's managerial style is not radically different 111 years later.'[2]

Although the Holt management system will become clearer later, it is

9

useful at this stage to give a brief summary. Controlling the company, in India Buildings, were the 'Managers', or 'partners' (both terms echoing the firm's nineteenth-century origins). Their number might vary, but was usually a group of about five or six. The Managers were always shareholders, and, as in the case of the Holts and Hobhouses, might be very substantial shareholders. In theory all Managers were equal, and all were responsible for the entire range of Ocean's activities. In practice these principles were eroded in two respects. First, among the Managers would be a 'senior partner' who might be, as in the case of Richard Holt or John Nicholson, every bit as much a 'leader' as the Chairman of P&O or Cunard. An unofficial ranking might be the product of experience, or force of personality, or shareholding, but in any event not all Managers were equal all the time. Interestingly, they were not all paid the same sum. The total set aside for 'Managers' Renumeration' was divided by the senior partner, usually into sixteenths or twenty-fourths, and distributed largely according to length of service or age. In 1914, for example, Richard Holt received six twenty-fourths, William Stapledon and Henry Bell Wortley five twenty-fourths each, and Lawrence Holt and C. S. Jones four twenty-fourths each (the sum divided was £12 333 6s 8d).

Second, in practice, Managers always 'looked after' particular departments and geographic areas of particular trades, and were effectively 'managing directors' in their own spheres. Nevertheless, the underlying principles of equality and catholic involvement were precious and pervasive: they made for free exchange of views, constant debate, and an intensity and colour of argument which became one of the hallmarks of the 'quarterdeck' (the raised platform where the Managers sat together in the large open-plan office in India Buildings).

The Managers might be aided by Assistant Managers, (some of whom probably started as young 'crown princes', groomed for succession), while below the management were the various superintendents and departmental heads. The chief departments in turn had mostly made their appearances before 1914, the steamship, inward and outward freight departments, conference, passengers, the various technical sections, and a few others.

To describe adequately the Holt management style is not an easy task, for in some ways we are dealing with an ethos rather than an organisation. Graham Turner, in *The Observer* article already quoted, was even driven to coin the rather inelegant phrase 'Holtishness' to describe this style. The key elements, perhaps, were two. In the first place, structure and style were closely related. Thus the system of 'group management' which evolved in the nineteenth century gave a platform for the style which then flowed from the personalities of the Managers. And second, the management structure, which allowed wide-ranging powers and responsibilities to each individual Manager, gave full scope to the personalities, talents, and strengths as well as weaknesses, of the individual Managers. Thus it becomes impossible to discuss the style and functions of the management without discussing also the Managers themselves, Crompton and Cripps, Holt and Hobhouse, and the rest. And here we need to emphasise, as we focus on the Managers, the unashamed intellectual and social elitism of the Holt system. To establish and sustain in Liverpool's Water Street something which was a cross between an Oxbridge Senior Common Room and a Pall Mall Club, yet which managed to guide one of Britain's major shipping lines through tumultuous times with conspicuous success, was an achievement which has no match in the annals of Britain's business history.

Before we look further at the role of individual Managers, however, we should ask how the peculiar management structure evolved and why it proved so enduring. Six factors seem of prime significance. First and foremost the Ocean Company started as a private family partnership, and, as we have noted, remained for a century cast in this mould, albeit with a waning family voice and as an incorporated limited company rather than as an unlimited partnership. Second, the company remained until the 1960s very much a specialised dry-cargo shipping enterprise, a 'one-product company' as it were. Third, the company remained based in Liverpool, with roots firmly planted in the mercantile, non-conformist, London-hostile world which had nurtured the firm's founders. Fourth, the remarkably long service contributed by many of the Managers was itself a powerful element in tradition and conservatism. Fifth, we should note the nature of the shipping business itself, which essentially was selling space and services. To quote Graham Turner yet again, 'the managing directors' room at Ocean Steam Ship is only possible because shipping is a relatively simple business: it would hardly work in a multiproduct concern'. And finally the Holt system of management evidently worked splendidly: the company remained successful in business, retained the loyalty of its employees, and never lost its pre-eminent reputation for the quality of its fleet and service. Without doubt the binding factor which created 'Holt style' was an ethos of quality and excellence, a determination never to yield to second best, and a sense of pride in maintaining an enterprise governed by such principles. Such an ethos permeated the company, from Manager to midshipman, and gave to Ocean its unique stamp.

Let us look briefly at these points in turn. In 1865 the Ocean Steam Ship Company was launched as a private unlimited company. The shares were held by a small, close-knit, family-dominated group, most of the shares being held by Alfred and Philip themselves. The brothers were the original 'Managers' of the company. The term was a common one in shipping concerns, deriving from the normal practice of dividing shares in a ship into fractions (usually sixty-fourths) held, perhaps, by a number of individuals. From the shareholders a smaller group would 'manage' the ship; this group in turn might well be a regular shipping business which would similarly manage other ships.

Alfred and Philip, being the main shareholders, were owners as well as Managers of the Ocean Steam Ship Company. As shareholding widened and new Managers were appointed, the total control exercised by the founders lessened somewhat. But the new recruits were appointed by the Managers themselves and this, coupled with the immense prestige and experience of the founders, and the bonds of family ties, ensured that the senior partners remained in control of the company as long as they wished. It is interesting, for example, that in 1895 a shareholders' meeting actually voted in favour of changing the firm into a limited company, against the wishes of Alfred and Philip Holt. Yet the founders were easily able to resist the change on the grounds that the majority in favour was insufficiently large. There were no dissenting voices.

The year 1895 was a significant one in the evolution of Holts' management. Until then only one new Manager, Albert Crompton, had been appointed, although a number of promising younger men had been brought into the firm. Crompton had joined the firm in 1872 as a cashier when he was 29 years of age. Son of Sir Charles Crompton, his mother was a cousin of Philip Holt's wife, and he was educated at Harrow and Trinity, Cambridge. Crompton became Manager in 1882, retiring in 1901. Alfred Holt recalled in 1911 that 'the effect of his skilful management, honesty and kindliness, survives to this day', though Crompton seems to have been essentially a conservative figure, resisting John Swire's advice to improve the size and speed of Blue Funnel ships. From the ranks of the younger recruits three in 1895 were appointed Managers, Richard Durning Holt (the son of Alfred and Philip's brother, Robert), George Holt (Alfred's son), and Maurice Llewellyn Davies, a nephew of Philip Holt and Albert Crompton. The Holts were both 27 years of age, Davies four years older.

So began a pattern which later became characteristic of Holts: a small group of promising young men would be taken into the company and groomed to become Managers; they would include family members, but the family must prove its worth in competition with 'outsiders' (or at any rate those less closely related); the recruits were of a homogeneous social background (public school and Oxford or Cambridge), and would have some intellectual distinction. Once in harness the new recruits would gain experience in all the principal departments, and travel to the east to acquire on-the-spot knowledge of Holts' business there.

In the course of time this recruiting and training pattern was to become increasingly refined, although there were always significant exceptions as we will see. But it is interesting that as early as the 1890s a course had been charted which was still clearly recognisable in the early 1970s. Incidentally it is tempting to date Holts' 'intellectualism' from the appointment of Roland Thornton in 1919, for Thornton was without question an outstandingly brilliant individual. Yet Crompton, educated at Harrow and Trinity, had been called to the Bar before joining Holts in 1872. Maurice Llewellyn Davies was educated at Marlborough and Oxford, holding a classical scholarship at Balliol. We may see even in the Greek classical names of the ships a certain scholarly vision. Philip Holt may well have been influential here, for there exists in the Holt archives a beautifully handwritten account of the various names of the characters used for Blue Funnel vessels by Philip Holt and presented to Captain Turner Russell in 1908 'as a mark of respect after a friendship of 53 years'. At the same time we should perhaps, keep the managerial intellectualism in perspective. George Palmer Holt always felt his own rather average university degree to fall below Ocean standards. Yet he could well have reflected that when he joined the company in 1932 three of the six Managers held no university degree, and two of them, Lawrence Holt and Leonard Cripps had not been to university at all (the third, John Hobhouse, had completed one year of a zoology degree at Cambridge when he joined Alfred Holt and Co. in 1912).

We have laid stress on the Holt family tradition. It is of significance, however, that the connection was never maintained at the expense of competence. Alfred's son George was never pushed ahead of his more able and dynamic cousin Richard simply by virtue of being the founder's son. Similarly family outsiders like William Stapledon (who became a Manager in 1901) and Henry Bell Wortley (1908) were brought into the management because of their special abilities. On occasion, neither Richard Holt nor his successors shrank from refusing preferment to family-members where the company's severe standards (said to be even more exalted for the family) were not met. And this Darwinian trait must be given its full due in explaining Ocean's continued dynamism; the management was never allowed to suffer the fate which befell other Victorian enterprises where succeeding generations proved incapable of carrying on the businesses established by their forefathers.

Until 1902 Ocean was an unlimited company (and had remained so despite the wishes of John Swire, Robert Holt, and others in 1895), which meant that in the case of bankruptcy the shareholders were liable for the debts of the concern to the extent of their private fortunes. Here was an act of faith in the Managers and an impetus to effective management if ever there was one. In the 1870s it was common for one-ship ventures in Liverpool to be launched as limited liability companies, yet here were the Holts, with their large and growing business, committing themselves completely to their company. Even more than self-insurance in 1874, unlimited liability established a standard of fleet management and financial soundness which never departed. This was the origin of 'Holts class' standards, of the amassing of large reserves, and of other manifestations of high quality and conservatism which stamped Blue Funnel over the years.

There was significance, too, for Holts' management structure in the company's lack of diversification. This had a dual aspect. In the first place the shipping business remained relatively straightforward. The company could remain centralised in India Buildings, its general business lying within the competence of its non-technical managers. The second consequence was rather more subtle. It may well be asked why such brilliant Managers were content to run for so long a business which, for all its commercial success, remained in the shadows of the more glamorous shipping empires trailing their clouds of subsidiaries. Part of the answer is that the Ocean management structure allowed each Manager to play a role, if he wished, in various public arenas. And most did wish. Nearly all the Managers were prominent public figures in Liverpool, many took leading parts in local or national shipping organisations or other national bodies, Richard Holt was for some years a member of parliament (and had spent even more years trying unsuccessfully to become one), Lawrence Holt had his Outward Bound schools and other interests, and so the list could grow. The management structure, in other words, allowed for diversification of the Managers' interests outside the firm itself, and we can read an anxiety to preserve these roles into John Hobhouse's memorandum quoted at the beginning of this chapter.

The origins and century-long residence of the company in Liverpool was equally of significance. The Holts and Durnings were already part of Liverpool's non-conformist business 'establishment' long before the Ocean Steam Ship Company was founded. Liverpool tradition gave to Ocean a special character: a sense of superiority and disdain for London, for example; a dislike of ostentation; a belief in the values and virtues of hard work, thrift, and loyalty. In a typical passage in one of his last public speeches, Richard Holt reflected in 1939 on his fifty years with Ocean:

> It has been work that has been a real true pleasure, and I have enjoyed living in Liverpool. I was brought up to the idea that we were Liverpool people and that was where I was expected to work, and it was the place for me, and if I didn't like it I had got to make myself like it, and do it properly. There is a great text in the Bible very relevant and suitable to the occasion of those appointed to the lead in the big concerns of life. 'I dwell amongst my own people' and I believe it is a very sound doctrine.[3]

Liverpool contributed something else too. Holts were among Liverpool's most highly regarded employers, and always enjoyed within the city an outstanding reputation as a fair and conscientious employer. In consequence Holts, alongside the major banks and insurance companies, always attracted the most able school-leavers. Indeed, headmasters of Quarry Bank, Liverpool Institute, and other leading local schools would send some of their brightest prospects to India Buildings. Thus the calibre not only of the Managers (and graduation from the ranks to the management was *not* a characteristic of Ocean) but the entire staff at India Buildings was high, and this in turn reinforced Ocean's standing within Liverpool and safeguarded its effectiveness in the wider business environment.

Tradition of management style came also from the long service of the Managers themselves. More than any other individual, Richard Durning Holt was responsible for building up and maintaining Holts' peculiar brand of paternal management. He had joined the company in 1889, became a Manager in 1895, and was the dominant voice in the company from the early years of the twentieth century almost until his death in 1941. But there were others, too, with lengthy service. Lawrence Holt, Richard's younger brother, had joined the firm in 1904, became a Manager in 1908, and retired only in 1953. Apart from the founders, William Stapledon, John Hobhouse, and Sydney Jones were all Managers for thirty years or more, while prior to 1970 it was common for Managers to serve forty years or so with the company. Young recruits in the 1920s or 1930s must have felt the older managers had been there for ever. And so they had. William Stapledon had joined his father's agency in Port Said at the age of 18 in 1876, and retired from Blue Funnel at the age of 73 in 1931. Richard Holt could, and did, talk familiarly of 'Uncle Alfred' and 'Uncle Philip' and became in turn 'Uncle Dick' to John Nicholson. Indeed, the presence of these founders seemed to increase with the passage of time. Lawrence Holt, too, never tired of quoting his illustrious uncles.

Continuity of service had a further impact on the style of management. The retirement or death of a senior Manager did not create the hiatus it might have done under another system. Already other Managers had grown up in the firm, and could readily assume additional responsibilities. Succession was never a problem, elevation to 'senior partner' not such a radical step. And in any case the dominant force in management need not necessarily come from the senior partner, as happened after the Second World War when, with Hobhouse and Lawrence Holt still in office, the firm was guided to a considerable extent by Roland Thornton; or a little later, when John Nicholson became the dominant force after 1953.

Another element we have noted in explanation of Holts' enduring form of management was the structure of the shipping industry. Given that Ocean diversified to only a limited extent, and remained centralised in India Buildings, it was possible for the growing complexities of the modern world, as it were, to be absorbed by the periphery. Thus the management of a ship at sea, as it adapted to new forms of locomotion, communications and design, changed considerably; so did the work and structure of some of the large agencies and shipbuilding concerns, and of other attendant businesses. Possibly it was more serious for Holts' management that the same process could be seen within India Buildings itself. Thus some of the more technical departments (for example, ship design), for all the intellectual capacity and occasional outstanding contributions by the Managers, were inevitably able to function at some distance from close management control. This was particularly the case after the untimely death of Henry Bell Wortley in 1919, the last technically trained Manager for a generation. Some technical skills were unrepresented either at management or departmental level and the absence of qualified accountancy staff remained an enduring lacuna until the mid-1960s.

Yet if there were weaknesses in the management structure these were effectively masked by the evident commercial success of the company. Throughout the difficult inter-war years, and for long afterwards, Ocean outperformed most of its British rivals, maintained its routes and trades, and reinforced its reputation for the quality and efficiency of its services. This reduced any pressures which might have built up for changes in the management structure, for the existing system was self-evidently a success.

The larger question, of course, is whether Ocean's style of management was responsible for success, or whether success, through largely exogenous factors (such as monopoly of the western berths for outward trades, conference quotas, prosperity of particular trades, and so on) sustained the management. We may argue, of course, that there were no exogenous factors. Management should be credited with creating and maintaining the conditions for successful business operations. The contributions of individual Managers in this way will become clearer in the following chapters, when we will examine the roles of some of the more prominent of them. Here, though, it is appropriate to reflect particularly on the career and influence of one of the Managers, Richard Holt, since his role was to prove so lengthy and so decisive.

Richard Durning Holt was born in 1868, the first of eight children born to Robert Durning Holt and his wife, the former Lawrencina Potter. Their youngest child, Lawrence Durning, was born in 1882 and he, like Richard, was destined to play a long and major role in the affairs of the Ocean Steam Ship Company. But the two sons cannot really be compared. Richard Holt became the firm's senior partner in 1904 upon

4 Sir Richard Durning Holt, Manager, 1895–1941

the retirement of Alfred Holt, but he had become the dominant force before that, and the reins of the company remained firmly in his hands almost until his death in 1941.

In many ways the Holt management structure for nearly a century was established the moment Richard Holt entered India Buildings as a young employee of 20 years of age in the spring of 1889, for Richard's 'career' became something of an ideal for later Managers. Richard, unlike his father, had received an excellent formal education at Winchester and New College, Oxford. After a spell in India Buildings learning the business in general routine jobs, he was sent on an extensive tour of the Far East. There he obtained a thorough grounding in the firm's business in Singapore with Mansfields, and with Butterfield and Swire in Shanghai and Hong Kong. On return, he soon became a full partner, or Manager, in 1895, along with two contemporaries, George Holt and Maurice Llewellyn Davies. In this way a soon familiar pattern was emerging: good formal education, youthful appointment and rapid advance to managerial position, sound grasp of the firm's overseas business acquired through travel, and recruitment alongside others which injected a competitive edge among the newcomers. The pattern, of course, may be viewed as that typical of a family firm, where the heir-apparent is brought in for obvious grooming for future inheritance. But in the Holt case we should re-emphasise a significant feature. This was that the senior Holt partners were willing to go beyond the immediate family in their search for talent, and to allow succession to be decided upon merit rather than upon family ties. And this principle was later extended to recruits from outside the family altogether, so that 'crown princes' could be drawn into a family business structure though of alien blood.

Circumstances, including both world wars, combined to make recruitment of new 'crown princes' and Managers by no means so orderly as their seniors might have wished, but it is nonetheless striking how similar was the career pattern of many of the firm's leading Managers before the great changes of the 1960s and 1970s.

Before looking further at Richard Holt's influence on the Ocean Steam Ship Company, it is interesting to recall the sort of firm which he entered in 1889. The whole of the management offices then were on the second floor of old India Buildings, the linen department, kitchen, and dining rooms were on the third (top) floor, while George Holt & Co. (the cotton-broking business established by Alfred and Philip's father) had most of the ground floor. The three Managers, Alfred and Philip Holt and Albert Crompton had their desks on a raised platform in full view of the heads of most of the very limited number of departments which then existed (Steamship and Inward Freight; Outward Freight and Passengers; Cashiers; Bookkeepers; Bills of Lading; a Medical Superintendent; and the Post Desk). In all, apart from the Managers themselves, there were only about twelve office staff and some ten women clerks who wrote letters, ships' manifests, and paid the seamen's allotments. Interestingly Holts were among the pioneers in the recruitment of female employees at a time when the male clerk reigned supreme.

Ocean's office arrangements were hardly forward-looking in other matters. India Buildings had no lift, no typewriter (the three Managers commonly press-copied their own letters), and the gaslit offices had no telephone except for a single private line to the loading berth at Morpeth Branch Dock. Not until after 1895 did the office staff and office facilities increase significantly.

This, then, was the environment into which the young Richard Holt stepped when he first entered India Buildings, and where he was to remain for the rest of his life. How shall we assess Richard Holt, as a man and as a business leader? A biography has yet to be written, but such a biography would certainly emphasise themes which can be of only tangential interest here. One, certainly, would be Richard Holt's strong Unitarian beliefs and his adherence to the Liverpool Unitarian Liberal 'establishment'. This triple-pillared edifice of provincialism, religion and politics does much to sum up not only Richard Holt's own outlook but also the outlook he imparted to Ocean. Lawrence Holt, too, was a practising Unitarian, as was Charles Sydney Jones who became a

Manager in 1912. Both came from Liverpool, and both were Liberals. Membership of the Unitarian 'gentry' or 'plutocracy' as it was sometimes called, defined social attitudes as much as it did religious belief. Like the Rathbones or Mellys, the Holts were not ashamed of their wealth (though they were not ostentatious in its enjoyment), but they saw wealth as a social responsibility. They were individualists and staunch free-traders, the epitome of Gladstonian Liberalism. But it was individualism tempered with a social conscience and great personal integrity.

Not that Richard Holt was a man of the people. His attitudes were paternal and even snobbish. He found his first (and probably last) visit to a football match at Anfield in 1901 'a most extraordinary sight – galleries for spectators all round the ground and these quite crowded. On the whole not a good development of modern city life, reminding one of a Roman circus, but free from brutality.' Socialism appalled him, and he certainly had no time for the radical views of his aunt Beatrice Webb (née Potter). The Webbs, he noted in his diary in 1903 'are very Imperialistic, with a great idea of spending money so as to please the working classes'. Of the turbulent industrial strife which struck the nation in 1911 he wrote at the close of the year 'Last summer the manual labourers, misnamed the working classes, got quite beyond themselves.' Yet he (and even more his brother Lawrence) were opposed to the degrading system of casual labour in the docks, and he showed little sympathy with the harsh views of some shipowners. In August 1912 he recorded, 'there has been a very bad dock strike in London, stupidly . conceived for inadequate reasons, and rather cruelly repressed by the employers against A. H. Co. and others' advice'.

Another theme in a biography of Richard Holt would certainly be his Liberal politics and his prominent part in both local and national affairs. In his Liberalism, as in his Unitarianism, Richard was following very much in his father's footsteps. His father, Robert Durning Holt (1832–1908) was the youngest of George Holt's five sons. Robert entered his father's cotton-broking business and played a most prominent role in Liverpool's political and charitable affairs, though unlike his son, he was too unsure of his educational attainments to seek a national stage. He was leader of the Liverpool Liberals for a decade in the 1880s, a City Councillor for 27 years, Liverpool's first Lord Mayor (as distinct from simply 'Mayor') in 1893, and he received the freedom of the City in 1904. Despite considerable social ambition Robert refused the baronetcy offered him in 1895 by Lord Rosebery's government, largely through the influence of his wife Lawrencina's democratic principles. Lawrencina ('Lallie') was the eldest of the nine remarkable Potter sisters. Among the many children of these sisters were Leonard Cripps and John Hobhouse, so that in time the Potter line proved a fertile source of recruitment into Alfred Holt and Co.

Richard Holt was and, despite the upheavals of the Great War and world depression, remained an old-fashioned Liberal free-trader. He first stood for Parliament – unsuccessfully – in 1903, and entered the House of Commons as Liberal Member for the Northumberland constituency of Hexham in 1907, remaining a Liberal Member of Parliament, representing Hexham, until 1918. What Liberalism meant to Richard Holt shines through his attitudes to some of the great events and personalities of the time. He was one of the handful of MP's to oppose conscription during the Great War, bravely argued on behalf of the rights of conscientious objectors, and advocated moderation in the terms of the peace settlement which was imposed upon Germany when the war was won. He abhorred Lloyd George's wartime system of government controls and wage boards, believing in freedom of contract even in wartime. He also abhorred Lloyd George, whom he considered 'a scoundrel'. He was also unimpressed with Winston Churchill. 'I don't like the clever Home Secretary Winston Churchill', he recorded in his diary in June 1911, 'He has a bad face and is needlessly provocative to the Tories.'

Like nearly all the Asquith Liberals who remained opposed to Lloyd George during the Great War Richard Holt lost his seat in 1918 and never regained it. He was then 50 years of age. In truth the world to which Richard Holt belonged, with its wealth and privilege, and its Unitarian Liberal values, had vanished forever. Although he became president of Liverpool Liberals between 1920 and 1928, he presided over a sinking ship still firing signals of an earlier era: free trade, low taxation, low government expenditure, 'compelling the unemployed to work in exchange for their keep', and the like. The themes of a biography would have to emphasise decline: decline and withering away of a class and way of life which had flourished brilliantly but fragilely before 1914. The historian P. J. Waller has aptly said of Richard Holt's diaries:

> Through Richard Holt's eyes it is possible to witness the erasure of a traditional way of life, particularly as he made such great efforts to withstand the process. This is also true of politics. In Richard Holt's career the extinction of the historic Liberal party is embossed, like a memento of Pompeii trapped in Vesuvius's lava.[4]

Richard Holt's failure on the national political scene could not but benefit the Ocean Steam Ship Company, to which he now devoted himself wholeheartedly (aided by an exceptionally happy marriage to Eliza Wells, an American lady, with whom he enjoyed a contented family life with their three daughters). In 1904 Richard Holt had been uncertain of his future career, for both he and Eliza were attracted by public life. 'But there is also business . . . my family have so much money invested in the steamers which really ought to be looked after.' Richard Holt also felt a sense of obligation towards his numerous local commitments, especially his major role in the Mersey Docks and Harbour Board, of which he had been an active member since 1896. 'On the other hand', he wrote, 'there is some ambition, some wish to move on a larger stage and an honest wish to help in crushing the miserable Jingo and protectionist movements.' Richard Holt's diary is curiously silent about business matters, but this should not be taken to reflect the part the company played either in his life or affections. In February 1926 he noted revealingly:

> During the week after the previous entry there was nothing of any note except the annual meeting of the Ocean Steam Ship Company. It is really amusing to note how it is possible for the most important event of the year to appear unimportant. After all, the OSS Co. does mean to me and mine worldly prosperity, bread and butter, all the pleasures of life – Abernethy, politics and so on – and yet in one sense it appears a very unimportant thing and we take its affairs like a rainstorm.

To summarise Richard Holt's character and personality is no easy matter. Certainly he had a strong personality. He was able to command and hold together a powerful management team of varied and often warring individuals who gave their complete loyalty to the company and never seemed to begrudge the hard work and expectation that they would be 'on call' at any time should an emergency occur. Richard Holt led by example, and during seamen's strikes in South Africa and Australia in the summer of 1925 returned from his country home at Abernethy because 'I don't like leaving my partners alone to face all these disturbances.' He was a punctual and meticulous man, arriving promptly at 9.30 each morning, collecting the conference reports, and taking them to the barber's shop where he would read them while being shaved. Then he would return and dictate replies. He expected his staff and agents to be fully conversant with even minute details of voyages and freights. Typical was the opening of a letter to Warren Swire in 1921: 'I assume you have a copy of the letter on the subject of freight on potato starch per *Ajax* . . .' He was also a dedicated non-smoker and forbade smoking in the India Buildings offices between 9.30 in the morning and 5.30 in the evening. The policy was continued by his successors until Sir John Nicholson's retirement in 1971 when the ban was less rigidly enforced, a curious example of swimming against the popular tide in both directions.

Within India Buildings Richard Holt seems to have been regarded with awe and respect rather than with affection. Some, like Roland

Thornton, resented the wealth of Richard Holt and the Holt 'clan', some of whom were substantial shareholders in Ocean, while the ordinary Manager's remuneration, remaining at the discretion of the senior partner, was kept relatively low. This remuneration, incidentally, was paid quarterly in arrears, and Richard Holt was apparently unaware how this might embarrass an impecunious newcomer. The system remained until Lindsay Alexander, shortly after his arrival in 1947, approached John Nicholson, who initiated more convenient arrangement.

Holt's House of Commons speeches show him as thorough rather than incisive, pedantic rather than brilliant, yet with a carefully rational approach to every subject. Roland Thornton, admittedly at an after-dinner speech long after Richard Holt's death, remembered him as 'the dominating, outstanding personality of it all, shrewd, vigorous and witty, and a very warm heart, unhappily marred by a most disconcerting rudeness. A man of extremely high commercial principle, dogmatic mind, ungracious in argument, invariably loyal in defeat.'[5] The reference to rudeness is interesting; others recalled a sudden temper in argument, when he would take off his glasses and plonk them heavily on the table in front of him as a well-known sign of displeasure. He could be aggressive and intolerant and would brook no criticisms of Ocean affairs at the docile shareholders' meetings. When one intrepid individual did raise a question at the 1939 meeting, Sir Richard fixed him with a stare and said nothing. Yet to Sir John Nicholson, Richard Holt 'will stand forever as a giant without whom the company could never have attained a fraction of its later stature'.[6]

John Nicholson was a disciple of Richard Holt. Beatrice Webb was not, and her views were less complimentary (though we would hardly turn to her for an objective view of a member of the Liberal plutocracy). In 1899, on a visit to Robert and Lawrencina Holt at 54 Ullet Road, she found the family 'common place in outlook and mediocre in ability'.[7] In 1913 she was back at the same house, which had passed to Richard and Eliza after the death of Robert Holt in 1908 (Lawrencina died in tragic circumstances two years earlier). Beatrice Webb now thought that Richard Holt, MP:

> destined at no distant date to become Sir Richard Holt, is a stout, straightforward businessman ... he is just now smarting at the reception of his Report on the Post Office, a poor shambling document which got the government into hot water and thereby injured Dick's chance of becoming an Under Secretary. Dick feels himself to be a man of importance. He is said to be an able business man. As the senior partner of the Blue Funnel Line he has the status, if not the capacity, of the great business organiser. He is at any rate the

sounding board of the first class business tradition of the Holt-Booth brothers. He represents the wide experience of the office filtered through an honest, fairly shrewd but narrow mind of no scientific or literary culture.

The Webb diary outlived Richard Holt. In March 1941 Beatrice Webb was able to look back on the life of her nephew:

> he was a failure as Liberal MP, a commonplace citizen with conventional capitalist dogmatically held opinions in politics and economics – conforming to the old fashioned unitarian Christian creed, so typical of the nineteenth century . . . on the whole the Holt family have gone down in the social scale – they are fading out in quantity and quality.

Beatrice Webb's perspective, of course, was clouded by her jaundiced view of business and businessmen. A more objective view would certainly emphasise Richard Holt's commercial astuteness and the significant responsibilities he took on at Ocean at an early age. He had a distinct entrepreneurial flair, and became a respected national figure in one of Britain's great industries. Already in July 1901, Richard, still only 32 years of age, recorded in his diary, 'Mr Albert Crompton retired from business leaving us younger men, Maurice Davies, George (Holt) and self as virtually managers of Ocean S.S. Co.' The crucially important takeover of China Mutual in 1902 was entirely at the instigation of Richard Holt, while the development of the Australian, Pacific and American trades, the expansion of property holdings in the Far East, and in fact most of the major initiatives from this time stemmed from him. Within Liverpool Richard Holt was soon, through his work in Liberal politics and with the Mersey Docks and Harbour Board, a significant figure. He was later, in 1927, to become Chairman of the Harbour Board, and his diary comment is interesting:

> It is a great compliment, not only to myself but also to the family and the business. This is probably the most important public office out of London and the highest post which I could have hoped to attain after losing a career in Parliament. In the world at large the Chairmanship of the Dock Board may not be a big thing, but in the vision of Liverpool as I knew it when I started on my career in business it was a very big thing.

Later, in 1937, he was to become President of the Chamber of Shipping, the first Liverpool shipowner to achieve this honour and a comment both on the esteem in which Richard Holt was held and the provincialism of the British shipping community, when Liverpool loyalties were traditionally focused on a Liverpool Steam Ship Owners' Association. By the 1930s Richard Holt had become a nationally respected figure in the shipping world, recognised when he was asked by the government to help in the rescue of Kylsant's Royal Mail group in 1932 (which we will discuss later) and recognised also by his belated baronetcy in 1935.

Richard Holt's great and enduring contribution to Ocean is beyond all dispute. In some ways man and firm were rather similar, respectable, unbending, orthodox, and, with reason, self-satisfied. Change did not come easily to Richard Holt, and in some ways we can see weaknesses in the very virtues of paternal management which Holt so exemplified. Rather revealing is Holt's diary comment on an offer made in 1926 for Ocean to manage a new enterprise planning to build six fast 22 knot steamers for the Australian trade. 'It is impossible to believe in the scheme financially ... moreover the whole scheme of our business life would have to be revolutionised.' Holt was an autocrat, and kept many important facets of company affairs in his own hands. These included the company's investments, the size and structure of reserves, and the system of accounts. Holts employed no qualified accountant, and the system of voyage accounts was left very much as it had been instituted by Philip Holt around 1880 until Leonard Cripps was able to introduce some long-needed improvements in the 1930s. There was no secretariat, and one of the junior partners would keep notes made of discussions and decisions reached during Managers' meetings. There was also a reluctance to bring technically qualified engineers or naval architects into senior management, and virtually no chance of a promising individual being promoted to top positions 'through the ranks'. The result was that Managers could be dominated by the heads of their technical departments.

Richard Holt remained at the helm of the Company almost until his death in 1941 at the age of 72, although for the last two years his health was failing. Rather characteristically the Annual Meeting of Shareholders in February 1942 (held at Sir Richard's old home, 54 Ullet Road, due to the bombing of India Buildings) did no more than record 'with deep regret' Richard Holt's death. The Company was rarely given to effusion in public. The private Managers' Book was more forthcoming. At a special meeting on the day of his death the Managers recorded that

> the death of Sir Richard Durning Holt, Baronet, took place on 22 March 1941. He was the senior manager of the Company. He joined the Company in April 1889 and was appointed a Manager in 1895. Throughout this long period he gave to the Company able and devoted service and by his energy and upright example laid the foundations ever more and more strongly upon the rock of business efficiency and human confidence. From the beginning to the end his

deep personal interest in the Company was unbroken. His fellow managers and all the staff found in him a leader of independence and insight and a friend of unfailing loyalty and human sympathy.

The words sound like those of Lawrence Holt who now, at the age of 59, became senior partner.

Circumstances seemed to combine to perpetuate the influence of Richard Holt long after his death. Lawrence Holt was certainly not one to seek change; the war made day-to-day survival the all-embracing priority; the post-war environment made rebuilding and recreating the former position a matter of policy; the mantle of senior managership which fell on Sir John Nicholson in 1957 fell on a man very much a disciple of Richard Holt and who saw himself as the upholder of traditions laid down during Holt's long leadership. Richard Holt had become a Manager in the days of Queen Victoria, and he guided the firm through five reigns until the wartime years of George VI. But in many ways his influence on the Ocean Steam Ship Company was to last well into the new Elizabethan age.

3

Sailings and Services

We came to an inhospitable shore and found a people embalmed in custom and the wisdom of the ages. We gave succour to their needs and, in return, took our just reward.

(John Samuel Swire)

At the very core of a shipping line's business stand the routes and ports served. Such services do much to shape the character of the enterprise, for they determine the cargoes to be carried, the types of ships required, the conditions under which ships served and were manned, and the agency business needed. A feature of Blue Funnel operations has always been the enduring and relatively concentrated character of its services. Virtually without exception the Company confined itself to operating regular liners; only rarely were Blue Funnel ships chartered to other lines or were alien vessels chartered by Blue Funnel. These liners carried cargo, with only a limited involvement in passenger traffic, and they traded primarily between the United Kingdom and the Far East via the Suez Canal. Endeavours in other directions, such as the Australian and American trades and the various local and feeder trades in which Ocean became involved never played more than supporting roles to the leading Far Eastern business.

For outward sailings, United Kingdom loadings were confined to west coast ports, with the home port Liverpool (or rather Birkenhead) remaining overwhelmingly the principal outward berth. Loadings were also made at various continental ports, especially Dutch and, after the First World War, North German ports. For homeward traffic London and Liverpool were always the main ports of discharge for the Far Eastern services, Amsterdam for the Java trade, and supplemented by various other continental and United Kingdom ports as circumstances dictated.

The purpose of this chapter is to explore the growth and extension of these Blue Funnel operations, so that the subsequent discussion of agencies, conferences, and the more detailed aspects of Company history may be viewed in the overall perspective of Blue Funnel's routine shipping operations. We may emphasise once more, though, that Holts' reputation rested firmly on the continued high standards of these operations. Frequency of services, quality of time-keeping (there was a saying in Amsterdam that you could set your watch by the arrival and departure of Blue Funnelers), and meticulous care of cargo all gave to shippers a quality of service unmatched by rival lines.

Broadly speaking the First World War brought to an end a long period of expansion. After 1918, because of difficult world trading conditions, the emphasis had to be upon consolidation rather than upon any new initiatives. Following the Second World War, wartime shipping losses and political uncertainties again largely precluded the development of services beyond those operated and enshrined in conference agreements before 1939. Thus in tracing the growth of Blue Funnel services we must necessarily focus largely on the years before 1918. And here we can see three distinct phases. First came a period of some twenty years following the opening of the Suez Canal in 1869 when the emphasis was almost exclusively on the Liverpool–Far East trade, supplemented with a small involvement in the local Sumatra–Singapore tobacco trade. In this phase the principal developments were the overall growth of services – the fleet growing from 5 in 1869 to 30 by 1890 – and the gradual extension of ports of call in the Far East. Second, in the 1890s, came a more diversified involvement in the Far East. Local services in South-east Asia were built up, a regular line between Singapore and Java (Dutch East Indies) and Western Australia commenced, and a direct service between Java and Europe started. Third, after 1900, came a notable expansion of horizons. A direct service between the United Kingdom and Australia was established, which by 1910 had developed to include regular passenger ships, while from 1902 some Far Eastern sailings were extended across the Pacific to the western ports of North America. The American services were later modified and extended as a result of two wartime developments, the opening of the Panama Canal in 1914 and entry into the New York–Far East trade the following year.

Before we discuss these developments in any detail it is helpful to reflect on a few general background points. In the early days of the Holt enterprise there were few restrictions, beyond commercial considerations and technical feasibility, on the routes a shipowner could operate, the type of cargoes carried, or the frequency of services. True, there were some countries (notably Holland and France) which regulated the carrying trade from their colonies to their home ports in favour of national flag carriers, but for the most part shipowners were free to trade where they pleased. It is worth emphasising that the 'unequal' treaties signed in the mid-nineteenty century by European powers with China, Japan, and Siam, prevented these countries from discriminating against foreign vessels. If Alfred Holt wished to send a vessel to Yokohama he could do so. Moreover, the relatively small steamship of the 1860s and 1870s could visit a great number of ports which would have been impenetrable to the larger vessels of a later era (though some ports, in turn, were later developed to take ocean-going ships).

Gradually, though, constraints on the freedom of shipowners to trade where they pleased grew. One important factor here was the development of the conference system after about 1880, a development in which Holts played a major role and which we will examine in a later chapter. By the opening of the twentieth century various kinds of conference arrangements covered most of the routes in which Holts were concerned, and to a greater or lesser degree regulated freight rates, types of freight to be carried, frequency of services, and ports of call. In this way the conference system was an integral part not simply of the financial side of Blue Funnel operations, but of the very size and structure of the services provided.

Until the opening of the twentieth century Holts' impressive growth (there were 41 ships in the Blue Funnel fleet in 1900 and the fleet's tonnage had doubled in the previous five years) had been achieved entirely through internal growth and financed through ploughed-back profits. Indeed a feature of Blue Funnel history has been the relatively small number of acquisitions of other enterprises or their ships. However in 1902 came Ocean Steam Ship's first major and its single most important acquisition before the 1970s (notwithstanding the Glen Line purchase in 1935), when the Company took over the China Mutual Steam Navigation Company. With the China Mutual came not only its fleet but the routes and conference allocations. During the First World War further acquisitions extended Blue Funnel services in this manner. Thus the development of Holts' services over the years was partly the result of expansion from an established base, and partly stemmed from taking over routes and ships operated by other companies.

The original China service established in 1866 soon grew from its initial simple schedule, taking in additional ports and adding more diverse cargoes, both outwards and homewards. Holt ships were quick to use the Suez Canal route, and by the mid-1870s the 14 Blue Funnel vessels were carrying goods 'right through to Penang, Singapore, Hong Kong, and Shanghai, calling at Galle and Amoy, or other ports in the Eastern seas when required', as the shipping historian W. S. Lindsay wrote at the time.[1] Point de Galle, incidentally, was Ceylon's main port until replaced by Colombo in the 1880s. The development of the main line service to China saw an improved frequency of service, which could be maintained as the fleet grew and as vessels became faster; the extension of the service to Japan and North China; the loading at additional ports in the United Kingdom and unloading at continental ports; and additional regular ports of call in the Far East.

The evolution of the main Far Eastern service was a gradual one, dictated by the needs of trades and seasons, and never fixed in an adamantine mould of inflexibility. Thus W. S. Lindsay's remark that Blue Funnel ships would call at eastern ports 'when required' long remained the case. Nevertheless the basic structure of services, though with a growing number of ports of call at both ends, soon settled into a coherent pattern determined by the need for regular schedules, conference arrangements, and the work of agency representatives.

Although it would be tedious to detail every small alteration in the main line services (some of which were short-lived), we must take note of some of the major landmarks. As far as outward sailings were concerned the Mersey remained the hub of Blue Funnel operations. As early as 1872 regular sailings were commenced from Birkenhead, where berths were less congested than at Liverpool Docks, and soon thereafter all loadings were confined to Birkenhead while unloading remained across the Mersey at Liverpool where Holts used Queen's Dock. Loading for the subsequent Australian services, incidentally, used Liverpool berths. The first Birkenhead loading berth was in the Morpeth Branch Dock, but so quickly did Blue Funnel services outgrow the facilities that the Company leased a new berth in the Birkenhead Egerton Dock in the early 1890s and, a few years later, an additional berth for the larger ships was taken, known as 'Cathcart Street Berth'. Just prior to the First World War Holts moved again to the new Vittoria Dock (opened in 1909), where, along with the Cathcart Street berth which was retained, the Company had four loading berths in a line.

In 1880 Holts began loading from London also, but this service was discontinued after 1887 and London thereafter remained a port of discharge only. For nearly a decade further, though, Holts retained conference rights for three annual London loadings, and Blue Funnel ships therefore continued to make their sporadic appearances until 1896. Another even more short-lived venture was the loading of Blue Funnel ships in Manchester in the mid-1890s, making use of the newly-opened Manchester Ship Canal. Manchester was, of course, the main provider of Britain's exports of cotton goods, and Manchester merchants saw the Canal as an opportunity to gain lower freight rates from enhanced competition among shipowners. Holts were not keen to use the Canal, but in order to forestall possible rivals the Managers despatched *Titan* to Manchester in January 1895 to load with goods for China, and thereafter regular loadings were continued for about eighteen months. However, due to the extra costs to Holts from using the Ship Canal, the service resulted in higher total charges for the Manchester shippers, and the experiment was allowed to lapse.

In the late 1880s Holts improved their outward service to the Far East by starting regular loadings at Glasgow, where the agents were C. W. Scott (a personal friend of Alfred Holt). The decision to load in Glasgow was made in 1887 in response to competition from the new China

Mutual Line, which loaded both at Glasgow and Liverpool. Sailings from Glasgow developed rapidly, the majority of ships going on to Liverpool, though by the late 1890s some vessels were regularly sailing directly from Glasgow to Java or China and Japan as well. From the mid-1890s, too, prior loadings were regularly made at a number of western industrial ports, especially Swansea and Newport (which became regular ports of loading after 1900), Cardiff and Barry, and also on occasions at Barrow, Belfast, Avonmouth, or Maryport. At this stage, however, only very occasionally were outward loadings for China and Japan made at continental ports, such services not becoming a feature of Blue Funnel operations until after the First World War. For homeward sailings the position was different. As cargoes and ports of provenance became more diverse, so Blue Funnel ships increasingly unloaded part of their cargoes at European ports before continuing to London and Liverpool. From the late 1870s calls at Amsterdam became regular, while in the 1880s and 1890s discharging at Marseilles, Havre, Antwerp, Rotterdam, Bremen, and various other European ports occurred with growing frequency. Vessels unloading at Marseilles sometimes carried Japanese raw silk, raw silk being Japan's principal export and Marseilles then being the major destination. This reminds us how important were British shipping lines in developing Japan's trades in the days before Japan had a merchant fleet of her own. British lines were similarly important in developing Japan's trades across the Pacific to the west coast of the United States.

The destinations and ports of loading in the east also saw some significant changes. The core service remained as it had begun, Liverpool via the Straits to China, with regular calls at Jeddah to carry Moslem pilgrims (a trade which will be discussed later). Already by 1883 the Holt fleet had grown sufficiently to allow a weekly service from Birkenhead to China, and by the eve of the First World War the service had improved to seven sailings every four weeks with occasional additional voyages. By this time the types of cargoes carried, and hence the types and capacity of ships required, had changed out of all recognition since the start of the Blue Funnel Line, and this involved additional ports of call both for the delivery and receipt of cargoes. The main developments were four: the extension of regular services to Japan and the Philippines; the growing significance of South-east Asia as a source of homeward cargoes; the growth of trade with the Dutch East Indies; and the development of new trades with Northern China, Manchuria, Korea, and as far north as Vladivostock.

The earliest Holt vessel to visit Japan was the *Ajax*, which had sailed to Yokohama in 1870 at a time when Japan had only recently been opened up to international commerce and was still largely unknown. And in December 1872 came the first call at Nagasaki, by the *Achilles* with Henry

Gribble and Co. acting as agents. In succeeding years spasmodic visits were sometimes made, but Holts were slow to realise Japan's potential, and both the Shire and Glen lines were much more active pioneers. Nonetheless, by 1881 regular calls were being made by Holt vessels at Nagasaki to collect tea and silk, and in 1883 a regular weekly service to that port was established. From 1888 Japanese ports became regular terminals for the main Far Eastern service, Philip Holt having written to John Swire shortly before, 'We *must* tackle the Japan trade; it is interwoven with that of China and we cannot let others have a monopoly.'

At the same time there was a marked increase in the ports visited in the Far East, both on outward and homeward voyages. Colombo became a regular port of call after 1884, and during the same decade occasional calls were made at Manila in the Philippines for homeward cargoes. After 1900 Manila became a regular port of call for ships returning from Japan, and the Philippines also became a regular destination for the trade across the North Pacific which was developed a few years afterwards. In the continued search for homeward cargoes Blue Funnel services to China were regularly extended from 1902 to the North China coast ports and Korea, with calls at Chefoo, Taku Bar (for Tientsin), Dalny (Dairen), Fusan, Chemulpo, and as far north as Vladivostock. Singapore remained the great gathering ground for the tobacco, rubber, and other produce of South-east Asia, but with the rapid growth of Malaya and French Indochina Blue Funnel ships were also regular callers at Saigon after 1907 and at Port Swettenham (now Port Klang), from 1909. The growing commerce of the Dutch East Indies was largely confined to a separate service between Java and Europe, as we will see later, and here too there was a widening of operations with Samarang, Sourabaya, Macassar and Cheribon as well as Batavia being regular ports of call by the early twentieth century.

I

We may at this point reflect on some very general trends which characterised Ocean's major far eastern trades over the years. The original inspiration was the search for tea cargoes in China, and until around the middle of the 1880s this remained the driving force of the trade. In these early years the outward trade consisted almost entirely of Lancashire cotton textiles, destined principally for Shanghai. In volume the trades were unbalanced, imports of tea providing full homeward cargoes, whereas ships departing from Liverpool were rarely fully loaded. Then as outward cargoes diversified and expanded, and as tea imports from China dwindled, there was a significant change. Richard

Holt, in evidence to the Royal Commission on Shipping Rings in 1907, put the matter succinctly: 'Originally, you did your best to find outward cargo in order to load the ship homewards, but it has exactly turned over.' He added that twenty or thirty years before 'when the ships were going out for the homeward business . . . they were going out frequently empty, or half-loaded, expecting to make their earnings on the homeward trade. Nowadays that is entirely reversed in this particular trade, and you expect to make your earnings on the outward trade.'

Even as Richard Holt was giving evidence there were significant changes which were reversing the pattern yet again. Increasingly the ships leaving China and Japan, usually with around one-third of cargo space occupied, were finding growing cargoes in 'the Straits'. Rubber exports from Malaya were in their infancy in 1907, but by the outbreak of the First World War total homeward cargoes were once more beginning to exceed outward cargoes in terms of volume, and this remained the pattern throughout the inter-war years.

The significance of the Straits for homeward cargoes can be seen starkly in Table 3.1 which shows the Conference allocations (which covered the major traded commodities) given to British shipping companies in the Far East to Europe trade, excluding Japan, for the year 1938.

We may note from this table three points of particular interest. One, certainly, is the strong position of British lines, with just over half the total tonnage allocations from the countries covered. Second, Holts' shares in the main trades were substantial, amounting to one quarter overall. And third, of particular relevance to present discussion is the very large volume of homeward tonnage from the Straits, 67 per cent of British cargoes and no less than 79 per cent of Holts' allocation.

Thus in broad terms we may see three phases in the pattern of Holts' main Far Eastern trade: (i) an early period, when homeward carryings exceeded outward, based principally on tea and cotton goods; (ii) a more diversified pattern, when outward cargoes, with textiles still very significant, exceeded in volume homeward cargoes; and (iii) from around the beginning of the First World War, a phase in which the growing imports of a growing range of raw materials and other primary products made the trade, to echo Richard Holt's word, once more 'a homeward one'.

We have touched at various points upon the cargoes carried in Blue Funnel ships to and from the varied destinations, and it is convenient here to say a little more about the trades in which Blue Funnel ships were engaged. At the broadest level of generalisation they consisted of outward cargoes of British (and European) manufactured goods with return cargoes of primary products, but naturally over the years the composition of trade in both directions underwent some fundamental

Table 3.1 Far East–Europe trade: Conference allocations, 1938

Lines	Shanghai 000 tons	%	Hong Kong 000 tons	%	Philippines 000 tons	%	Straits 000 tons	%	Total 000 tons	%
P & O	20.3	15.9	32.8	17.7	–	–	72.1	6.7	125.2	6.9
A. Holt and Co.	19.1	15.0	38.4	20.8	38.5	9.4	360.0	33.3	456.0	25.3
Glen	6.4	5.0	16.2	8.8	12.5	3.0	81.0	7.5	116.1	6.4
Ben	–	–	6.0	3.2	43.4	10.5	43.1	4.0	92.5	5.1
Ellermans	3.8	3.0	9.6	5.2	55.0	13.4	55.0	5.1	123.4	6.9
Foreign lines	77.9	61.1	82.0	44.3	262.1	63.7	468.8	43.4	890.8	49.4
Total	127.5	100.0	185.0	100.0	411.5	100.0	1080.0	100.0	1804.0	100.0

Note: tons = space tons of 40 cu. ft.
Source: Ocean Archives: Conference documents

changes. Textiles remained a major outward cargo, and Holts built up a near-monopoly in the carriage of cotton piece-goods from Liverpool to China. In 1898 Albert Crompton estimated that Blue Funnel ships were carrying around 40 per cent of all cotton goods exported to China from the United Kingdom (Holts' position on the Mersey was impregnable but there was growing competition from lines using Southampton). As time went by, capital goods of various kinds, especially machinery, locomotives and locomotive parts, steel pipes, cables, and others, were of increasing importance as Holt cargoes. To textiles were added other consumer goods, especially food products and drinks, while as the twentieth century advanced chemicals and motor-cars became large carryings. Homewards came wool and refrigerated meat from Australia, edible oils and egg products from China (which gradually replaced in importance early cargoes of tea and silk), and tin, rubber, coconut products (including copra), tea, tobacco, sugar, coffee and other plantation products from South-east Asia. Rubber was a particular mainstay of Blue Funnel homeward sailings, loadings coming principally from Singapore and Port Swettenham. Blue Funnel ships started calling at the latter port in 1909, and loadings of rubber there grew from 1300 tons in that year to over 12 000 tons by 1914. Total exports of Malayan rubber, meanwhile, grew from around 100 tons in 1905, when the trade began, to 47 000 tons in 1914, and over 400 000 tons by the late 1920s. Much of the rubber destined for the United Kingdom was carried in Blue Funnel ships.

After the Second World War increasing homeward cargoes of manufactured goods became a feature of the Far Eastern trade, especially textiles, canned foods, and 'shop goods' from Hong Kong and Japan. Bulk liquids too became a major homeward cargo. Some carriage of liquids in bulk, in special tanks rather than in drums, had begun in the 1920s, but a significant development came in 1933 when the first shipment of bulk latex (a solution of rubber in ammonia) was made on behalf of Dunlops, who had invented the process in 1928–9. Bulking the latex involved major technical problems, and Holts were pioneers in the field. The first ship to carry the latex was *Eumaeus*, which sailed from Singapore in December 1933 and pumped the latex into Dunlops' installation at Gladstone Dock the following month. Following the Second World War the trade rapidly became a major one, with tanks in some ships capable of carrying 250 000 gallons each (the three small tanks in the *Eumaeus* had carried a total of 50 000 gallons). Palm oil, too, became a major bulk import from South-east Asia, growing from small beginnings between the wars to a principal cargo after 1960.

Another significant inter-war development was the carriage of chilled beef from Australia, a trade which Blue Funnel pioneered. Chilled beef was a superior, and higher value, product to beef which had been frozen, but chilled meat could be kept for a much shorter time than frozen. While chilled beef could be brought to Europe from South America, the longer journey from Australia made this unsatisfactory. However, it was discovered that gas-chilled beef (air mixed with carbon dioxide) would last much longer, and in 1934 an insulated chamber was installed in the *Idomeneus* and a consignment of chilled beef made a first, and successful, voyage to Liverpool. Five vessels on the Australian run were at once adapted for this purpose.

Over time the cargoes carried in both directions became far greater both in numbers of individual commodities carried and in the range and complexity of goods. This called for increasing sophistication in the ships' designs, to cater for the chilled products, bulk liquids, chemical products, and other cargoes which they were called upon to carry. At the end of the 1950s the *Cargo Notes* issued by the company to each officer listed no fewer than 200 commodities under classifications 'cargo liable to damage', 'cargo likely to cause damage', 'taint and infestation', and 'valuable cargo', alone. The list ranged from lizard skins from Java to 'human hair in bales', and by this time a single cargo might well consist of over 1000 separate items.

Loading such a varied cargo was a complex business. Special care had to be given to valuable, dangerous, or fragile cargoes, while some cargoes could be damaged by proximity to others (for example, chemicals). In planning loadings the chief officer also had to take account of the order in which cargoes would be discharged at various ports so as to minimise unloading times, while the placement of cargoes in the various holds had also to preserve the stability of the ship. A complex stowage plan was drawn up for every sailing by the Chief Officer. To take at random the homeward voyage of the *Diomed* in 1969. On the outward journey *Diomed* had discharged some 70 separate commodity 'parcels', 5321 tons weight, in Hong Kong, 383 tons (4 parcels) in Shanghai, and 1964 tons (20 parcels) in Manila. The cargo, all from Birkenhead, ranged from chemicals to branded food and drinks, soap and toothpaste, steel plates and diesel engines. For the return journey *Diomed* brought cargoes from Manila, Shanghai, Singapore, Port Swettenham, Penang, and Colombo, discharging in Portugal before completing unloading in Liverpool. The cargoes included bulk liquids (palm oil and latex) in special tanks, refrigerated products, sheet rubber, tea, timber, canned foods and tapicoa (all typical carryings) as well as a variety of other commodities.

Even in an earlier age the range of commodities which Blue Funnel ships were expected to carry was considerable. Table 3.2 gives a list of main cargoes carried homewards from the Straits to London in 1907, together with the rates charged (to Glasgow an additional 7s 6d was levied, and to Liverpool, 5s).

Table 3.2   Homeward Straits rates (per ton), 26 February 1907

20/-	Lead Ore
22/6	Kapok Seeds (14 cwt)
23/-	Tin, Sago Flour, Tapioca Flour, Sugar Cane & Sugar Cane Refuse per 20 cwt
27/6	Rattans (cargo), Sugar, Oilcake, Mangrove Bark (10 cwt)
30/-	Gutta Jelutong, Groundnuts in shells (12 cwt), shelled (16 cwt), Bamboo Scraps, Pepper shells (7 cwt), Cutch in bags (20 cwt), Cocoanut oil per 60 cft
31/3	Kapok in bales (50 cft), Wild Cinnamon (30 ft)
31/3	Gambier, Fl. Gambier (12 cwt)
35/-	Antimony & Tin Ore, Metals, Cutch in boxes, Gum Copal in bags and baskets, Illipenuts, Teelseed, Canes, Cube Gambier, Planks of low value (not Teak), Paraffin Wax (17 cwt) Asphalt in cases
37/6	Teak Keys
38/9	Cocos Triage (15 cwt)
40/-	Black Pepper, Pines
42/6	Coffee, Pearl Sago, Peral Tapioca, Flake Tapioca, Tallow, Arrowroot, Hides in bales, Horns, Sapanwood, Sandalwood, Shells in baskets except Mother of Pearl Shells, Planks, Long Pepper, Sesamun Seeds (14 cwt), Tapioca Siftings (18 cwt), Ipecacuanha Roots (10 cwt)
45/-	Old Gutta Percha Core (20 cwt)
47/6	Loose Hides, White Pepper
50/-	Gutta Percha, Gum Benjamin, Arrack, Cardamums, Cassia, Cigars, Cloves, Cordage, Cotton, Cubebs, Dragon's Blood, Elephant's Teeth, Gamboruge, Gum Arabic, Gum Copal in Cases, Gum Damar, India & Borneo Rubber in cases & baskets, Mace, Nutmegs, Piece Goods, Rum, Sticlac, Sugar Candy, Tea, Fishmaws, Essential Oil, Patchouli, Chillies, Mother of Pearl Shells in cases, Silk, Logs, Tortoise Shells, & other measurements not otherwise specified
57/-	Cocoa (15 cwt)
60/-	Dunnage Rattans (20 cwt)
62/6	Plants (50 cft)

Note: Spelling as original.
Source: Ocean Archives.

These were the main commodities, but often there were less mundane articles. For example in 1905 Mansfields arranged for the *Patroclus* to ship tigers, at £20 each, 'Orang Outans' at £7 10s each, and monkeys, snakes, birds and ants at 100s per 50 cu.ft. And there were the regular, if macabre, carryings of coffins from the Straits to China, as deceased Chinese emigrants were returned for burial to their ancestral home-lands. In 1905 Mansfields was quoting a rate of Straits $100 from Singapore to Hong Kong, and $125 to Swatow or Amoy, an apparently astonishingly high rate (£11 or £12) for the period, until we realise that the coffins were large containers carrying numbers of corpses. This was learned in a rather ghoulish way by Captain E. W. C. Beggs on his first voyage in the *Titan* in 1887.[2] He later recalled how after leaving Penang:

we took a number of Chinese deck passengers returning to Hong Kong, as well as several very large cases as deck cargo. These were also for Hong Kong. After leaving, my watch companion and myself found the heat very oppressive in our room, and we thought it would be much cooler on deck. So we decided that during our watch below (or off duty) we would sleep on top of the cases taken in at Penang . . . so we slept for the remaining days or nights on top of these cases. On arrival at Hong Kong, two large lighters came alongside. The awnings which covered the cases were taken in, and derricks got ready. The cases were at once discharged. After the last one had landed in the lighters, the chief officer made this remark, 'Well, we have got rid of those fellows.' On my asking him what the cases contained he answered, 'They are full of dead Chinamen.' Had we known this, we might not have slept so soundly.

Live passengers were carried too. The early ships sometimes had cabin space for a limited number of travellers, but all vessels were planned to passenger certificate standards and could take a large number of deck passengers. Before the First World War there was a constant emigration of Chinese from their homelands to South-east Asia, and also a steady flow of those returning. Already by 1875, as the *Chinese Customs Trade Report* for that year recorded, Holt steamers had taken over the passenger traffic between Amoy and the Straits, 'taking cargo to Shanghai (and) passengers from Amoy to Singapore in their 'tween decks and often also cargo for the Straits and London'. By the end of the decade Holt steamers were taking one-third to one-half of the Chinese emigrants from Swatow to Singapore, and though their interest subsequently lessened, they were running three emigrant ships between 1899 and 1914 engaged in the trade between southern China and the Straits.

Another significant, and much longer lasting, passenger interest was in the pilgrim trade. Holts' interest in the carriage of Moslem pilgrims dated from the earliest days when the company's steamers began using the Suez Canal. The passage past Jeddah on the Saudi Arabian coast made the opportunistic carriage of these pilgrims to Mecca an obvious and profitable one, and soon became a regular feature of Blue Funnel services. Pilgrims were carried primarily between Jeddah and the Moslem territories of Malaya and Indonesia, and some from North China were carried too. Like many Holt trades, the traffic was a seasonal one, varying according to the Moslem year and moving roughly ten days earlier with each succeeding year. Pilgrims were carried in the bare-'tween deck compartments as deck passengers, below the upper deck (and later in the centrecastle). An ordinary Holt cargo vessel could carry 1000 or more pilgrims on a single voyage, and before the First World War were carried in what must have been very primitive conditions, although they were certainly an improvement on the notorious local sailing craft used hitherto. Even in the mid-1920s a Master recalled how dysentry 'nearly always occurred with resultant mortality', and many accounts survive of the simple funeral rituals on board ship performed with dignity and acceptance by the pious pilgrims making their once-in-a-lifetime journey. No food was provided for them, this being their own responsibility. Areas with stoves were set aside for cooking, and each pair of pilgrims, before 1914, was provided with three pieces of firewood for cooking and with three gallons of water a day.

Slowly the trade became more regulated by government authorities in the Straits and in Java, and the Managers and Masters took considerable care to carry the pilgrims in a responsible manner. In July 1913 we find Stapeldon, who looked after the business from India Buildings, writing to Mansfields about the *Jason*. The agents were not to risk trouble with the Singapore authorities by taking too many pilgrim passengers, and 'on no account is space to be taken for cargo when it has once been measured for pilgrims'.

A typical pilgrim seasonal voyage was that of the *Peleus*, which carried the following numbers from Jeddah to Java in December 1914:

Adults	1217
Children (half price)	72
Infants (carried free)	9

That infants were carried free was perhaps a necessary provision, since nearly every voyage ended with more infants aboard than had commenced the journey! Table 3.3 gives some details of the pilgrims carried between the Straits and Jeddah in the years before 1914.

The table shows not only that substantial numbers were carried (with similar numbers carried from Java, and on the return legs from Jeddah

Table 3.3   Pilgrim traffic from the Straits, 1905–14

Season	From Singapore		From Penang	
	Adults	Children	Adults	Children
1905–6	4896	453	919	106
1906–7	5964	560	932	113
1907–8	5482	422	1026	133
1908–9	4322	358	799	117
1909–10	4808	457	855	109
1910–11	6528	619	1196	162
1911–12	10673	1049	1649	181
1912–13	7808	935	1747	246
1913–14	10086	1157	2174	278

Source: Ocean Archives.

also), but it shows, too, the considerable fluctuations in the numbers carried from year to year. These fluctuations remained characteristic of the trade throughout the inter-war years too. The main reason was the variable prosperity of the peasant pilgrims, dependent largely on the state of the harvest. Pilgrims were invariably classified in the passenger lists as 'farmers', or 'gardeners'. If the harvest failed, poverty would be widespread and travellers correspondingly few; in good years large numbers would make the pilgrimage. Religious factors, too, influenced numbers, some years being particularly auspicious. In the inter-war years political disturbances in the Arabian peninsula on occasions caused a drop in numbers.

Fares, of course, were as modest as the conditions of travel. In 1913 the passage money from Singapore to Jeddah was Straits $110, and from Penang Straits $115, at a time when the Straits dollar was worth around 2s 4d. £12–£13 might represent a life's savings for the typical peasant who made the trip.

Until after the Second World War the pilgrim trade continued to be undertaken in the normal Holt liners, specially adapted to the carriage of pilgrims during the season. There were, though, substantial improvements over time in the conditions under which pilgrims were carried, and there was stricter control by both the British authorities in the Straits and by the Dutch in the East Indies. In the 1920s permanent wash places and lavatories were installed, rice boilers provided, and food and a canteen available. In addition, side portholes were cut in the pilgrim ships. Incidentally, during the First World War itself the pilgrim traffic largely disappeared. A particular problem was that although the British government allowed Holt ships to carry Moslem pilgrims 'who were British subjects', the danger arose that Turkey, an enemy country, would fire on the ships – a particular irony since Turkey, too, was a Moslem country.

II

While the Far Eastern service via the Straits remained at the centre of Blue Funnel operations, by the close of the 1890s there had been significant expansion in three new directions. Two were related enterprises which strengthened Ocean's position in the Malayan and East Indies archipelago. The third was a local service running between Singapore and Western Australia. Singapore changed from being largely an intermediate port on the route to and from China to being a major hub in the Ocean network, the centre of feeder services which provided tobacco, tin, tropical timber, sugar, coffee, tea, and other plantation products, and a wide variety of other goods for transhipment to the main line services. With Singapore's growing importance to Ocean, Mansfields' role as Holt agents, which will be discussed in the next chapter, was strengthened accordingly.

The move into local services in South-east Asia based on Singapore began on a small scale in 1879 with the first incursion into the Sumatran tobacco trade; later, in 1882, Holts entered the Bangkok–Singapore rice trade, and by the end of the decade small vessels owned either wholly or partly by Holts were trading regularly between Singapore and Borneo and to various ports beyond.

Holts' decision to enter the Sumatran tobacco trade came as a result of Philip Holt's visit to the Far East at the end of 1877, at a time when tobacco exports from Sumatra were growing rapidly. Commercial production had begun there only in the mid-1860s when an enterprising Dutchman, Jacob Nienhuys, began to export tobacco leaf to Amsterdam. By the late 1870s considerable quantities of tobacco were being shipped to Europe from the Dutch East Indies, mainly in Dutch and German vessels, and this was the trade brought to Philip Holt's attention by George Mansfield and Theodore Bogaardt when Holt visited his Singapore agents.

The time was propitious for some diversification of Holts' interests. They had built up considerable financial reserves from previous years of success, but were now encountering a period of some difficulty in their traditional business, with growing competition and falling freight rates. At the same time the dynamic Theodore Bogaardt was actively expanding business in South-east Asia, sometimes on his own account, sometimes through Mansfields, and often with the support of Holts in Liverpool. In 1878 Bogaardt had moved to Penang to establish an office of Mansfields named Mansfield, Bogaardt, and Co. He resided there until 1882 when he returned to Singapore to take over Walter Mansfield and Co. upon George Mansfield's retirement the year following. Bogaardt, with his Dutch connections, was able to secure tobacco cargoes for Blue Funnel vessels.

Accordingly in 1879 Ocean built the tiny (405 tons) *Ganymede* for trade between Deli and Singapore, and later placed a storage hulk at Belawan Deli. Another small steamer, the *Ascanius* (107 tons, and locally known as 'the flying coconut') was added in 1880 to run between Sumatra and the Straits, and a storage hulk was also placed at Penang. During the 1880s the tobacco trade was developed substantially with the acquisition of a number of other small steamers, so that Holts' main line vessels, taking cargo from Singapore and Penang (mostly from Singapore) became major carriers of tobacco from the Dutch East Indies (Indonesia) to Europe. Indeed, by 1890 Holts' vessels were carrying around 20 000 tons of Sumatran tobacco to Britain out of a total export from Sumatra of 32 000 tons.

By the 1870s tobacco estates were also being developed by Dutch planters in North Borneo (Sabah), a territory which came under British rule in 1877 and was administered by the British North Borneo Company after 1881. Holts' interests in this trade developed from 1886, when Bogaardt persuaded the Liverpool company to take a financial stake in several small ships trading between Singapore and Borneo. Mansfields established an office in Sandakan in 1890, by which time partly owned Holt ships were also trading to Ternate, Menado and the Moluccas, and Bogaardt had introduced a service between Borneo and Hong Kong.

In addition to the tobacco trade Holts also became involved in the fast-growing Bangkok–Singapore rice trade, though unlike tobacco, rice was not normally transhipped to Europe, but was consumed by the expanding Chinese communities in South-east Asia and Hong Kong. The first move, again through the promptings of Bogaardt as soon as he had returned to Singapore from Penang, came in 1882 with the purchase of the *Hecuba*. Rice was a profitable, if highly competitive, trade, with the *Hecuba* returning profits of around £10 000 a year in the late 1880s. By 1889 a further three ships were engaged in the trade, Windsor & Co. acting as Holts' agents in Bangkok. These vessels took a considerable share, around two-thirds, of Bangkok's mounting rice exports to Singapore between 1885 and 1899. The rice steamers were not wholly owned by Holts, however. The Ocean Steam Ship Company put up rather more than half the capital, but the remaining shares were taken by Bogaardt himself, by Mansfields, and also by the Ocean Manager Albert Crompton.

So in the 1880s the Ocean Steam Ship Company, working in close partnership with Mansfields and Bogaardt, had become involved in important new trades, tobacco and rice; new areas, Sumatra and Borneo; and had developed significant regional trades from Singapore with Bangkok, Borneo, and Sumatra, and also betwen Penang and Sumatra. These trades, especially the Deli and Bangkok trades, brought considerable profits to Holts. But success inevitably brought also growing competition from rival concerns; and it was the growth of Dutch, and to a lesser extent German, competition which led to further, and more permanent, changes in the operation of Ocean's trading in the Straits in the 1890s.

Holts' position both in its main line link carrying tobacco to Europe and also its regional feeder services based on Singapore and Penang had been threatened by increased Dutch activity from the 1880s. In 1880 a new Dutch line, the KPM (Koninklijke Paketvaart Maatschappij), had been formed under the auspices of the subsidised Dutch mail lines, the Nederland (SMN) and the Rotterdam Lloyd. The KPM established a regional line in 1888 and began to provide a feeder service for Sumatran tobacco and other archipelago products to the main line fleets of the

Dutch mail lines based on Batavia and other Indonesian ports. Holts therefore found their lucrative tobacco monopoly under threat, and determined to enter the direct Java–European trade in response. Under Dutch law, non-Dutch registered vessels could not trade in Dutch colonial ports on equal terms with ships of the Dutch flag, and Holts therefore decided to form a new company registered in Amsterdam. Such an expedient, incidentally, had already been adopted by the British India Steam Navigation Company in 1865.

In 1891, therefore, Holts established the NSMO, or 'Oceaan' (correctly, Nederlandsche Stoomvart Maatschappij Oceaan), in partnership with the Dutch agency of J. B. Meyer. The new service began on 1 January 1891, and the NSMO was incorporated a few days later. The Ocean Managers told their shareholders at the Annual Meeting that year that it was intended to load the vessels homewards for Amsterdam if possible, and that the service would load from 'all the more important Netherlands Indies ports'. The homeward voyages were both to Amsterdam direct, and to London via Singapore. The company had a registered capital of 2 million guilders – three-quarters of which was subscribed by Ocean – and two Liverpool Managers, Crompton and Llewellen Davies, became directors. NSMO was a unique institution, Dutch run, with ships crewed by Dutch officers, yet completely integrated into the Blue Funnel network with Blue Funnel traditions and standards. The Dutch company retained its highly individual identity, and relations between Meyers, who managed the service, and India Buildings were always cordial and cooperative. The original fleet consisted of ten older Holt ships (including the 1866 *Agamemnon* – the *Ajax* and *Achilles* were to join later), and the normal route by the end of the decade was from Amsterdam to Liverpool, then on to Padang (in western Sumatra) and Java via Jeddah (in the pilgrim season); then returning to Amsterdam with tobacco, sugar, and other estate products. By 1900 the NSMO fleet had been strengthened by the acquisition of some newer vessels (still about twelve years old on average), but not until 1924 did the Dutch company obtain its first new ships.

The establishment of the NSMO allowed Holts to maintain their preponderant share of the tobacco trade, and, although the transhipment trade at Singapore inevitably suffered, by 1914 Blue Funnel vessels were controlling some three-quarters of the tobacco sent from the Dutch East Indies to Europe. The direct Java–Europe trade added a significant new string to Blue Funnel's bow, the trade being conducted both by the Dutch-flagged ships and, as Dutch regulations were relaxed, some from the Ocean and China Mutual fleets.

Holts' local feeder services in the archipelago were less successful. In the face of competition from the KPM and German lines Holts in November 1892 established a separate local company, the East India

Ocean Steamship Company. This company had a registered capital of £98 000 and the fleet, initially numbering 14, consisted primarily of the small local ships which had been engaged already on the Mansfield-Holt feeder services, and in addition several new ships were specially built for the fleet. Most of the company's capital was held by Ocean, and the ships of the fleet were registered in Liverpool, but day-to-day management was largely in the hands of Mansfields in Singapore.

Holts doubtless hoped that a successful local fleet could be built up along the lines of the KPM in the archipelago or Swire's China Navigation Company, which had been established in 1872 in Shanghai. But the competitive world of the 1890s proved overwhelming, while the difficulties of organising management and control at such distances were also severe. The departure of Theodore Bogaardt from Mansfields in 1895 (under what circumstances we do not know) must have added to the difficulties, and by that time Bogaardt had formed the Straits Steamship Company in January 1890, to provide a service between Singapore, Penang, and the ports of the Malayan peninsula. The Straits Steamship Company must have been a considerable competitor in certain local trades, and seems to have had a close connection with the Straits Trading Company, formed in 1887 to operate a modern tin smelting works in Singapore. Despite the close Mansfield connection (Mansfields were managers of the Straits Steamship Company between 1893–5) Holts did not invest in the new concern, doubtless because of their plans for their own regional line. But the Straits Steamship Company was destined long to outlive the East India OSS Co., and eventually, after 1914, to form an integral part of the Ocean fleets.

The main competition to the Holt East India OSS Co. came, as always, from the Dutch, especially on the Padang–Penang–Singapore run. Eventually Holts were driven out of this trade, while German competition similarly gained the upper hand in Borneo. By the late 1890s the regional trades had become unprofitable, and in 1899 the Ocean Managers decided to sell the East India OSSC to North German Lloyd (NDL) 'on terms which have allowed the capital to be returned in full to the shareholders'.

The disposal of the East India Steamship vessels to North German Lloyd was based on sound commercial considerations; the trades were uncertain and fluctuating, the management problems severe, and since Holts were able to reach an agreement with NDL to safeguard the transhipment of East Indies products to the main line Blue Funnel vessels at Penang and Singapore, the original purpose of the local service had been secured. Yet it is surprising to find Holts selling its strategically placed fleet to a German concern at a time when German commercial and diplomatic rivalry was causing widespread concern in Britain. In Bangkok, for example, the consequence of Holts' action was a fall in Britains' share of the carrying trade from around 70 per cent to 20 per cent, with corresponding increases for the German flag. Mansfields, certainly, were shocked, and the rumour at the time was that the Ocean management did not know whom the ultimate purchasers were until the deal was clinched.[3]

Holts still retained their interests in the Singapore–Deli tobacco trade, which remained an important branch of Holts' activities in the region. Three ships were engaged in this trade, and in 1913 new vessels, the *Circe* and *Medusa* replaced two of the older ships. After the First World War this service became redundant when Belawan was developed as an ocean-going port so that main-line steamers could call there, and the vessels were sold in 1925 to the China Navigation Co. Holts also retained their involvement in the traffic in Chinese migrants, three Ocean vessels operating before the First World War between the Straits and southern China.

Another significant development in the early 1890s was the West Australian service, connecting Singapore and Fremantle. During the early 1880s there was a considerable development of trade between Western Australia and the Far East, and in 1884 an Australian firm, Trinder, Andersons and Bethell, had started a service between Fremantle and Singapore. This firm formed the West Australian Steam Navigation Company in 1886, with two ships maintaining a regular service, and it was on one of these ships, the *Australind*, that Alfred Holt sent Captain Frank Pitts from Singapore to Fremantle in 1889 to investigate the possibilities of Blue Funnel entering the growing trade. As a result of Pitts' favourable report and concessions promised from potential shippers and government officials, Holts made an arrangement with the West Australia company for a joint monthly service which commenced in 1891. Holts committed a specially built vessel, the *Saladin*, (1874 tons gross), constructed in 1890 by Thomas Royden and Co. of Liverpool, which was registered in Fremantle. The Managers had considered expanding the service to Adelaide and Melbourne, but were dissuaded from doing so by John Swire, who replied such a move would conflict with the interests of the British India Steam Navigation Company and other lines.

The West Australia service, running between Singapore–Batavia (Djakarta)–Fremantle, was destined to become a useful adjunct to the main Holt business, and lasted until well after the Second World War. Initially the main cargoes to Australia were British and Far Eastern goods transhipped from Singapore, and Australian demand developed rapidly after gold was discovered in Coolgardie in 1892. Return cargoes were predominantly cattle, sheep, and wool. In 1894 Holts built another ship for the trade, the *Sultan* (2063 tons) the first ship built for Ocean by Workman Clark and Co. of Belfast. Both the *Sultan* and *Saladin* were

WESTERN AUSTRALIA.

Derby, Pearling Grounds, Broome (Roebuck Bay),
Cossack (for Roebourne), Onslow (Ashburton),
Carnarvon (Gascoyne), Sharks Bay,
Geraldton (Champion Bay),

AND

FREMANTLE

Via Suez Canal and Singapore.

By Ocean Steam Ship Co.'s Steamers and by s.s. "AUSTRALIND"
or s.s. "SALADIN."

Passengers by this route have the advantage of
visiting Singapore and the various Northern Ports of
Western Australia.

Passage Money to Western Australia:

Cabin .... **40 Guineas.**

Intermediate .... **24 Guineas.**

(Bedding and Cabin Requisites are provided.)

Embarking from Liverpool by the s.s. " ," to sail
about , and transhipping at Singapore to the
s.s. "AUSTRALIND" or "SALADIN," to sail about

The above dates are estimated so that Passengers can go from
one Steamer to the other at Singapore, but should there be any
unforeseen delay, Saloon Passengers must pay any hotel expenses;
Intermediate Passengers are kept at the expense of the Steamer.

*The Ocean Steamships from Liverpool to Singapore carry
both a Surgeon and a Stewardess.*

Apply to—
ALFRED HOLT,
1, India Buildings, Water Street, Liverpool;
TRINDER, ANDERSON & Co.,
4, St. Mary Axe, E.C.;
C. BETHELL & Co,
110, Fenchurch Street, London, E.C.

5 Sailing notice for the West Australia trade, 1892

capable of carrying about 100 head of cattle and 750 sheep each, the shipping of live animals being a permanent and particular feature of the trade. In Western Australia most of the agency work for the joint service was undertaken by Dalgety and Co., who had opened a branch in Fremantle in 1889, and handled the business from the arrival of the *Saladin* in 1891, although in some ports of call (such as Derby, Broome, and Shark Bay) other agents were appointed. In Singapore Mansfields, of course, acted for the Blue Funnel vessels, while Boustead and Co. were agents for the West Australia Company.

A characteristic of the West Australia service for many years was the large number of ports at which the little ships would call in both directions. From Singapore they would call at numerous ports in Java and then in Australia at any or all ports between Wyndham in the north to Fremantle in the south. Elements in the trade included mother-of-pearl shell from Shark Bay and Broome, migrants for the gold fields, and islanders from Koepang and Cocos for the pearling industry in Broome. The main staples of the trade, though, remained cattle and wool taken to Singapore for transhipment.

The various extensions of Ocean enterprise in the 1890s centred on Singapore were both profitable and timely. The Java trade, for example, could be profitably served by older boats which, in the words of the Managers 'have become too small and otherwise out of date for the China trade', while, as we have seen, the lucrative tobacco trade was retained largely in Holt hands down to the First World War. At this stage Blue Funnel vessels were carrying around half the rapidly expanding rubber and tin exports which were sent from the Straits to Europe. By the late 1890s the combined profits of the West Australian, Java, and Deli–Singapore tobacco trade were contributing roughly one-third of total profits, and in the subsequent decade diversification from the original China trade brought a still further shift of emphasis.

III

The annual meeting of shareholders held in January 1892 had been the first to receive a report distinguishing the various company interests, 'West Australia', and 'Local fleet at Singapore', in addition to the main line vessels. A year later the Report detailed 'China and Japan Line', 'Java Line', 'Local Services', and 'West Australia'. This regional approach set a permanent pattern not only for the presentation of reports but, over time, for the division of responsibility within the management. By the close of the 1890s each Manager 'looked after' specific regions, and corresponded with, and built up close relationships with, the agencies operating there. There was no hard-and-fast rule; all

Managers were free to involve themselves in any part of the business and the principle of 'collective responsibility' remained. But the growing complexity of business necessitated some form of divided interest among the Managers and staff, and in the years between 1900 and 1914, 'America', 'Australia', and 'Passengers' made their appearance as new areas of responsibility as the company developed and widened its interest further.

Apart from the Fremantle–Singapore cross trade, Holts had, before 1900, sent the occasional ship to Australia to load with wool cargoes, returning to Europe via Suez or the Cape. The close of the nineteenth century saw a rapidly growing trade betwen Australia and Europe, and at the same time the increasingly efficient steamships with triple-expansion engines at last invaded the final stronghold of the sailing vessels. Circumstances were dictating a more permanent interest, although the initial move came not from Holts but from an Australian agent. In 1898 Holts were approached in Liverpool by a Mr F. W. Braund, of the Adelaide firm of George Wills and Co., the London representative of 'The Syndicate', a group of three of the most powerful shipping agents in Australia. These three, Wills of Adelaide, John Sanderson of Melbourne, and Gilchrist Watt of Sydney, had formed a permanent association in 1892 and were thereafter known as 'The Syndicate', and the group was anxious to expand trade links between Australia and England. As a result of the Liverpool visit, Holts sent the *Glaucus* to Australia in 1898, and in the following year five further vessels from China. These ventures were successful, and in 1900 Richard Holt (who was already becoming a dominant force in the affairs of the company) put forward plans to run a regular monthly service between Liverpool and Australia, outwards via the Cape, and returning via Suez. The ships would carry Australian wool in the season, but otherwise would load with miscellaneous goods, including fruit for which refrigeration compartments would be necessary. In Australia the ports of call would be Adelaide, Melbourne, Sydney and Brisbane.

For the Australian service Holts committed six vessels, three of them, *Sarpedon*, *Nestor*, *Orestes*, being the first Blue Funnel ships to have refrigerated holds fitted for fruit and other perishable commodities. The service was commenced in February 1901, not from Liverpool initially, as Richard Holt had hoped, but from Glasgow (where competing interests were fewer and friendly arrangements could be made with other companies for available berths). The Liverpool service was started the following year. The service was initially monthly, and proved so popular that by 1903 half the outward cargo from Glasgow to Australia was being carried in Blue Funnel steamers.

In 1909 the Managers decided to extend their operations further by entering the Australian passenger service, for this was a period of rapidly expanding emigration from the United Kingdom. In 1910 three specially designed ships were launched, the *Aeneus, Ascanius*, and *Anchises*, which were each capable of carrying 288 first-class passengers (reduced to 180 after the First World War). In addition each ship had space for 450 000 cu.ft. of cargo, one-sixth of which was in refrigerated holds. In 1913 the company launched two much larger vessels, the *Nestor* and *Ulysses*, both of around 14 500 tons gross and carrying some 350 first class passengers (reduced in stages to half that number by the mid-1930s). These vessels cost about £250 000 each (the entire China Mutual fleet had cost only a little more), could take 600 000 cu.ft. of cargo, one quarter being refrigerated space, and remained the largest conventional ships ever built for the Blue Funnel Line.

The fares on these pre-war passenger services make interesting reading. Perhaps only a cargo carrier as meticulous as Holts could have arranged children's fares and luggage space with quite the passion for detail and complexity (Table 3.4)

The ships built for the Australia passenger service (which also served South Africa) marked Ocean's first significant venture into the carriage of cabin passengers (there had always been the pilgrim and Asian migrant traffic, of course). Early vessels had carried a few passengers, usually with cabin accommodation for 12 to 24, although from the 1890s some ships were built for cargo only. After the First World War Holts

Table 3.4   Passenger fares, Australian service, 1914 (£) *Nestor* and *Ulysses*

	Luggage allowance (cubic feet)	Cape Town Single	Cape Town Return	Adelaide/Melbourne/Sydney Single	Adelaide/Melbourne/Sydney Return
Adults (single berth)	–	35	63	55	99
Servants	–	24	44	36	64
Children					
under 1 Year	nil	free	–	free	–
4 Years	10	5	10	10	20
5 Years	10	7	13	11	21
6 Years	12½	19	17	14	26
7 Years	15	11	21	17	31
8 Years	17½	13	24	20	36
9 Years	20	15	27	23	41
10 Years	22½	17	31	25	46
11 years	25	19	34	28	51
12 Years	27½	21	38	31	56
13 Years	30	23	41	34	61
14 Years	32½	25	44	37	66
15 Years	35	27	48	40	71
16 Years	37½	29	51	43	76
over 16 Years	40	full	full	full	full

Source: Blue Funnel Line, publicity brochure, 1915.

6 *Nestor*, built 1913. The ship was in service until 1950, and for many years boasted the world's largest funnel (*photograph courtesy of the Liverpool Museum*)

began the regular carriage of passengers to and from the Far East, but apart from this service many of the new generation of Holt ships had accommodation for twelve first class passengers. This service, long preserved to maintain goodwill among the 'planter and government classes' became a font of treasured memories in the more leisured days of steam and empire.

The other major development at the opening of the twentieth century was the service provided across the Pacific to the Puget Sound ports of the United States and Canada. This trade had been developed by the China Mutual Company, which regularly sent Far Eastern steamers across the North Pacific. When China Mutual was taken over by Blue Funnel in 1902 this trade was brought within the orbit of Ocean services, and linked well with the established services between Britain and the Far East.

Prior to 1914 the trade across the Pacific was very profitable. A staple of this trade was timber carried from the American Pacific ports, and specially designed ships were developed to deal with the heavy logs. The first of these ships was the 9000 ton *Bellerophon*, Holts' largest ship to date, launched in 1906; the last, the *Tyndareus*, in 1916. These were so-called 'goal-posters' because of their powerful wide-reaching derricks to handle the large logs. In addition to the timber trade, the ships also carried large numbers of Chinese migrants and temporary workers between China and the Pacific coast of the United States, carrying up to 200 passengers in rather spartan open areas. These ships were the first Holt vessels to be fitted with twin-screws. Hitherto all Blue Funnel ships had been driven with a single propeller only, such was Alfred Holt's confidence in the reliability of his engines. The larger vessels, though, and those carrying more passengers, were now fitted with twin screws.

IV

By the outbreak of the First World War, therefore the company's operations were well defined, with a network encompassing the Far East, the Straits, the Pacific, the Dutch East Indies, and Australia. By 1913 the fleet numbered 77, with a gross tonnage of 467 815. In 1909 the figure had been 62, totalling 350 853 tons; in 1902 (after the acquisition of the China Mutual), 55, with 259 000 tons; and in 1900 41 vessels with a gross tonnage of 165 646. In other words in the short space of thirteen years the number of ships had almost doubled and the tonnage almost trebled. Although the China trade still remained the core of operations, accounting for nearly two-thirds of the voyages, the steady increase in the 'subsidiary' trades meant that these latter were contributing nearly one half of gross earnings in 1913. By this time it was increasingly difficult to distinguish between 'Straits trade' and 'China trade' since, with the decline of tea imports to Europe from China in the 1880s, vessels going to China and Japan normally returned to the Straits with part loads for further loading. The 'China trade' was in fact a single main line service, later supplemented by the Java and Australian trades.

The period of the First World War, disrupting and damaging as it was, saw some significant new additions to the company's services. One major change was the acquisition of the Straits Steamship Company which, it will be recalled, had been founded in 1890 by Theodore Bogaardt and had been managed by Mansfield and Co., until Bogaardt's resignation from the agency in 1895. The outbreak of war in 1914 brought the immediate end of the German lines, which had established a strong position throughout South-east Asia. The North German Lloyd line, in particular, carried a great deal of the exports and imports of Siam and Borneo. When the German company had purchased Holts' East India Ocean Company in 1899 the terms included NDL providing through bills of lading for Blue Funnel cargoes between Singapore and Siam and Borneo. Thus Blue Funnel cargoes destined for Bangkok or Borneo would be taken onwards from Singapore in NDL ships, and freight from these countries would similarly be transhipped to Blue Funnel steamers at Singapore. Such trade was considerable. In 1913, for example, Holt vessels had landed at Singapore 23 286 tons of cargo destined for Bangkok and Borneo, and shipped 6046 tons from these places.

The sudden departure of NDL in 1914 naturally meant new arrangements, and on 5 September a few weeks after the outbreak of war, Richard Holt wrote to E. Anderson, Managing Director of Mansfields in Singapore:

> Owing to the war and the consequent termination of our agreement with the NDL, it is necessary to provide for services between Singapore and Borneo and Bangkok capable of dealing with the Ocean Steam Ship Company's through traffic creditably under our general control. We are agreed that this result can best be obtained by cooperating with the Straits Steamship Company, thus providing for local management and securing local goodwill and interest.

Richard Holt's initial proposal was that Blue Funnel should take shares in the Straits Steamship Company in return for providing ships to replace those of the NDL. Straits Steamship Company would obtain all Blue Funnel traffic to Bangkok and Borneo, while the Straits Steamship Company in turn would ensure that all homeward traffic from these places would be shipped in Holt vessels. Richard Holt also demanded that the Ocean Company would nominate two directors to the Straits Board.

The resulting negotiations did not bring all Richard Holt had hoped. The Straits Company refused to allow Ocean to nominate directors or to agree to send all homeward cargoes on Blue Funnel ships, but by October 1914 satisfactory agreement was concluded. The Straits Company obtained three new ships in return for shares taken by Holts, and the *Kajang, Kamuning,* and *Kepong* were built for Straits in Hong Kong in 1916 at the Taikoo Dockyard. These vessels were placed on the Borneo service.

Ocean was now the largest single shareholder in the Straits Steamship Company, with roughly one-third of the total number of shares, Straits Steamship then being in the capable hands of H. E. Somerville, Managing Director from 1914 and Chairman from 1919. From the outset Holts' relations with Straits were completely cordial. With the successful conclusion of the agreement between the two companies, Somerville had written to Richard Holt that his Board 'wish to place on record their high appreciation of the very liberal way you have dealt with the matter'; for their part, Holts were careful to ensure that, in Professor Tregonning's words, 'their strong position did not become a dominating one'.

As a result of wartime expediency, therefore, Holts had enlarged their interests in Malayan waters, and achieved something they had failed to do in the 1890s – acquire a local fleet to serve their main line steamers in South-east Asia.

A second consequence of the war was the development of a service with the east coast ports of the United States. This came about as a result of a piece of opportunism by Richard Holt, who in August 1915 acquired the seven ships of T. B. Royden's small Indra Line. The fleet itself was of little consequence, but the acquisition gave Holts a seat on the China–New York Conference, so enabling Blue Funnel vessels to serve regularly the eastern ports of the United States for the first time.

The Indra purchase and the near contemporaneous opening of the Panama Canal (which, like the war, started in August 1914), enabled some modifications to Holts' existing American services. The first Blue Funnel ship to use the new canal was the *Astyanax*, which passed through in February 1915. In 1915 the Pacific journeys of the main line Far Eastern ships were abandoned, and the Trans-Pacific trade between the Far East (Japan, Hong Kong, and the Philippines) and the Puget Sound ports (Vancouver, Seattle, Tacoma, and Victoria) became a local cross trade. The Trans-Pacific linked with a service between the United Kingdom to the Puget Sound ports via the Panama Canal. Other Blue Funnel ships provided a 'round-the-world' service based on New York. Outward services went from New York via Panama to Japan and China, linking there with homeward services to New York via Suez. There were further services through the Suez Canal between New York and Hong Kong, via the Straits and Manila, and also direct services between New York and Java.

The pattern of routes and services built up before 1918 remained largely unaltered throughout the inter-war years. Indeed, in the changed world trading conditions between the wars it became an object of policy to maintain the *status quo* rather than expand. As quickly as possible the major services and agencies were re-started and already by 1920 the main Far Eastern, Java, and Australian services had been established, the Java trades in cooperation with the Dutch Mail Lines (that is, the KPM, Nederland, and Rotterdam Lloyd), the Australian services jointly with Hamburg–Amerika and North German Lloyd. During the inter-war years some significant changes nonetheless took place. A major development was the extension of both main line Far Eastern and the Java services to incorporate prior loadings from the North German ports of Bremen and Hamburg. Since Britain's traditional exports, like textiles, faced sluggish world markets after 1920 the additional sources of cargo were of substantial value. Thus from 1921 monthly sailing from Birkenhead loaded first at Bremen, Hamburg, and also Rotterdam, while the fortnightly Amsterdam service to Java also loaded first at the two German ports. The Australian services, too, were soon revived. A three-weekly service between Liverpool and Glasgow and the eastern ports of Australia via the Cape was reintroduced in 1920, maintained principally by the five passenger steamers, *Ulysses, Nestor, Aeneas, Anchises*, and *Ascanius*. In addition, a new monthly cargo service was started jointly with North German Lloyd and DADG (the German–Australian Line) in 1922.

Another post-war change was an attempt to develop further Ocean's passenger business by extending the service to the Far East. Previously, as we have seen, the main line ships to China and Japan each carried only a limited number of passengers, some none at all. At the conclusion of the war the government requested Holts to provide temporary additional passenger accommodation on some of their Far Eastern ships, and the popularity of the service encouraged the Managers not only to maintain the additional accommodation on the *Mentor, Teiresias,* and *Pyrrhus,* but to order four 15 knot passenger steamers capable of carrying 155 first-class passengers. These ships came into service in 1923 and 1924.

The great depression of the early 1930s had relatively little impact on the overall pattern of Blue Funnel routes and services although, as we will see in a subsequent chapter, profits fell sharply, the American services were pared, and the Company attempted to utilise its spare shipping capacity by promoting passenger excursions and cruises. A new service was introduced in 1933 when Far Eastern services began loading at Gdynia, in Poland, Blue Funnel ships being the first and only liners to maintain a direct regular service between Poland and the Far East. The first ship to go to Gdynia was the *Helenus,* with a freight of under £1000, but by 1938 the volume had grown to such an extent that chartered vessels were sometimes used. The most significant change in the 1930s, and a direct consequence of the depression, was the purchase by Holts of the old-established Glen Line in 1935. The Glen fleet, whose acquisition will be discussed later, was not integrated with Blue Funnel (though conditions of sea-service were brought into line), Glen maintaining their organisation in London and the Glen ships continuing to sail under its own flag and colours. Glen ships also traded to the Far East, from 'east coast' ports, and the acquisition thus gave Holts an interest once more in outward loadings from London. The Glen purchase, especially after the line had been strengthened by the building of new modern ships in the later 1930s, greatly improved the overall balance of services offered by Holts. If with the round-the-world New York service Holts had become truly international, with the purchase of the Glen Line the Liverpool company became truly national as well.

4

Representatives Abroad

You find the cargo and I will find the ships.

(R. D. Holt to Killick Martin and Co.)

The history of the Blue Funnel Line has been intertwined with that of its agents. Characteristic of the Company's development has always been its close and long-standing relations with its main agents. Indeed, in the case of two, Mansfields in Singapore, and Stapledons in Suez, Holts were directly responsible for the foundation of the agents' business in the first place, and in the case of Butterfield and Swire were indirectly responsible. Relations were built up over many years. In all the main ports of call agents appointed in the very early days were retained well into the post-Second World War era, and in many cases the connection has remained unbroken. Table 4.1 below gives a list of Blue Funnel's principal agents in 1923, though most of them had been appointed at the

Table 4.1   Principal Blue Funnel agents, 1923

Agent	*Principal ports*
Butterfield & Swire	Shanghai, Yokohama, Hong Kong, Kobe, Dairen
Mansfield & Co.	Singapore, Penang
Smith, Bell & Co.	Manila
Boustead & Co.	Port Swettenham
William Stapledon & Sons	Port Said, Port Tewfik
Fraser, Eaton & Co.	Sourabaya
The Borneo Co.	Bangkok
Mitchell Cotts	Capetown, Johannesburg
Australian Syndicate	Perth, Adelaide, Melbourne, Sydney, Brisbane
Dalgety & Co.	Perth, Fremantle (West Australia Service)
Booth American	New York
Maclaine, Watson	Batavia
Dodwell & Co.	Seattle, San Francisco, Los Angeles
Delmege, Forsyth & Co.	Colombo
Meyer & Co.	Rotterdam, Amsterdam
Harrisons and Crosfield	Belawan, Medan, Sandakan
Haacke & Co.	Padang
McNeill & Co.	Samarang
Michael Stephens & Co.	Macassar

Source: 'The Blue Funnel Line: Alfred Holt & Co.', *The Manchester Guardian Commercial*, 24 January 1924.

inception of the particular trade. The principal agents in 1923 were virtually identical with those in 1939, while if we compare agents, or 'correspondents' as they were then called, for the same ports in 1902 prior to the absorption of China Mutual, we find still a very similar array of names. The full list of 1902 'correspondents' is given in Appendix III, and it can be seen that Butterfield and Swire represented Blue Funnel at 18 ports (though not yet at Dairen); Port Swettenham had yet to become an established deep-sea port; Windsor and Co., a German firm, was the Bangkok agent (replaced later that year by the Borneo Co. upon Holts' acquisition of China Mutual); and the Pacific trade (and hence Dodwell's representation) had not yet commenced.

The main agents played a full role both in furthering Blue Funnel business and developing the standards of service achieved by Blue Funnel ships. Holts in turn always considered their relations with the various agencies to be of the utmost importance. Each Holt Manager had an area of trade designated for particular responsibility, and made it a major task to cultivate mutual confidence with the appropriate agencies in their region. Over many years firms such as Dodwells, representing Blue Funnel on the Pacific coast of North America, Booth American in New York, Maclaine Watson in Java, Smith Bell in the Philippines, the Australian agents, Meyers in Holland, and many others became a part of Holts' history; Holts was similarly a not-insignificant part in the histories of their agents. Contacts with shippers and customers were the life-blood of Blue Funnel operations, and it was the agents and brokers who completed the chain. Especially important were agents such as Boustead in Port Swettenham (Guthries there also after the Second World War), Harrisons and Crosfield in Sumatra, Smith Bell in the Philippines, and similar representatives in the great cargo-gathering regions of Asia. Nor should the role of agents who acted for the associated companies be forgotten. Straits Steamship operations, for example, were closely bound up with Harper, Gilfillan and Co. at Port Swettenham, Harrisons and Crosfield at Borneo, Bousteads in Kelantan, Sime Derby at Malacca, and the Borneo Company in Bangkok. Mansfields, as managers of Straits Steamship, and Meyers, as manager of NSMO, were also deeply involved in the shipping side of Ocean affairs.

Holts similarly retained close and equally long-standing relations with their United Kingdom agents and brokers. Much of Blue Funnel's homeward cargoes was destined for London, where the company's business was handled by John Swire and Sons. For outward brokerage, Killick Martin canvassed cargo from London-based merchants and manufacturers for Blue Funnel's west-coast sailings. These agencies lasted from the days of Alfred Holt until the Second World War. Glasgow Far Eastern trades were represented from 1870 by Colin Scott, later Roxburgh Scott, and the connection lasted even longer. In 1901 began representation with another Glasgow firm, Aitken Lilburn, who handled Blue Funnel's new Australian services. The following year brought Holts into contact with the firm of John Roxburgh, also in Glasgow, agents for the newly acquired China Mutual Company. The dual representation was continued until 1923 when the firms were amalgamated as Roxburgh, Colin Scott and Co. Following the Second World War Ocean Steam Ship acquired the shares of both Glasgow companies and the businesses were combined in January 1947.

Before turning to the eastern agencies we may first consider Holts' close links with Meyer & Co. in Holland, links which extended back to the formation of the NSMO as we saw in the previous chapter. Meyers were managers of NSMO and principal continental agents for Blue Funnel ships, and since the parts of Meyers and NSMO in Ocean history are inseparable, in this chapter on agencies we should consider also the development of NSMO.

'Oceaan', founded jointly by Meyer and Co. and Holts in 1891, was a small but significant part of Blue Funnel operations. The company's place rested on its unique status as a Dutch subsidiary, with close and cordial relations established between Liverpool and Amsterdam; on the long-standing link established between Ocean and Meyers; and upon the integral part played by NSMO in the important 'Java trade' side of Ocean business. Meyers acted as Dutch agents for NSMO and Ocean (and also for Glen and Elder Dempster through the Blue Funnel links). Meyer's Amsterdam office handled the Java trade (that is, Blue Funnel services between Indonesia and Europe), while cargo-broking business was conducted both from Amsterdam and Rotterdam.

As we saw earlier, NSMO was established originally to run ships under the Dutch flag between Amsterdam and the Dutch East Indies. The first NSMO Board in 1891 consisted of Albert Crompton and Maurice Llewellyn Davies from Liverpool, J. B. Meyer and B. R. Hyde from Meyers in Amsterdam, and W. N. Cool from Meyer's Rotterdam office. For a time NSMO operated ships under the Dutch flag only for the required part of the voyage, then transferring them to the British flag. Later, around 1905, after agreement had been reached with the existing Dutch companies involved in the Europe-Java trade, SMN (Stoomvaart Maatschappij Nederland) and Rotterdam Lloyd, NSMO ran their ships permanently under the Dutch flag. Under this arrangement British-flagged Blue Funnel ships could also enter the trade. Initially the NSMO fleet consisted of four ships, but this had been increased to five by the First World War. The ships were always manned by Dutch officers, with Chinese crews recruited in Hong Kong and Liverpool.

Relations between Liverpool and Amsterdam were always close, and were naturally concerned with the Indonesian operations in general. Ocean Managers were therefore very much involved in what an NSMO

and Meyer Director, David Rahusen, described in 1960 as 'this extraordinary Anglo-Dutch cooperation, where no fifty-fifty financial investment keeps the balance, but where it proves to be possible in a 100 per
cent British concern to maintain a Dutch branch which flourishes in a
real sense of independence'. Between the wars the NSMO–Meyer link
was a leading interest of John Hobhouse and Leonard Cripps. Following
1945 George Holt played a notable part in the revival and extension of
both the Dutch and Indonesian operations, taking over from Hobhouse
in 1955 formal 'first string' managerial responsibility for Meyers and the
NSMO.

Highlights of NSMO operations included both world wars, when the
Dutch-flagged ships had an eventful history. NSMO ships played a
heroic role in both conflicts, and losses of men and ships were grievous,
as is related in the appropriate chapters. In the First World War Oceaan
vessels operated their customary services as far as possible, though they
were requisitioned by the Dutch government from time to time to
transport grain. However, the period of unrestricted submarine warfare
brought great peril for such ships, and the last two ships to pass through
the Channel with government grain were Oceaan's *Tantalus* and *Dardanus*.
The new phase of naval warfare induced the British government in the
summer of 1917 to seize all foreign-registered vessels owned substantially
by British owners. Thus all NSMO vessels except the *Dardanus* (which
the Dutch government refused to allow to leave Amsterdam) passed
temporarily under British government control. The Managers reported
to shareholders that they 'must record their regret at an occurrence
which was probably inevitable in view of the German refusal to respect
the laws of maritime warfare and the Dutch Government's inability to
give protection to vessels flying their flag'.

In the Second World War all NSMO ships escaped immediate seizure
and served under requisition. They participated in many theatres of
war, playing a notable part in convoy operations. Three of the six vessels
were lost, and of special note was the fate of the *Polydorus*, torpedoed off
West Africa in November 1942. In recognition of the long and gallant 36
hour battle against the German U-boat a 'Royal Mention in the Orders of
the Day' was granted in 1950 to the lost ship by Queen Juliana of the
Netherlands. The Order spoke of the 'excellent moral strength and
discipline' of the crew 'from high to low'. In 1951 a bronze plaque
commemorating the incident was presented to NSMO and kept in the
Amsterdam office.

Among the individuals who shaped NSMO and guided its development in Amsterdam were P. C. Adrian, who was the main executive
director throughout much of the inter-war period, and Baron Taets van
Amerongen, whose long service as director ended in 1946, 'hastened by
his work under occupation', as Ocean shareholders were told. Another

prominent NSMO director, D. Rahusen, escaped to Britain during the war. He had joined NSMO in 1921 and took over the number one position in Amsterdam until his retirement in 1960, when he was succeeded by T. G. Gleichman until the latter's retirement in 1974. Gleichman, like Rahusen, was a major figure both in NSMO and Meyers, and was Ocean's chief agent on the continent especially in conference dealings with the German lines.

As cargo-brokers for Blue Funnel vessels in Amsterdam and Rotterdam, Meyers were always a significant cog in Ocean's commercial wheel. A legendary cargo-broker in Rotterdam was J. Tieman, who originally entered Meyer's Rotterdam office as a boy of 15 in 1901 and became a director of NSMO and Meyers in 1938. He retired in 1950, having played a big part in the post-war revival of trade especially in the success of Blue Funnel's *Peleus*-class service to the Far East. Prior to the war he was also closely involved in Ocean's new service from Gdynia in the 1930s.

Blue Funnel ships belonging to NSMO were long a familiar sight in Amsterdam. At first, until 1906, the ships berthed at the Handelskade, but in that year they became the first to use the new Javakade. In 1925 berthing was moved to the Borneokade, with greatly improved facilities, and this remained the hub of operations until the 1970s, when the rundown of conventional services forced the closure of NSMO after a span of 87 years. A special ceremony marked the sad occasion, the flag of the last ship, the *Patroclus*, being hauled down at Liverpool by Captain A. A. Duif and his Chief Engineer, J. Blaase, on 4 October 1978.

Holt's relations with others of their principal agencies went far beyond the normal commercial one, and was sometimes intimate to the point of being incestuous. Stapledons were founded by Holts, the Stapledons were substantial shareholders in Ocean, and in 1901 William Clibbett Stapledon returned from Egypt to become a full partner in the Ocean Steam Ship Company. Mansfields became a wholly owned subsidiary of Holts in 1903, and ten years later we find Stapledon writing to one of the Mansfield Managers, 'We wish your young men to regard themselves as fully belonging to the OSS Co.'. And the relationship with Swires was the most remarkable of all, both through John Swire and Sons in England and Butterfield and Swire in the Far East. John Swire may almost be considered as one of the founders, and certainly one of the architects, of the Ocean Steam Ship Company. Moreover Butterfield and Swire, from their office in Hong Kong, controlled the entire homeward trade of Blue Funnel vessels from the nineteenth century until the Second World War. Actual agency work was, in fact, secondary. 'Butterfield and Swire and the well-known and ubiquitous Blue Funnel line are the two great British firms engaged in the commerce of the Far East', wrote C. A. Middleton Smith, Taikoo

Professor of Engineering at Hong Kong University in 1919, and 'their ships carry the red ensign into all latitudes between Singapore and Vladivostock'.[1]

In view both of the commercial significance of the Straits and China trades for Holts, and of the unique relationships built up by Holts in their conduct of these trades, we shall for the rest of this chapter concentrate upon the early history of the Swire and Mansfield connections. Without such a perspective we can appreciate neither the expansion of Ocean's own business, nor the close personal ties with these agencies which remained a permanent feature of Ocean's operations.

To understand these unique relationships, four points need to be emphasised. First, without the early dynamism of Swires and Mansfields, in their very individual ways, Holts could not have achieved either the prosperity or range of operations which had developed before 1900. Holts' success is therefore part and parcel of a wider story which also embraces that of the main agents. John Swire's role in the formation of the conference system, as we will see, was largely due to the Blue Funnel connection, and the conference in turn was an enduring and fundamental influence on Blue Funnel's fortunes. Second, the Blue Funnel agencies were the first agencies held by Mansfields and by Butterfield and Swire, and both were established within months of Alfred Holt's initial steamer sailings in 1866. Both agencies, moreover, followed from long-standing personal contacts. Thus the three firms more or less began together, and it was the Holt contracts which gave the agencies their start in life.

Third, from the earliest stages Holts and Swires built up strong business connections beyond normal agency business. Thus the Swire brothers were initial and substantial shareholders in the Ocean Steam Ship Company itself, while members of the Holt family invested in various Swire enterprises, including being original shareholders in the China Navigation Company, while Alfred Holt put his engineering skills at the disposal of Swire's shipping ventures on many occasions. Fourth, and following from these earlier points, came a close and long-lasting personal involvement of the firms at all levels. Especially was this true of the remarkable relationship between John Swire and Alfred and Philip Holt, for John Swire remained as loyal and devoted to the interests of the Ocean Steam Ship Company as any partner could possibly be. Personal links between the firms evolved by the interchange of personnel, by the visits of Holt managers to the east, and by the initial 'tours' of the 'crown princes' who would typically spend time either in the Far East under the aegis of Butterfield and Swire, or in the Straits in the Mansfield offices. John Hobhouse, for example, worked in Mansfield's offices after the First World War, and became responsible later for many matters connected with South-east Asia, the Straits

Steamship Company, and Mansfields itself. And at the close of 1921 we find Warren Swire writing to Richard Holt to agree to take the young Roland Thornton for a period 'in the Blue Funnel department in Hong Kong or Shanghai ... we will tell B & S to give him all the experience possible, including, if necessary and desirable, say Hankow during the season, or Japan'. Similarly in 1935 Swires sent Holts a detailed programme covering John Nicholson's projected eighteen months in the east, from autumn 1935 to April 1937, a programme which included stays in Hong Kong, Shanghai, Hankow, Tientsin, and Japan.

In sum, therefore, a very special relationship existed between Holts and the two main agents forged out of personal friendships, mutual debt, and mutual esteem. In many respects it is far more appropriate to consider Alfred Holt & Co. and the relevant activities of Swires and Mansfields as elements of a single firm than the terms 'principal' and 'agent' imply.

I

John Samuel Swire was four years older than Alfred Holt, and the two Liverpool men had already formed bonds of business connections and personal friendship before Alfred launched Ocean Steam Ship in 1865. John Swire was destined to become one of the great commercial figures of the nineteenth century. He laid the foundations of a firm which was to play a dominant role in the commercial affairs of the Far East and he may be considered the founder and organising spirit behind the system of shipping conferences. Richard Holt thought John Swire the most able businessman he had ever met, and the impact of John Swire on the Ocean company was fundamental. There was only a small element of exaggeration when Robert Holt (Alfred and Philip's brother and father of Richard) wrote to John Swire in 1894 'You are the moving and ruling genius of the OSS Co.'

In the mid-1860s the fortunes of the Liverpool merchants John Swire and Sons were in a state of flux. The first John Swire had died in 1847 leaving the firm he had established in 1832 to his two young sons, John Samuel, 'The Senior', then 22 years of age, and William Hudson, only 18. The pair were involved in a variety of enterprises, John for a time even having sought his fortune in the Australian goldfields. By the outbreak of the American Civil War (1861–5) Swires were engaged in the import of American cotton, a trade which was severed abruptly during the war. As a result the brothers increasingly turned their attention towards the eastern trades, consigning woollen and cotton textiles to two Shanghai firms, Augustine Heard and Co., and Preston, Bruell and Co. The arrangement did not work satisfactorily, for in 1866 John Swire

7 J. S. Swire (1825–98) founder of John Swire
and Sons, and originator of the first China
Conference (*photograph courtesy of John Swire &*
Sons)

felt compelled to send a representative to Shanghai to work in the
Preston, Bruell office. This, too, failed to satisfy John Swire, who came
to the conclusion he should open his own firm in Shanghai.

At this juncture the fortunes of both Holt's and Swire's new ventures
became linked. John Swire had been quick to see the potential of Alfred
Holt's China steamers when most were sceptical. He was one of the
original subscribers to the Ocean Steamship Company (the shares being
in the name of W. H. Swire) – a great act of faith when it is recalled that it
was an unlimited private company. For the initial sailings of the
Agamemnon and *Achilles* in 1866 Alfred Holt had used old-established
agents in each of the main ports: Preston, Bruell in Shanghai, Syme and
Co. in Singapore, and Birley and Co., in Hong Kong. Like Swire, Alfred
Holt was quickly disillusioned with the service given by these representatives and he decided to change his agents.

In October 1866 John Swire, together with William Lang (who was to
stay in China as Shanghai manager) sailed for Shanghai on his first visit
to China in the P&O ship *Aden*. Their object was to set up in Shanghai a
trading and agency business to be run as a joint enterprise with Richard
Shackleton Butterfield. Butterfield was a Yorkshire textile manufacturer

with whom Swire had done business over a number of years. The new firm of Butterfield and Swire was set up in Shanghai in December 1866 and started trading on New Year's Day, 1867. In the event, Swire's partnership with Butterfield lasted less than two years, for the partners disagreed and dissolved the partnership in August 1868. This left the firm of Butterfield and Swire entirely in the hands of John Swire and Sons, though the name of the firm remained unchanged. Never can a businessman have achieved such widespread and enduring recognition for so minimal a contribution as has Richard Butterfield.

While there must have been several motives for the establishment of Butterfield and Swire, it appears that a principal factor was the acquisition of the Blue Funnel agency, and Swire's consequent determination to have a company to look after Holts' China interests on the spot. Swire was aware that Butterfield and Swire, as importers of textile goods, would be able to provide cargo for Blue Funnel ships on their outward journeys, and in going to China in 1866, John Swire must already have reached some arrangement with Alfred Holt. Indeed, a large quantity of cotton goods was consigned to Butterfield and Swire in the *Achilles* on her first voyage, which arrived in Shanghai on 24 December 1866. On this ship, incidentally, also sailed one James Henry Scott, third son of C. C. Scott, head of the Greenock shipbuilders with whom Alfred Holt had close connections. Young Scot, then 21, was generously given a free passage by Alfred Holt, together with an introduction to John Swire. Scott was taken on by Butterfield and Swire and was made a partner in 1874. Scott recalling in 1909 how 'Mr Holt offered me a passage in one of his boats coming East to see if I could find something . . . On the way out I tried for employment everywhere but was unsuccessful until I came to Shanghai, where I found the best friend a man could find.' Scott eventually became senior partner. Butterfield and Swire were able to furnish the *Achilles* with a return load of raw cotton for her homeward trip, since no tea was forthcoming from the Shanghai merchants most of whom were tied to the clipper owners. The *Achilles*, departing from Shanghai on 16 January 1867, was thus the first steamer cleared by Butterfield and Swire for the Ocean Steam Ship Company.

The Swire brothers had opened a London office as early as 1864, and they decided in 1871 to move their headquarters from Liverpool to London. From this time John Swire and Sons were appointed as London homeward agents for Blue Funnel, and handled the tea and other china goods which were the major part of the early homeward trades.

The Holt brothers quickly recast their other main eastern agencies. Singapore, as we will see later, was given to Mansfields early in 1868. In Hong Kong, too, Alfred Holt was not satisfied with the existing arrangements. Birley and Co., according to J. H. Scott, always showed 'a certain amount of unwillingness to take orders from Butterfield and

Swire, the chief agents of the Ocean Steam Ship Company'. In consequence, Alfred Holt asked John Swire in 1869 to open a branch of Butterfield and Swire in Hong Kong. In September 1869 John Swire wrote to Lang in Shanghai:

> Mr Holt, anxious to augment his fleet of China Ocean Steamers to commence the coasting trade, and dissatisfied with the management of Birley & Co. has requested us to open in Hong Kong, and, after mature consideration, we have consented, firmly believing that his agency alone authorizes this important step, besides which other business to and from England and the Australian Colonies will add largely to our commission account. The Singapore Agent is to be under your authority. We think that Holt's Line is only in its infancy, and that Hong Kong is far the most important station in the East, for the future of the Steam trade – also that the specie, passenger and intermediate business, is nothing to what it will be.

This remarkable prophesy was made two months before the Suez Canal transformed eastern shipping.

So began a period of expansion in which Butterfield and Swire opened a large number of branches throughout China and Japan, representing Blue Funnel interests in all the major ports, and maintaining overall control of Blue Funnel's entire far eastern trade from Shanghai and Hong Kong. The first additional branch was opened in Yokohama in 1867, a pioneering venture at the time of the Meiji Revolution when Japan was still largely unknown to the outside world. The Hong Kong office was opened at the beginning of 1870, while further branches were established at Foochow in 1872, Swatow in 1882, Hankow in 1885, Tientsin in 1886, Kobe in 1887, and at various other ports in China, Manchuria, and Japan, even as far north as Vladivostock in Russia, where a branch was maintained until 1923 (six years after the Bolshevik revolution). There were in all 15 branches of Butterfield and Swire by the mid-1890s and 21 by 1913.

A living relic of the early visits of Holt ships to Japanese waters is continued in the still-sung 'Lullaby of Shimbara', a region of Japan's far south. There Holt vessels often called at the port of Tsunokuchi to carry coal to Shanghai, but sometimes they carried too *kara-yuki-san*: young girls sold by their destitute families to 'work' in far-away places. The lullaby asks 'Where have our girls gone?', with the reply '*ao-entotsu no Battanfuru* (by Blue Funnel boats of Butterfield, far, far beyond the sea to China)'.

In 1872 John Swire had founded a China coast and river fleet unique in shipping annals. The China Navigation Company, formed with a paid-up capital of £300 000 was also destined to have a long-standing role in

Holts' affairs. Already on his visit to China in 1866 John Swire's fertile imagination had been struck with the possibilities of the Yangtze river trade. This trade, opened to foreigners after 1860, was then very much the preserve of the famous American company, Russell and Co, which also ran the Shanghai Steam Navigation Company. Swire at first envisaged an enterprise which would act as a branch and feeder line for Blue Funnel steamers, and he therefore proposed to Alfred Holt in 1867 that the two firms form a joint company to exploit the river trade. Holt declined, doubtless with his hands full running his new enterprise, and for a time the project lapsed. A few years later Swire was ready to embark on the project on his own, and he wrote to his eastern managers, 'we shall have 3 ships on the Yangtze next year. We are going to run the River and don't intend to be bought off – or frightened off', referring to possible moves by his great rivals, Russells. Later, in 1883, he recalled in another letter to the east:

> *In 1867* I urged Holt to go on the Yangtze, and two or three years before we bought *Foochow* and *Swatow*, as well as when we bought them (November 1874) I pointed out to him the fine opening for steamers on the China coasts, and the advantages of such an invest-ment to the OSS Co., but he distinctly stated that nothing would induce him to own steamers not under his personal supervision, which shut me up.

Alfred and other members of the Holt family were among the original shareholders in the China Navigation Company, and Alfred advised on the designs of some of the early ships. The complex history of the China Navigation Company, and how it overcame vigorous opposition from Russells, Jardine Matheson, and other established interests, cannot be detailed here. The first decade of the Company's existence was charac-terised by Marriner and Hyde, in their excellent biography of John Samuel Swire, as 'hectic and stormy' but by 1882 'Swires had succeeded in establishing the China Navigation Company as a major shipping interest on China's local routes'.[2]

We may note, though, that in the early years of the China Navigation Company's existence, Swire confined his operations almost exclusively to the Yangtze river. The object was in part to avoid competing with Blue Funnel vessels which in those days picked up considerable cargoes from 'wayside' trade along the China coast. Later, from the mid-1870s, Swire extended his operations beyond the river and became more and more involved in the coastal trades. Swire largely pioneered what became known as the 'beancake trade', a trade in beancake fertilisers carried in the China Navigation's 'beancakers' from Manchuria to south China. In 1883 the Company ran 5 vessels on the river and 15 on the coast, by 1900 the numbers were 7 and 41, and trade was carried on as far afield as Australia, the Philippines, Japan and Siam. The centre of gravity, though, remained the China coast and the Yangtze river.

The reference to Swire's paramount concern to protect Blue Funnel interests is a recurring theme throughout the first thirty years of Ocean's existence. As we will see in a subsequent chapter, John Swire's inception of the first China Conference in 1879 was largely a defensive measure to protect Holts from the consequences of falling freight rates and the growing uncompetitiveness of Blue Funnel vessels. Whenever Swire's China Navigation Company ships were likely to find themselves in conflict with Blue Funnel John Swire was adamant that Holts' affairs should take precedence. In 1880 he wrote to Alfred Holt 'I have ever declined business that could by any chance interfere with the OSS, and actually, our other lines are feeders to, and very much benefit your steamers'. When in 1881 Butterfield and Swire suggested to London that China Navigation ships should enter the Swatow–Straits emigrant trade, John Swire replied firmly, 'I will not go in for special steamers for the Swatow/Straits trade, nor put your proposal to A. H. who might consider it to interfere with OSS Co earnings.' Swire's reluctance was due to the already firm position Blue Funnel had developed in carrying Chinese emigrants from Swatow and Amoy to the Straits. Again in 1883 he reassured Alfred Holt, who was concerned that China Navigation was taking Blue Funnel business, 'My orders have always been to give OSS Co. the preference. On the contrary the CN Co helps OSS Co., as its parent line, and that it would do so was one of my strongest arguments to you before the CN Co., was formed and when I wanted it to be a branch of OSS Co.' Swires continued to give priority to Holts in various spheres long after John Swire's death in 1898. One instance was the wartime acquisition of the Shire Line's Vladivostock agency in 1917. Swires at once informed India Buildings, 'it is, of course, understood that if the Shire agency prejudice your position at this point after the war it will be given up'.

Swire's concern and constant protection of Blue Funnel interests in the time of John Samuel Swire cannot by any means be explained solely in terms of the profit the Holt agency brought. Indeed, the agency was for long not particularly profitable, and at times completely unrewarding. Yet John Swire never wavered in what Marriner and Hyde aptly term 'his stormy devotion to the interest of the Ocean Steam Ship Company'. It is worth quoting at length these authors' assessment of John Swire's motives in his dealings with Holts. They write:

In all his disagreements with the Holt brothers, disagreements which, over fairly lengthy periods, verged on the acrimonious he never forgot that it was Alfred Holt who gave him his first real opportunity to establish himself in the Far East. The purely financial rewards from the Blue Funnel agency could certainly not be said to have generated such loyalty. As we have seen the true assessment of his reward from the handling of Holt business more often than not was a debit, rather than a credit item in his firm's accounts. In years of depression his commissions from such business often failed even to cover overheads but despite this, he always gave prior consideration to the Ocean Steam Ship Company's interests. Such loyalty from a man whose whole life had been governed by the strictest of business principles, is pleasurable to record but difficult to understand. In modern psychological jargon it could, no doubt, be explained away in terms of a love–hate relationship. In the uninhibited world in which John Swire lived and worked, there is little room for such an explanation; his loyalty, however much it may have conflicted with his business interests, sprang from guiding principles which controlled the whole of his behaviour. In this case, the close connection between Holts, Butterfield & Swire and the China Navigation Company was engendered by something more powerful than considerations of profit and loss. All three firms had grown out of a common identity of purpose; their individual driving force originated in their own mutual well-being. To have abandoned the Holts during the critical years of the 1880s might have assuaged John Swire's business conscience, but would have been morally indefensible when judged by his own ethical standards and as Philip Holt himself wrote to him 'I never knew a man with a stronger sense of justice in business'.[3]

Interestingly, even after John Swire's death the handling of Holt's China agency did not bring particularly large returns. Between 1908 and 1913 Swires were paid a management fee of £2000 annually, while the additional annual profits from the main agencies in China and Japan, averaged over the years 1908–13 were from Hong Kong, £1358; from Shanghai, £5985; from Kobe, £680; and from Yokohama there was an average annual loss of £365.

Holts, too, were protective of China Navigation interests. In 1891, a dispute arose with Mansfields (also China Navigation's Straits agents) over transhipment of Blue Funnel Manila cargoes at Singapore rather than at Hong Kong. After an angry exchange of correspondence with Mansfields (whose relations with Swires were often turbulent in this period) John Swire approached Philip Holt, and some months later Swire was able to write to his Hong Kong manager 'M & Co. never replied to mine of 14th January, but their correspondence since then has been very deferential. I am certain that P. H. H. gave it them hot!'

Holts were also involved as shareholders in various other Swire enterprises. One such was the Taikoo Sugar Refinery, established in 1881 and constructed between 1882 and 1884. The word *Taiku*, incidentally, means in Mandarin Chinese "great and ancient"; it was the Chinese name chosen by Swire for his new Shanghai enterprise at an early stage, and, amply fulfilling the founder's rather immodest confidence, came to be the generally used Chinese name for Butterfield and Swire and all Swire enterprises. Holts were among the original subscribers to the large initial capital (£98 000) of the Taikoo Sugar Refinery. Alfred and Philip subscribed for 10 shares of £1000 each and a further two bonds of £5000 each. Later, during the 1890s, they increased their investments in what became a substantial and flourishing concern. Another major Swire enterprise in Hong Kong was the Taikoo Dockyard and Engineering Co. In 1900, two years after John Swire's death, John Swire and Sons decided to construct a large graving dock, building berths and engine works on a large 52 acre site east of the Taikoo Sugar Refinery. Not until 1908 did the first ship dock at the new yard, and in 1910 the yard built a small steam tug for Holts. During the First World War Holts further contracted with the Taikoo yard for the *Autolycus*, delivered in 1916, a vessel of 5806 gross tons and the largest yet built at the yard. The *Laertes* followed in 1919 at a cost of £226 000, and the turbine steamer *Rhexenor*, 7957 tons, delivered in 1923.

A more modest concern, the Tientsin Lighter Co., was promoted in 1904 jointly by the China Navigation Company and Ocean. Such a company had been envisaged by John Swire since the 1880s, existing lighter facilities for the growing Tientsin trade of the China Navigation Co. being at that time hopelessly inadequate, with long delays at Taku Bar. But the promised scheme did not come about until after Blue Funnel vessels had started calling at Taku. The first Blue Funnel ship to do so, the *Prometheus*, arrived there in 1899 and lifted around 8000 tons, 'a cargo beyond expectation'. With Holts now anxious to develop trade through Tientsin, and to develop trade with North China and Korea, plans to form a lighter company went ahead, and in 1904 the two companies launched the Tientsin Lighter Company with Ocean contributing initially £5000.

The areas of cooperation between Holts and Swires remained substantial, and continued into the inter-war period when for a long time Warren Swire was the leading influence on Swire operations. Throughout this time, and until 1941, the homeward loading programme from the whole of the Far East except the Dutch East Indies, was controlled and coordinated by Butterfield and Swire in Hong Kong, and was the major occupation of two senior staff. In addition came the normal agency work.

This led to friction on several occasions. For example Richard Holt in 1925 complained that Butterfield and Swire in Hong Kong 'have completely lost their heads; they are obsessed with the idea of a weekly

8 At Hazelwood (the Butterfield & Swire Shanghai Manager's mansion) first week in April 1920. *left to right*: E. F. Mackay (B&S Shanghai Manager); J. K. Swire (in spats); Mrs Mackay; Grace Holt; Mrs R. D. Holt; S. Timms (B&S); and R. D. Holt (*photograph courtesy of John Swire & Sons*)

London service which, in my opinion, is at present quite impossible except from the Straits'. Another cause of friction was Swire's disinterest in Japan, an area where Richard Holt and John Hobhouse were anxious to expand. The problem was that Butterfield and Swire, as coordinators of the whole eastern trade, were responsible for Japan too. On several occasions Holts threatened to establish direct representation in Japan, bringing on one occasion, in 1925, the response from one of the Swire partners, 'B and S *are* the Blue Funnel in the Far East and we consider it essential to have the control centralised in Hong Kong.'

Cooperation, though, continued to be the hallmark of relations, and a few examples will illustrate the continuing and developing connections. In 1918 the Straits Steamship Company, the China Navigation Company, and the Borneo Company (Holts' agents in Bangkok) formed the Bangkok Wharf Syndicate and in 1921 Ocean and the China Navigation Company jointly financed a warehouse property development in Hankow. In 1922 Swires agreed to give 'private' discounts to Holts for work done on Blue Funnel ships at Taikoo Dockyard 'as an earnest of our desire to secure for Taikoo the bulk of your work'. On several occasions in the 1930s Holts provided engineers familiar with diesel engines for China Navigation Company vessels making their maiden voyages from British shipyards, and in December 1934 Lawrence Holt agreed to give one year's training in Blue Funnel ships to young Chinese cadets being trained by China Navigation. Many other examples of close cooperation could be given, while there were also avenues of possible joint working arrangements which did not come to fruition. In 1933, for example, Swires offered to arrange joint cooperation between Ocean and the China National Airways (a Sino-American concern), which Richard Holt declined. The following year Swires turned down Holt's suggestion of a joint loan to the Ho Hong Steam Ship Company, a company in which China Navigation and Straits Steamship had taken a stake. Moreover, as we have seen, the two companies did not always see eye to eye, and Warren Swire resented various attempts by Roland Thornton to change arrangements with shippers in the eastern trades and to tread in areas thought by Swire to be the preserve of the agency. Indeed, while we should rightly stress the excellent working relationship which existed between India Buildings and principal agencies, it would be surprising indeed if there were not occasional disagreement or even friction.

Criticisms could, of course, go both ways. In 1921 Warren Swire complained to India Buildings, not for the last time, that Holts were very poor at giving instructions to their agents. Instead of communicating directly with Butterfield and Swire, or with other agents, instructions from India Buildings arrived 'via masters and other subordinates'. This comment serves to emphasise the very strong support and reliance placed by Holts on their Masters, and the Masters in turn were fully aware of Liverpool backing (an enduring memory at Butterfield and Swire was the rather ominous statement from dissatisfied Masters, 'we will have to see what Liverpool thinks about that'). Only very rarely did Masters fail to live up to the Managers' high expectations, although when Butterfield and Swire dismissed a Master in 1919 for drunkenness Richard Holt fully supported the agents, writing to them that 'in such cases we are bound to leave the decision to our agents on the spot'.

For much of the inter-war period Holt–Swire relations were dominated by the powerful personalities of Richard Holt and Warren Swire. Despite occasional disagreements and surface tension, mostly resulting from pressures generated by the great depression, both personal and commercial contacts between the two firms remained close and cordial. Both firms were family concerns (John Swire and Sons remaining under family leadership to this day under a series of exceptionally able chairmen), and both concerns retained financial involvement in the enterprises of each other. Swires remained a major shareholder in Ocean. In 1927 they held around £30 000 of a total capitalisation of £425 337, while in 1952 out of the same capital, John Swire and Sons held £16 745 worth of stock and other members of the Swire family £9400. In both periods the Swire holding was easily the largest single holding outside the Holt family itself.

After the war there were some significant changes. In 1945 the long-standing agency agreement with John Swire and Sons was reluctantly severed, the firm having acted as Holts' inward London agents since

1871. The move was caused by Holts' acquisition of the Glen Line before the war, commonsense dictating that the London agency should pass to Glen. By the same token the long link with Killick Martin and Co., who had acted as London outward brokers since 1887, was also broken, with the outward brokerage passing to Glen's McGregor, Gow and Holland. In their statement to shareholders the Managers recorded that 'this decision is not unmixed with regret that these long and valued relationships with Swires and Killick Martin should come to an end', and they reassured shareholders that the moves would have no effect on Ocean's relationships with Butterfield and Swire 'or indeed with John Swire and Sons personally'. In 1954 an even older connection was lost when Butterfield and Swire were obliged to close down their remaining offices in mainland China after representing Holts there for 87 years. The Blue Funnel agencies then passed to the government-controlled China Ocean Shipping Agencies. Swires continued to be involved closely in the affairs of Ocean. In 1970, with a holding of £2.2 million, John Swire and Sons remained principal shareholders, and in 1977 John Anthony Swire, Chairman of John Swire and Sons, became a non-executive Director of Ocean.

Until the age of containerisation in the 1970s Butterfield and Swire continued to represent Blue Funnel ships in Hong Kong and Japan, while relations between what increasingly became the Holt and Swire 'groups' developed in a number of different directions. Among other post-war developments may be mentioned the taking over in 1953 by Swire's China Navigation Company of Blue Funnel's carriage of Moslem pilgrims between the Straits and Jeddah. And in June 1967, a century after John Samuel Swire had first mooted the idea with a reluctant Alfred Holt, Ocean and Swires became joint partners in the China Navigation Company, with Ocean purchasing a half share and Swires retaining the other half and continuing to manage the line. A further co-operative development between the two groups was the formation in January 1968 of McGregor, Swire Air Services (MSAS), a joint venture between Ocean Steam Ship and John Swire and Sons. Neither joint venture proved enduring, China Navigation passing once more into exclusive Swire ownership in 1975, while Ocean became sole owners of MSAS in 1982. One of the original Swire directors of MSAS, incidentally, was James Hinton Scott, grandson of the James Henry Scott who, over a hundred years before, had set off on the *Achilles* for China. Eventually came the end of Butterfield and Swire's agency representation. The passing of Blue Funnel ships in the 1970s and 1980s inevitably drew to a close the old arrangements and finally, in 1988, with the sale of last Barber Blue Sea ships which were still represented in Japan by Butterfield and Swire, came the final curtain on an era which had lasted more than 120 years.

II

In a number of ways Holts' connections with Mansfields, though vastly different, were even closer than with Swires. W. Mansfield and Co. became Holts' Singapore agents almost from the earliest sailing, and from this time there developed a close and mutually supportive relationship between the two companies. Business bonds were strengthened by personal ties. After the First World War, for example, as with Butterfield and Swire, it became common practice for young 'crown princes' at India Buildings to work in Mansfields to gain experience of the eastern trades at first hand, and a number of prominent figures gave distinguished service to both firms. Furthermore Mansfields also developed, through Holt connection, a close relationship with Swires, and acted as Singapore agents for the China Navigation Company.

It could well be said that the agency business of W. Mansfield and Co. was a creation of Holts, although Mansfields' presence in Singapore predated the launching of the Ocean Steam Ship Company by some years. The connection once established, Blue Funnel, and Blue Funnel-related, business remained overwhelmingly the major activity of the Singapore firm, notwithstanding some significant diversification in the 1930s such as the move into airline agency business. Following the Second World War Mansfields also expanded through the agency business of certain Holt joint ventures, for example, Overseas Containers Ltd, Panocean Shipping and Terminals, and Blue Sea Line in the 1960s. There can be no doubt that Mansfields and Holts benefited each other, and if Holts brought business for Mansfields, Mansfields also opened up new avenues for Holts. This was true especially of the expansive period under Bogaardt's leadership in the 1870s and 1880s: Holts' interest in the Sumatra and Borneo tobacco trades, the Bangkok rice trade, even the Singapore–Fremantle service, and most important of all, the Straits Steamship Company, all had their origins directly or indirectly with Mansfields. In 1903 the link between Liverpool and Singapore became closer, and formal; for Mansfields now became a limited liability company with nearly all the shares owned by Ocean. A few years later, in 1914, Ocean became the major shareholder in the Straits Steamship Co., and in 1922 Mansfields became general managers of Straits Steamship.

The origins of the Mansfield enterprise lay in a ship-chandler's business set up by Captain George Mansfield in 1857. This George was a brother of Walter and father of George John (born 1850), both later to play a prominent role in the firm's formative years. We know from Captain Mansfield's notes of his voyages that he first visited Singapore in 1851, but did not return there until 1857. On this latter occasion he came from Melbourne via Batavia in the *Enterprise*, his own ship, sold both vessel and cargo, and joined the well-established ship-chandler's business of G. J. Dare and Co.

Captain Mansfield was evidently a remarkable man. Hailing from a Devon seafaring family he had gone to sea at the age of 12 and by 1857, when still only 30, had travelled extensively in Europe and the Far East, and spent two years in the Melbourne pilot service. For his part in a heroic rescue of treasure and passengers who were wrecked off Melbourne he was awarded £5000, doubtless the capital with which he financed his shipping and financial interests.

The association with Dare and Co. in Singapore seems to have prospered, and George Mansfield became a full partner in the concern on 1 January 1858. At this stage his main business seems to have been with Siam which had been opened to European traders only in 1856. In 1859 George went to Bangkok to set up the firm of Orr, Dare and Co. (which was sold to Orr the following year). This was a period when George Mansfield was constantly on the move, sailing to Hong Kong, Shanghai, Amoy, Swatow, and numerous other Chinese ports, and rarely was he in any one place more than a week or two. From England, and from his wife and son, he was absent nine years. In October 1861 he was in Singapore again, and now set up the firm of George Mansfield and Co., 'general commission agents and auctioneers'. The ship-chandler's business was behind him, and his firm, with its small godown (warehouse) in Commercial Square – later Raffles Place – became one of many European agency houses founded in Singapore at this busy time. The foundation-stone of the future Mansfield and Co. had been laid.

Almost at once tragedy struck. Captain Mansfield at last journeyed to England in 1863 to see his wife and son (who did not know him), and to purchase a ship. While in Liverpool on the latter task he was suddenly taken ill and died at the age of only 37. An irony that a man who had survived the tropics and the sea so vigorously should fail to survive Liverpool. George's brother Walter now went to Singapore to attend to the business there and wind up the firm. But instead of selling up, Walter decided to remain in partnership with Mr R. J. Wright, and early in 1865 arranged for young George John to join him there as a clerk in the firm.

George John was only 13 when his father died. Some fragments of his youthful diary survive, and we know he was capable of feeling sorry for himself: 'My poor father taken away from me. No brother. No sister. No one except my mother to love me', he wrote a few months after the tragedy. But he was also capable of quiet determination and recorded in September 1864 'my uncle John has explained to me that I must take care of my money as I shall not be left so well off as he expected'. We can possibly read into these remarks that Walter had not found the Singapore business in a flourishing state, certainly not sufficient to provide adequately both for widow and son.

Young George arrived in Singapore in March 1866, after a journey of 110 days (a relatively quick passage) in the 55 ton barque *Belle of Southesk*; he was 15 years of age. The diaries tell us little about the day-to-day commercial business of Mansfields, though it appears that trade was not particularly flourishing and there are several references to the lack of activity. But the diaries also capture something of the novelty and excitement of the arrival of the first Blue Funnel steamers. Less than four months after settling in Singapore, George's entry for 20 June 1866, noted 'the first of [a] new line of Steamers from Liverpool round by Cape calling at Mauritius, Singapore, Hong Kong and Shanghai and back again. Also calls at Penang on way out. She is a new boat called the *"Agammenon"* [*sic*], Capt. Middleton, of about 2000 tons.' George noted with admiration that the *Agamemnon* had made the journey from Liverpool in only 61 days, including a stop in Mauritius. 'The Manager of the Company is Alfred Holt who was formerly agent for the West India Navigation Coy. Fortunately both uncle and Mr Wright knew the Capt. so we had him up with us.' This connection with Captain Middleton was to prove significant both for Mansfields and Holts.

Many years later, in a letter written in 1923, George recalled this first call of the Holt vessel. He wrote:

> my recollections take me back to 1867 [he meant 1866, of course] when I saw the old *Agamemnon* just arrived in Singapore, lying in the Harbour. No one had any idea in those days of the probable growth of the Blue Funnel Line, except my uncle, Walter Mansfield, who as an old Liverpudlian recognised who A. Holt was! A great friend (in fact a partner of my uncle's) thought otherwise and told me, a youngster, "Have a look at that ship for she won't be here again – she can't pay or stand up against P&O – and about 18 months afterwards he was dead and my uncle was agent."

The first arrivals of the other Holt steamers, the *Ajax* under Captain Kidd, and the *Achilles* under Captain Turner Russell, were also faithfully recorded in George's diary. On 8 November 1866, 'Ajax came in considerably surprised me' (presumably because of its speed). Then, on Monday, 25 November, the *Achilles* arrived in the fastest time of the trio, '57 days from home – splendid passage – 53 days fair steaming and sailing', with some news for Mansfields: 'Middleton had arrived home and his owners had given orders to all vessels coming here, that they should give their orders to us.' In consequence Captain Turner Russell gave orders 'to his stewards and officers to take everything from us and get enough to last to China and back'.

In effect, the firm of Mansfields had secured the Blue Funnel agency, had there been any firm to undertake it. For only two weeks or so before Captain Turner Russell arrived with his welcome news, the Singapore company had gone bankrupt, and already uncle, nephew, and partners were involved in the melancholy tasks of chasing debtors, settling with creditors, and disposing of stock. The root cause of the failure may well have lain in the great commercial crisis of 1866 which set in with the collapse of the London banking firm of Overend Gurney in May of that year. The crisis brought failure to many concerns, in Europe and throughout the world. The great China house of Dent and Co. was a victim; so was a branch of the Scott shipbuilding company in France. And George was given the task of making about twenty copies of the following note to be sent to the various creditors, dated 13 November 1866: 'Dear Sirs, Owing to the severe depression of our business we are reluctantly compelled to inform you we are unable to fulfil our engagements – and have to ask your kind attendance at our Godowns tomorrow at 3 o'clock when a statement of affairs will be laid before you with a view of making arrangements. We are, yours obediently, Mansfield and Co.' The letter was signed 'R. J. W.' (R. J. Wright) whose death was referred to in George Mansfield's 1923 letter. Despite frantic attempts to salvage the business (and with considerable friction between Wright and Walter Mansfield) there was no alternative to closure, and on 17 November formalities were concluded 'and Mansfield and Coy was no more', as George recorded. Walter then, sometime in 1867, returned to England, and George was able to find work as a bookkeeper with the firm publishing the famous newspaper *The Straits Times*.

About Walter Mansfield's activities in England in 1867 we know very little. Did he meet Alfred Holt and Captain Middleton in Liverpool? Did he travel to England explicitly to seek the Blue Funnel agency? Given the previous good relations between Mansfield and Holts, and the arrangements made just on the eve of Mansfields' bankruptcy, it seems highly likely. It seems also probable that Middleton himself was responsible for furthering Mansfield's cause, being a personal friend and business contact as we have seen. This at any rate, has been a traditional tale within Mansfields.

By the beginning of 1868, in any event, arrangements had been made. Back in Singapore, George (now 17 years of age) wrote:

I then received a letter from uncle conveying the good news that he had been offered the appointment of Holt's agency for their steamers in this place. He offers to take me with him if I like. He will be out in a very short time if all goes well. It gave me a great deal of pleasure but excited me *much*.

And a few days after George received his exciting news Alfred Holt recorded in his diary the first departure of his new steamer the *Diomed*:

By her went to act as our Agent at Singapore Mr Walter Mansfield under a two-year engagement. We have found it necessary to send Mr Mansfield out, our previous agents Messrs. Symes & Co. conducting our affairs in a manner very prejudicial to the success of the line.

On 29 May 1868, the *Diomed* arrived in Singapore bearing Walter Mansfield and his wife, and with the Holt agency in his pocket. His new responsibilities began formally on the departure of the *Diomed*, and from then on Walter Mansfield and Co., as the new firm was styled, took on full responsibility in Singapore for the Holt vessels.

Naturally the success of the Mansfield enterprise was largely bound up with the success of Holts' steamships, while Holts in turn depended upon their agent's diligence and integrity. Establishing an agency cannot have been an easy task, but Alfred Holt evidently made a very good decision to entrust his vital Singapore agency to Walter Mansfield. Here is yet another example of Alfred Holt's enormous perception of individuals; time and time again he made an unorthodox decision based on his own estimation of an individual's worth, and then backed his decision to the full. Rarely was his trust misplaced.

We do not, unfortunately, know a great deal about Walter Mansfield. From such glimpses as we have through the pages of George Mansfield's diary and a few other sources, we can visualise a large robust man, governed by strong religious principles, with a wide range of personal and commercial contacts, a hospitable nature, a capacity for hard and conscientious work, but with a fiery temper and sharp tongue. There were many periods of coolness between uncle and nephew, George recording after a few months of the new business, 'if ever I get the chance of getting a good situation I should jump at it. So much for working with relatives.' Fortunately for both Holts and Mansfields this particular quarrel was soon forgotten.

Thanks to Walter's hard work the Mansfield firm quickly established itself. No sooner had the *Diomed* left Singapore than a godown was rented (possibly, it is thought, in the Fleet-Street–Battery-Road area), negotiations undertaken to get adequate wharfage at Tanjong Pagar dock, and commercial business sought. George was initially pessimistic, writing on 5 June 1868, 'I hope we shall get on all right with this agency business but prospects at moment look very gloomy to me.' But despondency was short-lived, and on 20 June letters were received from Captain Middleton of *Agamemnon* saying that he was loading teas at Hangkow for £8 8s and from Captain Turner Russell who was about to load at Foochow for £6 6s a ton. 'This was indeed splendid news for us', wrote George. 'Between the two therefore there should be about £60 000 freight on the voyage home, a splendid freight all in all.' Three days later George proudly noted booking his first passenger for Holts: 'In the office today I booked Mr Cateaux for Mauritius, $230.77 which was at once *paid*.' And on 30 June came the *Achilles*, the first Holt ship fully entrusted to the new agency.

Mansfield's work in these early months was both varied and spasmodic. Exact arrival times of the various Holt steamers were unknown of course, and the faster mail-ships would bring letters with estimated dates. Then Walter and his nephew would anxiously watch the flagstaff on Government Hill, waiting for the Holt vessels to be signalled, fearing delay or disaster, and satisfied only when the ship came finally into view. With a ship's arrival imminent, activity would mount. Freight would be canvassed, coals purchased, passengers booked, and various arrangements made for the vessel's arrival and clearance. Then, once the ship had docked, would come a few days quite feverish activity: cargo loaded and unloaded, coal deliveries made, passengers received, captain and officers taken care of, light duties and pilotage paid, and so on. Then any repairs and stores for the onward journey had to be attended to while there were usually seamen to be taken on and discharged (mostly Asian crews). All this involved a great deal of bookwork, of course, and manifests had to be prepared of the amounts of freight loaded and the rates charged.

When the *Achilles* arrived as Mansfield's first Holt ship George noted 'we of course hurried down as soon as possible to the docks and had everything in readiness for her'. This vessel brought passengers from China and took on some 630 tons of coal, four passengers, but no cargo. Later vessels took on tin, gutta-percha, and a wide variety of miscellaneous cargo, while there were always migrants arriving with the boats from China. On 15 November 1868, for example, the *Nestor* brought 'a lot of Chinese passengers among whom were a lot of women. There were also 80 Naval Invalids for Europe. Nice Mixture.' Not all Mansfield business was done on Holt's account, of course, and there are several references in George's diary to such independent business. But the Holts business remained overwhelmingly the foundation of Mansfield's success, and Mansfields in turn proved themselves worthy agents for the Blue Funnel steamers.

Walter Mansfield successfully guided the firm for five years, a period which, encompassing the opening of the Suez Canal in 1869, saw a quite remarkable boom in Singapore's fortunes. Oddly, many Singapore merchants had feared the Canal's opening, believing that the entrepôt trade would be ruined as new direct lines could be established. But such pessimism was short-lived, and the volume of Singapore's trade more than doubled in the 1870s. The impact of the Canal on Singapore's fortunes and on the development of steam shipping in South-east Asia was, indeed, dramatic. In 1869 sailing tonnage at Singapore still exceeded steam. The figures were 347 596 tons and 264 790 tons respectively. But by only 1872 748 322 tons of steam shipping were cleared at the port, while sailing tonnage was stagnant, with only 313 018. Four years later, in 1876, sailing tonnage had halved to 163 385, while steam tonnage had grown to 1 291 304. The Canal was not the only new expansionary force, for in 1870 came the extension of the European telegraph to Singapore via India. By the close of 1871 cables had been laid connecting Vladivostock, Nagasaki, Shanghai, and Hong Kong with Singapore, and from Singapore telegraphic communication went onwards round the world, via London to New York and San Francisco. Singapore and China became ever more closely drawn into the world economy. There was a striking development, too, in Singapore's trade with the Malayan peninsula, adding a new dimension to the traditional entrepôt trade of the archipelago. From 1874 British colonial rule was extended to Perak, Selangor, and other Malayan states, and Malayan tin and other local products (coffee in the 1880s and rubber after 1905) became of rapidly growing significance in Singapore's trade.

George Mansfield was taken into partnership by his uncle in 1872, the same year that an enterprising Dutchman, Theodore Cornelius Bogaardt, was brought into the firm. In 1873, however, the company received a great blow with the death of Walter Mansfield while on a visit to England. W. Mansfield and Co., as the firm remained, was now headed by George and by Bogaardt, and between them they laid the basis for a new period of expansion. For a decade George, as senior partner, proved himself as able and conscientious as his uncle, though Bogaardt was the more dynamic, and in these years Mansfield's reputation as one of Singapore's leading agency businesses was consolidated. Mansfields prospered, of course, *pari passu* with Holts. When the Suez Canal was opened the Blue Funnel line consisted of but 5 ships. In 1883, the year of George Mansfield's retirement, the fleet numbered 26.

George's departure heralded a period of twelve years in which the firm was guided by Bogaardt. It is worth quoting at length an extract from the *London and China Telegraph*, 28 August, 1883, on the final departure of George from Singapore, an event which marked the end of the Mansfield family connection with the firm:

Another of our old and esteemed residents, Mr G. J. Mansfield, for several years past the head of Messrs W. Mansfield and Co., agents of Holt's Line of steamers, took his departure from us by the steamer *Agamemnon* on the 15th July to enjoy a well earned rest at home. Mr Mansfield was for many years resident here before he rose to his present position, where his great business capacity has enabled him to keep pace with the largely increased work of his establishment, for whereas Holts' steamers formerly only arrived here once a fortnight between England and China and vice versa, there are now one or more weekly each way, besides three fine steamers engaged in the local trade, the successful establishment of which, as regular local lines, is due to Mr Mansfield's influence and able management. In all his business transactions, as well as in social life, Mr Mansfield held a high place in the esteem of everyone.

No one reading this extract could guess that George was only 33 years of age at the time.

Mansfield's new head was a powerful and forceful character who showed a flair for imaginative entrepreneurship both within the Mansfield organisation and through his own enterprises. In his own way Theodore Cornelius Bogaardt was as influential as John Swire in charting the direction in which the Ocean Steam Ship Company was to grow. At the beginning, Singapore was primarily a port of call for Holt ships on the way to and from China. But under Bogaardt's inspiration Singapore became a focus of Blue Funnel trade throughout the Southeast Asian region, with a series of trades radiating to the Dutch East Indies, to Malaya, Siam, and Borneo. These trades have been discussed in the previous chapter. Here we will look briefly at the formation of the Straits Steamship Company – Bogaardt's principal legacy – and at the parallel progress of Mansfields.

The Straits Steamship Company was incorporated in Singapore in 1890, basically a creation of Bogaardt in conjunction with the powerful Straits Chinese Tan family of Malacca. The initial capital was $500 000, and the subscribers included Bogaardt himself, Tan Jiak Kim, head of the important Kim Seng Company, Tan Keong Saik, and A. P. Adams, a Mansfield director. The Kim Seng Company provided three of Straits Steamship's first five ships, which at first participated chiefly in the fast-growing tin trade of the region. Straits Steamship's vessels carried tin ore from the mining regions of the western Malay states to the smelting works of the newly established (in 1887) Straits Trading Company in Singapore. And from Singapore there was a steady stream of Chinese coolie labourers, many of them arriving in Blue Funnel ships from the South China coast, destined for the mining districts and elsewhere in western Malaya and Siam.

The early history of the Straits Steamship Company is not well documented. It is known that in 1892 Mansfields took over the management, with Bogaardt then in control of both companies. The following year Ocean set up its own feeder line, the East Indian Ocean Steam Ship Company, with Mansfields also managing the line. But about the relations between the two lines and the numerous interconnected business dealings of Holts, Mansfields, and Bogaardt, little information has survived. In any event Bogaardt left Mansfields in 1895 and from that date the involvement of both Holts and Mansfields with the Straits Steamship Company was severed until the First World War.

In the two decades before the war, Mansfield's business appears to have grown considerably. Even with the loss of the Straits business in 1895, and that of Ocean's East India Company four years later (when the ships were sold to North German Lloyd), other activities were prospering. The main Blue Funnel business was growing all the time at this period, especially with the development of Malayan rubber plantations after 1905. Mansfields handled the Straits Moslem pilgrim traffic, the Blue Funnel West Australian service, the agency of the China Navigation Company, and a variety of other activities.

Early in the new century came a significant change in Mansfield's status. In 1903 the senior partner, A. P. Adams, retired, and at his instigation Ocean took the opportunity to acquire control of the partnership. A new company, W. Mansfield and Co., was incorporated in Liverpool in that year with an authorised capital of £10 000, the Ocean Steam Ship Company holding 135 out of the 200 shares issued. The former business of Mansfields in Singapore and Penang passed to the new company, and the directors were henceforth the Managers of the Ocean Steam Ship Company. However, the actual conduct of Mansfields' business was to be in the hands of managers appointed by the directors, and three managers were appointed for a term of seven years. Two of the managers, J. E. Romenij and J. G. Burkhuyzen, were Dutch appointees of Bogaardt. The third, Edward Anderson, had joined Mansfields in 1899 and, upon the retirement of the last of the Dutch managers in 1912, became the first Managing Director of Mansfields until his own departure in 1917. We can glimpse something of the sort of problems with which an agency had to cope (and the tartness of some of Richard Holt's letters) from a complaint to Anderson in 1910:

> Pilferage of cargo to Singapore still goes on in a methodical kind of way – a few dozen belts, singlets, or handkerchiefs being paid for by every steamer. We have gone to a good deal of personal pains at this end to do away with theft, and I think we have succeeded, for it seems to be only at Singapore that a regular weekly pilferage of the same class of goods continues to recur.

Meanwhile, external events were once more linking the fortunes of Holts and Mansfields with the Straits Steamship Company. As we saw earlier, the First World War meant the abrupt withdrawal of North German Lloyd from the feeder services they had inherited from Ocean's

9 New Ocean Buildings, Singapore, 1920s (*photograph courtesy of Liverpool Museum*)

East India Company in 1899, and Ocean took a majority interest in the Straits Steamship Company in return for the provision of new ships. In 1915 Edward Anderson of Mansfields joined the Straits Steamship board. Arrangements were formalised further after the war when in 1922 Mansfields were appointed as agents for Straits Steamship in succession to Adamson Gilfillan. In the same year Mansfields was registered in Singapore as a private company as Mansfield and Co., (without the 'W'), and H. E. Somerville, the outstanding chairman of Straits Steamship, joined the Mansfield board. In 1923 the staff of both Mansfields and Straits moved to Ocean Building, a fine new headquarters, and for the remainder of the decade Straits Steamship prospered through the expansion of its 'little white fleet' of 75-ton ships, built to comply with a law peculiar to Malayan waters.

The Straits Steamship fleet expanded both by new building and acquisition, the most significant purchase being that of the Eastern Shipping Company in 1922. This was a Penang-based company which had started life in the 1880s as the Koe Guan Shipping Company trading principally between Penang and the ports of western Siam and Malaya. From 1907, when a tin-smelting enterprise was developed in Penang, Koe Guan (which changed its name the following year) was able to expand considerably by carrying tin ore, tin, and Chinese coolies throughout the region. The Penang-based tin trade was an important acquisition for Straits Steamship in 1922, and the sphere of operations moved north of Penang as, in K. G. Tregonning's words, 'now at one bound its little ships were nosing round the corners, dodging between the islands of the Mergui Archipelago'.[4]

In addition to the Eastern shipping Company, Somerville also acquired a controlling interest in the Sarawak Steam Ship Company in 1931 and Straits Steamship became a majority shareholder in 1932 in the Ho Hong Steamship Company. Both brought new trades to Straits Steamship. Ho Hong, in which China Navigation also had an interest, was formed to take over an established line of three ocean-going ships providing a service between South China, Singapore, and Rangoon. Mansfields, incidentally, provided both a European manager and secretary for this essentially Chinese enterprise. The main significance of Ho Hong for Straits Steamship was that it brought the Burmese port of Rangoon within the company's sphere of operations. Rangoon had been barred to Straits under the 1925 Victoria Point Agreement signed between the British lines P&O, British India, Alfred Holt and Co., Straits Steamship, and China Navigation. The Agreement identified spheres of operation, Straits Steamship having an area which went southwards from the Burmese border to Bangkok, eastern Sumatra, Borneo and the South Philippines. Straits Steamship, rather surprisingly, could also trade in the China Navigation sphere, but Straits and China Navigation concluded a gentlemen's agreement not to invade each others' waters. At the same time, competition with Holt ships was removed by Straits' purchase in 1925 of the *Ayuthia, Circe,* and *Medusa,* the small Singapore–east Sumatra ships which were the remains of Holts' pre-war coastal operations. Table 4.2 summarises the growth of Straits Steamship before 1932.

Mansfields, as managers of Straits Steamship, played a full part in this progress. Mansfields negotiated the Eastern Shipping Company deal in Penang in 1922, and relationships between Holts, Mansfields and Straits remained close. In Liverpool, John Hobhouse maintained an active interest in Mansfields, and Blue Funnel Managers were frequent visitors to Singapore. Yet we should emphasise that Holts, having reorganised both agency and fleet management in 1922, and having strengthened Mansfield management with its own men, allowed both Singapore

Table 4.2   Straits Steamship, fleet and
tonnage, 1914–32

Year	Ships	Tonnage
1914	17	13745
1922	24	25446
1932	52	40591

Source: K. G. Tregonning, *Home Port Singapore: A History of Straits Steamship Company Limited, 1890–1965* (Singapore: Oxford University Press, 1967) pp. 43, 52, 70.

companies to flourish under their own direction. In Tregonning's words:

> although Mansfields was controlled by Holts, Liverpool was well aware that a maximum degree of independent operation was necessary for Mansfields to be efficient, and Holts therefore interfered very little. The same applied to Straits Steamship Company, whose agents, when accepting cargo, shipped it as directed, or on the first available ship. No pressure of any kind was applied for them to ship Blue Funnel.[5]

5

Eastward Ho! The Beginnings of Ocean Steam Ship

The end was that the 'Holt' line soon became firmly established, and, in spite of occasional bad time, has greatly enriched everyone concerned in it.

(Japan Herald, 1880)

In previous chapters we have seen how the Holt enterprise expanded after 1865 as new trades were steadily introduced, and we have seen, too, how critically important was the relationship built up between Blue Funnel and the various overseas agents. Beginning as a small-scale venture trading between Liverpool and China, the service widened at both ends, and by 1918 a world-wide network had come into being, served by a large modern fleet built to exacting standards.

In this chapter we will return to the beginnings of the Blue Funnel Line in the 1860s, examine the background to this unique venture, and, in particular, look at the circumstances which led to the development of the Far Eastern trade. A number of interesting questions suggest themselves. Why did Alfred Holt launch his China enterprise just when he did, and what accounts for its great success? Can we attribute success to Alfred Holt's genius, to the enterprise of his agents, or to good fortune? How was Holt able to stave off rivals in an increasingly competitive world? As we look at these questions three major points will emerge. First and foremost, Alfred Holt was very much a child of his time. The launch and success of the Ocean Steam Ship company in 1865 moved with strong currents already flowing. These included the extension of steamship navigation to increasingly long-distance trades as improvements in the marine engine and ship construction made this possible; the great increase in Britain's overseas trade with distant parts of the world as Britain's own wealth and commercial policies encouraged trade; and the spread of Britain's influence and the British empire which brought more and more territories within the orbit of world commerce. By the 1850s and 1860s political circumstances were fast opening the great Chinese empire, Japan, and other countries in the east. Economic circumstances, too, with the rise of Western industrialism, the spread of railways, steamships, electric telegraph and submarine cable, and the extension of free trade, encouraged the expansion of international trade.

Second, Alfred Holt was certainly a pioneer and his place in commercial history as founder of the first steamship line trading directly between England and China is forever assured. In turn this achievement depended on the design of efficient ships in which Alfred Holt himself played the leading part. But such achievements need to be placed in perspective. Holt was not the first to send cargo steamers directly between England and China (although it is sometimes stated that he was); already in 1863 such a ship had made a voyage from Hankow with a cargo of tea and others followed. Moreover Holt's first three ships were examples of perfecting and bringing together then-known technology rather than representing a major leap in the dark. Iron ships and compound engines had both proved their success by the 1860s. Holt brilliantly combined the two in a safe and commercially viable vessel, but he by no means possessed the visionary engineering genius of, say, a Brunel. But then neither were Brunel's achievements notable for their commercial success. The very speed with which Ocean's competitors made their mark from the early 1870s indicates clearly that Holt may well have been in advance of his time, but only just. And therein lay the secret of his success.

Third, it will become clear that the Ocean Steam Ship venture was by no means a suddenly conceived enterprise which then advanced steadily in size and prosperity. On the contrary, success was achieved in a series of phases, and not without struggle, nor, for a time, without the prospect of failure. Prior to 1914, by which time the foundations of a great national business had been well and truly laid, two critical phases had passed. The first, the opening of the Suez Canal in 1869, ushered in a period of great opportunity and prosperity for the Blue Funnel steamers, for suddenly Europe was brought closer to Asia and the economics of long-distance steamshipping transformed. We may suspect that a line of small steamers chugging round the Cape of Good Hope to Mauritius and on to China, not obviously quicker or more competitive than the best sailing ships, would have had an unspectacular future. Certainly improvements in marine engines made the ultimate demise of sail inevitable, but sailing ships, too, could benefit from iron hulls, the telegraph, and other developments. What the Suez Canal did was to compress into a few short years what otherwise might have taken a couple of decades, and with Holt already in the field the Ocean Company was to benefit accordingly from the shorter route.

The second critical phase occurred in the late 1880s and early 1890s. By this time the basis on which the Blue Funnel ships had built their early success was passing rapidly. The nature of trade was changing and, in the new commercial environment, Holt's vessels were too small and too slow. The venture might, at this stage, have come to a halt. Instead, as we will see, Holts took the courageous decision to rebuild and expand their fleet to meet changing circumstances. This was the point of no return. From the early 1890s came more than two decades of rapid growth, and of the penetration of the company into new regions and new markets.

Let us look at some of these points in more detail. Until well into the nineteenth century China, the mainstay of Holts' success for so long, had remained largely outside the scope of western trade. From 1757 until 1840 China's rulers forced Western merchants to trade only at one Chinese port, Canton. Canton, in southern China, was ill-situated to tap the populous areas of the centre and north, though it was conveniently situated far from Peking so that the Manchu rulers could have as little as possible to do with the Western barbarians. In Canton, Western merchants were obliged to deal only with officially licensed Chinese merchants (the monopolistic Cohong). As far as England was concerned China had only one product of great commercial significance to offer, tea; and tea was until 1833 monopolised by the venerable East India Company whose majestic East Indiamen made their stately voyages to and from the Thames over a period which was normally well over one year for a return voyage, and could well be eighteen months. As well as tea, other Chinese luxuries such as silks and porcelain were also in demand. At this stage China, despite her vast population, required few goods that England could produce, and before about 1800 financial settlements had usually to be made by the export of silver from England.

By 1860 the position in China was transformed. Large areas of China were now opened up to Western commerce and China was increasingly becoming a market for textiles and other products of Britain's industrial revolution. This process of Western penetration continued in the succeeding decades, but by 1860 the main steps in the process had been taken whereby China, once isolated, had become an integral part of the international economy.

The transformation involved several phases, of which the most significant was the Opium War which was fought between Britain and China between 1839 and 1841. The rather complex origins of the Opium War need not concern us. Sufficient to say that whatever the immediate causes of the conflict, an irresistible force was in any event building up, making China's traditional exclusiveness and ways of conducting trade with the West impossible to sustain. Britain took the lead for Britain was the pre-eminent commercial nation, and because it so happened that opium exports from British India to China had become a vital link to settle payments for Chinese tea exports to England. But underlying the visible conflict was a more general confrontation between rising Western industrial power on the one hand, and the declining strength and vigour of the Manchu Empire on the other.

The results of the Opium War were momentous. As confirmed in the 1842 Treaty of Nanking, four ports in addition to Canton were opened up to the British (and later to other Western merchants): Foochow, Amoy, Ningpo, and Shanghai. British consuls were placed in the five ports, with jurisdiction over their own subjects. The Cohong system of trade was abolished. Furthermore, strict limits were placed on China's ability to levy import and export duties, and Western merchants were permitted to reside and trade freely in the various treaty ports. In addition, Britain acquired Hong Kong, a tiny, virtually uninhabited island to which the British community in Canton had withdrawn at the outbreak of the troubles in 1839.

The partial opening of China as a result of the Opium War was only an initial step. Indeed, for a time the high hopes of British merchants remained unfulfilled, for none of the new treaty ports developed as anticipated. Most successful, unexpectedly, was the small port of Shanghai, most northerly of the treaty ports, and well situated to serve the tea-producing regions of central China. Shanghai soon attracted British and American merchant houses which dealt in tea and opium, Jardine Matheson opening a branch there in 1844. Already by 1850 Shanghai had become a thriving and bustling centre of overseas trade. But the remaining treaty ports were inadequate, too few in number, far from the main areas of population and production, and suffering from ambiguities in the treaty arrangements which left unclear the rights of Westerners to participate in the coastal trade. Above all, no Western trade was yet possible along the great Yangtze river, which gave direct access to the populous regions of central China and the main tea-producing regions.

A new, and for the future Holt enterprise, far more significant phase took place in the 1850s. This decade was one of great political turmoil in China, for following the outbreak of the Taiping Rebellion in 1851, civil war swept the country causing great slaughter and destruction. Central control by the Chinese rulers waned still further and the treaty ports, protected by their foreign communities and by their extra-territorial privileges, became havens of peace and security. Events in Shanghai in 1854 led to the creation of what was virtually a foreign merchant enclave there, with its own laws, taxation and administration. The customs too, passed under international control in 1854. In 1863 the British and American Settlements joined to form the International Settlement, with more than 2000 foreign residents. In the words of one writer, 'the future centre of China's modernisation at the entrance to the Yangtze Valley became a semi-foreign city run by the local foreign land-renters under the protection of their treaty rights of extraterritoriality'.[1] In the disturbed conditions, many hundreds of thousands of Chinese flocked to Shanghai and Hong Kong, and these migrations added greatly to the growth and enterprise of these ports which developed rapidly after 1853. Disturbances in the southern provinces diverted tea and silk from Canton to Shanghai and Foochow, a pattern which became permanent, and was further cemented when Shanghai and Hong Kong were linked in 1855 by regular steamship services pioneered by the well-known American firm of Russell and Co.

Another direct consequence of the Taiping rebellion was the mass emigration of Chinese beyond China's borders, to the goldfields of Australia and California, to the tin-mines of Perak and Selangor, to Bangkok, Singapore, Batavia, and other cities in South-east Asia. These Chinese migrants produced overseas communities which endeavoured to maintain their traditional life styles and consumption patterns, and so established important trading links with the Chinese mainland and elsewhere. For example rice exports from Siam and Indochina fed Chinese populations in Singapore, Malaya, Hong Kong, and elsewhere. Such new patterns of migration and trade were not insignificant elements in the commercial world which beckoned Alfred Holt and other carriers in the 1860s.

The Taiping rebellion also provided an opportunity for Britain and other Western powers to wrest further major concessions from the Chinese rulers. Such concessions had been angrily demanded by merchants in Shanghai and Hong Kong since the 1840s, and their achievement was of direct significance both for Alfred Holt and for John Swire.

Under the treaties of Tientsin and Peking signed in 1858 and 1860 respectively the whole of the Chinese empire was effectively opened up to the West, and the 1860s therefore saw a new era of opportunity for Western merchants. In addition to the existing five treaty ports a further twelve were opened to foreign trade, including Tientsin, Swatow, Hankow, Chingkiang, and two in Taiwan. Thus Westerners could now trade directly with the great tea centre of Hankow on the Yangtze river, and also carry on trade and navigation to the north in Gulf of Pechili. Of particular significance for the future of Swire, and so of Ocean, was the freedom given to foreigners to participate in coastal shipping on the Yangtze as far as Hankow. Western steamers could now compete with the multitude of small Chinese sailing vessels. The treaties also strengthened Britain's position in Hong Kong, adding to the colony the small mainland area of Kowloon (the much larger New Territories were added in 1898).

Surprisingly, in view of the political turmoil existing in China – the Taiping Rebellion was not finally put down until 1864 – China's foreign trade grew considerably in these years and Western merchants enjoyed

Table 5.1   British exports to China and
Hong Kong, 1854–70
(current values, £million)

1854	1.0
1860	5.4
1870	10.0

Source: British Parliamentary Papers, *Statistical Abstracts for the United Kingdom.*

a period of great prosperity. The basis of prosperity was the continued growth of opium imports from India, and the development of China as a market for British textiles and other industrial products. Table 5.1 shows the very rapid expansion of British exports to China after 1854 and hence indicates the sort of opportunities which must have attracted Alfred Holt. Incidentally it is worth mentioning that neither Alfred Holt nor John Swire were involved or interested in any way in the opium trade, believing that such trade was evil. Such could not be said of Jardine Matheson, Dents, Russells, P&O, or some other great China names of the period.

Western penetration of China was part of a wider British-led drive into Asia during the nineteenth century. The process involved the growth of European colonies, the opening up to trade of hitherto isolated countries like Japan and Siam through 'unequal treaties' similar to those imposed on China, the establishment of regular steamship communications, and the spread and enforcement of free trade under the watchful eyes of the British fleet.

Colonisation, certainly, was a potent force in the development of Europe's trade with Asia. In 1786 Britain, already firmly established in India through the East India Company, made its first acquisition in South-east Asia, the small island of Penang. Expanding Indian interests proved the cutting edge for further penetration after 1800. In 1819 Sir Stamford Raffles founded Singapore, a small almost uninhabited island, as a British colony; Malacca was obtained in 1824 by agreement with the Dutch, and certain Burmese provinces were acquired as a result of the First Burma War, 1824–6. A second war with Burma between 1852 and 1854 added the important region of Lower Burma, and the way was open for Rangoon to become a major international port. The remainder of Burma fell into British hands in the mid-1880s, by which time Britain had also taken control of the main tin-producing regions of Malaya, after 1874, and were establishing colonies in Borneo. Nor were other European powers inactive. Both the French and the Dutch had become major Asian powers by the last quarter of the nineteenth century, the Dutch consolidating their hold over the islands of the Dutch East Indies

(Indonesia), and the French establishing their empire in Indochina, acquiring Vietnam, Laos, and Cambodia between 1858 and 1893, and developing Saigon as an important trading port in the 1860s.

Some areas, as in the case of China, were obliged to open their doors to foreign traders by force or the threat of force. It was the Americans, rather than the British, who first broke down Japan's two-hundred year period of isolation from foreign contacts in 1853. An American naval force under Commodore Perry obliged the reluctant Japanese to concede 'unequal treaties' which opened the country to trade and virtually abolished Japan's right to impose import or export duties. Britain and other Western countries almost immediately signed similar treaties with Japan. Siam (Thailand) too was opened to Western trade at around the same time in much the same way though without an overt display of Western force, and a treaty with Britain was signed in 1855, coming into operation the following year. Bangkok soon developed as a major rice exporting port and began to attract many tens of thousands of Chinese migrants.

I

While there were many causes of Britishers' increasing attention to trade with Asia in the first half of the nineteenth century, two factors above all stand out for their relevance to Alfred Holt's shipping venture: the abolition of the East India Company's monopoly of the Far Eastern trade, and the rise of Singapore. In 1813 the Company's monopoly of trade with India was ended, partly as a result of pressure from Liverpool merchants and other free traders. In India, the East India Company became a political institution only. As a result, encouraged by the freedom of trade and the great commercial possibilities of India as a market for Lancashire's textiles, a great many agency houses established themselves in India in the succeeding years, purchasing goods on consignment, selling on commission, handling freight, chartering shipping, dealing with insurance, and so on. The East India Company still clung to its monopoly of the China trade, but by this time the Company's business there was virtually confined to the tea trade. Silks from China, and Indian exports of opium and cotton to China were handled by 'country traders', licensed by the Company, and dealing usually with British and American firms established in Canton.

As trade between China and India expanded, Singapore, too, became a significant transhipment centre from the mid-1820s, and began to attract agency houses, with links in London or Liverpool, and sometimes as branches of houses in India or Canton. Singapore was aided partly by its strategic location between the mainland of South-east Asia to the north and the archipelago of the East Indies to the south; but it benefited also as a transhipment centre of goods from China to Europe, for by such transhipments merchants were able to break the legal monopoly of the China trade held by the East India Company.

In 1833 the China monopoly of the East India Company was also abolished, as Britain's adherence to free trade became stronger, and as, in any case, the old system was in decline. A new era seemed promised: British houses, many with Liverpool connections, sprang up in India, Singapore and Canton, though, as we have discussed already, the high expectations were largely unfulfilled and there soon followed agitation for opening up more areas of China. George Holt, Alfred's father, was not immune from the excitement. He named his new office building opened in 1834, 'India Buildings' in honour of the abolition of the East India Company's monopoly. He might more properly have called it 'China Buildings' for it was in China where the monopoly was ended and where the future prosperity of Alfred's and Philip's business career was to lie.

Thus, by the 1850s and 1860s, the Asian world was fast changing, a world of which Liverpool merchants were fully aware. As yet steam-ships had made little impact on these developments. All long-distance trade between Asia and Europe went by sailing ships, while within Asia both oceanic and river traffic was overwhelmingly carried in such craft. Tea cargoes from China were carried in fast clippers, the Americans especially proving adept in designing these beautiful ships which could carry tea from China to the Thames in as little as 90 days. But steamships were increasingly making their appearance in the east. A few paddle steamers had been introduced in the 1820s and 1830s to undertake coastal and river work in India, the Dutch East Indies, and elsewhere. In Java, Maclaine Watson, later to be Holt agents, was the first to introduce a steam vessel as early as 1827. In 1840 a new era opened with the founding of the Peninsular and Oriental Steam Navigation Company, which was to run a subsidised mail and passenger service (and some high-value freight) in two sections: from England to Alexandria via the Mediterranean, and, after an overland journey to Suez, by steamer from there to India. The inaugural service between Calcutta and Suez took place in 1842, and three years later the P&O introduced a service to China running from Ceylon, Penang, Singapore, and on to Hong Kong. In 1848 the company started a feeder service between Hong Kong, Macao and Canton, and in 1853 commenced a new steamer service between Singapore and Australia. Nor was P&O alone. Prior to 1865 many steamers were plying their trades throughout the east, some belonging to large lines like the P&O, the British India Steam Navigation Company, or the French Messageries Maritimes, others the individual ventures of independent entrepreneurs.

Some unsuccessful attempts were made even before 1860 to trade directly by steam between England and the east. Most spectacular was Brunel's monster *Great Eastern*, designed to voyage between London and Ceylon (and link there with steamers to Australia), but although finally launched in 1858, proved never remotely capable of making an economical return on the route planned. Then in 1852 two lines of steamers were projected to run via the Cape of Good Hope respectively to India and Australia. Neither project came to fruition. But not until Alfred Holt started his three-vessel Blue Funnel Line in 1866 was the commercial and technical viability of sending steamers on these lengthy voyages proved beyond doubt.

We have mentioned the word 'technical', and it is appropriate to say a few words about the development of the marine engine, for this was to prove decisive in Alfred Holt's success. By the 1850s the commercial viability of iron screw-propelled steamers was firmly established for certain types of voyages. The great advantages of steam over sail were speed and reliability. The great disadvantages were cost, especially at high speeds, where fuel costs rose dramatically, and lack of cargo space, for where lengthy voyages had to be made stocks of coal might take up most of the space which might otherwise have carried cargo. Steam therefore achieved its best results on short voyages, on rivers, around coasts, or over short sea journeys, and when valuable space-saving cargo was carried such as passengers, subsidised mail, bullion, and high value freight.

The story of the advance of steam shipping is really the story of how better marine engines allowed such ships to voyage further and take lower value freight, while still returning a profit to the owners. Regular steam services from England were operating across the channel in the 1830s, conquered the Atlantic in the 1840s (Samuel Cunard commenced his line between Liverpool and Boston in 1840), and had spread to the West Indies and South America in the 1850s. In this latter trade, Alfred Holt and the shipping firm Lamport and Holt were deeply involved (the Holt here was George – Alfred and Philip's elder brother. Alfred was involved in various aspects of the firm's operations, while Philip joined the company as a boy, remaining until he left to found Ocean Steam Ship with Alfred). China and Australia remained the province of sail, but not for long.

The key development was the compound engine invented by John Elder in the early 1850s. In 1856 the Pacific Steam Navigation Company launched two ships with such engines developed by Elder. The principle of the engines, later adapted and improved by Alfred Holt, was to prove the answer to successful long-distance steam journeys directly between Europe and the Orient. The compound engine greatly increased the power generated by a given quantity of steam, and hence of coal, by passing steam through two cylinders instead of one. The steam was first admitted to a small cylinder where it attained only a portion of its expansion, and thence to a larger cylinder where expansion was completed.

II

Alfred Holt's promotion of the first cargo-liner service directly between Europe and the Far East in the 1860s grew from his interest in marine engines. It is time now to turn from the general background of events which were unfolding at this period and examine how Holt himself came to establish his company. Fortunately there is considerable material available on Alfred's early life and the circumstances which led him to the shipping business. We have, for example, some rather sketchy autobiographical notes written by Alfred himself, mostly dating from 1879 when he was 50. There are also scraps of a diary, various technical and professional papers, as well as assessments by others. We also know a fair amount about Alfred's rather formidable father, George, a prominent Liverpudlian and pillar of the Unitarian establishment in his day. The obverse side to these details is that we know far less of others who figured prominently in the early days of the Holt enterprise, and the danger is that we may exaggerate the role of Alfred himself, especially once the firm was established. We would certainly like to know more about Alfred's brother Philip, for example, who was joint founder of the Ocean Steam Ship Company, and remained a Manager until 1897. Many tributes were paid to Philip's wisdom and guidance, and Alfred certainly laid great store on his brother's advice, but about his background and detailed contributions we know little. We know even less about Albert Crompton, the first Manager to be added to the firm. Nevertheless, it is wholly appropriate that we should concentrate on Alfred Holt. His was the genius which first made the decisive step of trading directly with the east, he set up the original company and made all the necessary arrangements, and he left such an indelible personal stamp on the firm that his influence could still be seen a century later.

Alfred Holt first became interested in marine engineering shortly after finishing his apprenticeship as a railway engineer in 1851. There seems never to have been any question that Alfred would make his career in engineering, and from his boyhood days he was fascinated by the steam engine. Indeed, it might almost be said that he grew up with steam locomotion, for he was born in Liverpool in 1829, the same year that the famous Rainhill trials proved decisively the superiority of steam locomotives which were then adopted for the Liverpool and Manchester Railway the following year – the true beginning of the railway age.

Alfred seems to have had a sound, if rather general schooling, the years from 9 to 15 being spent in a school run by the Unitarian minister, Mr Green, at Heathfield, Knutsford. Even at school Alfred learned to make a model steam engine, and George Holt probably followed his son's wishes when he apprenticed Alfred to Edward Woods, engineer of the Liverpool and Manchester Railway. Alfred proved himself hard-working, able, and enthusiastic; he wrote:

> My apprenticeship was a happy, active, industrious one. Mr Woods was more of the scientific tutor than of the planning mechanical engineer. He would, and did, teach me anything I wanted to know, and I devoured everything I saw with immense avidity. I never could have enough of the Railway; early or late, at night time or Sunday, I was there if there was anything to be seen or done.

Outside his apprenticeship Alfred records how he had his private lessons in mathematics and literature, and also studied chemistry and freehand drawing ('I attained considerable skill as a mechanical drafts-man'). But his main delight was a small workshop with two lathes which he equipped with his own savings, and in which he made five steam engines and other machines.

Alfred was obviously cut out by predilection and training to be a noted railway engineer. Indeed, when he finished his apprenticeship in 1851 he records that 'I hardly knew one end of the ship from the other.' But fate would have otherwise. A short depression in the railway industry induced Alfred to join as a clerk in Lamport and Holt. Lamport and Holt were just beginning to develop cargo steamers, a venture still in its infancy, and Alfred was soon able to watch and assist in the installation of engines to the second such vessel, the *Orontes*, built by Alexander Denny of Dumbarton. He went on the vessel's first voyage to Egypt in 1851, as a voluntary (unpaid) engineer, 'and in this manner I got my first taste of marine engineering'.

Young Holt was not long in his brother's office, for the following year he went to India Buildings as consultant civil engineer, renting an office from his father for £40 a year (a family concession of £5 on the going rate!). This office Alfred was to occupy for the rest of his life. But Alfred's future was to lie not in civil but in marine engineering, for a certain Thomas Ainsworth asked Alfred's help with a small steamer, the *Alpha*, which he hoped to run between Whitehaven and Cardiff, but whose engines refused to go. Shortly thereafter, probably in 1852, Ainsworth appointed a Captain Middleton as master of the vessel, and so began an acquaintance with Alfred which was to have significant consequences for Blue Funnel's operations in South-east Asia.

Let us digress for a moment to consider Alfred's bond with Middleton, a personal link and attachment which was also evident in Alfred's relationships with other of his employees, both on ships and land. He wrote of Middleton that until his death on board the *Stentor* in 1878, 'he remained closely and constantly attached to me, and was pioneer Captain in all my undertakings. His opinions, advice, and assistance have had a very great influence over all my shipping life.' No hint here of master and servant, but rather of friends with mutual respect and trust. And this close personal bond between 'head office' and the ships' Masters became one of the hallmarks and great traditions of the Blue Funnel line.

Alfred's career edged further forwards. Pleased with the success of the *Alpha*, Thomas Ainsworth and George Holt jointly purchased a small steamer the *Dumbarton Youth*, also built by Denny. On board was a quantity of blue paint used by the new owners to paint the long funnel, a modest forebear for such famous progeny. Alfred was not only engineer of the *Dumbarton Youth*, but agent as well, learning at first hand the business of Custom House clearances and freight collections. The *Dumbarton Youth* also gave Alfred an insight into the possibilities of screw propulsion. It should be emphasised, perhaps, that iron screw-propelled steamers were very much in an experimental stage in the 1850s. Although there were various such ships constructed at the time (including Brunel's *Great Eastern*, launched in 1858, which also had paddles and some sails for good measure) they made few inroads into ocean-going traffic. Technical difficulties with the engine and design of the ship abounded, and Alfred Holt was to be one of the first to combine satisfactorily the three key elements: iron, screw, and compound engine.

The *Dumbarton Youth* venture was a profitable one. Alfred now (in 1853) persuaded Ainsworth and his father to advance money to buy another steamer, the *Cleator*. Fortuitously the Crimean War enabled this ship to earn enormous profits on charter to the French government, and Alfred now embarked on a larger vessel, the 535 ton *Saladin*, launched in 1855. With war at an end, this vessel traded between Liverpool and the West Indies. Alfred's own words are interesting: 'we resolved (for Philip, though partner with Lamport and Holt, assisted in all my plans even then) to put her into the West India trade, to which part of the world, at that time, no unsubsidised steamer plied'. Obviously Philip would have been well situated to learn the shipping business in his brother's firm, and it is clear that even at this stage there was a close business bond between Philip and Alfred. It is equally clear that, sure as he was of the technical quality of his vessel, Alfred was untroubled at the prospect of a pioneer venture.

In the event success in the West India trade was hard won, and Alfred even, for a time, thought of withdrawing from shipping. Eventually, though, the *Saladin* began to make good returns and Alfred decided to expand his West Indian interests. Again with support from his father and others he ordered the *Plantaganet* from the Greenock shipbuilders, Scott and Co., in 1857. The *Plantaganet* was soon followed by four other ships for the West Indies trade. Three points are worth making about the *Plantaganet*. She was the first to be built for Holt by Scotts, thus beginning a famous partnership which lasted until after the Second World War. Alfred had met John Scott through an introduction from W. S. Lindsay, a shipowner and shipping historian, and the two became firm friends. Alfred wrote of Scott in 1880 'our intercourse has ripened, through mutual services and esteem, and almost uninterrupted business connection, into one of the happiest and most satisfactory friendships I have ever made or probably ever shall make'. By 1880 Scotts had built some 25 ships for Holt. Second, the *Plantaganet* was constructed in a year of severe depression with prices very low and advantageous. According to Alfred his steamer was the only one being built at the time in the three great shipbuilding towns of Greenock, Dumbarton, and Port Glasgow, and in subsequent years Alfred frequently ordered his ships at depressed times. The policy benefited both himself, and, of course, the shipyard. Third, we may note that the *Plantaganet*'s maiden voyage was under the command of Captain Middleton, whom we have already seen had a long and close association with Alfred Holt. Another of Holt's West India ships, the *Talisman*, was commanded by Captain Kidd, another close associate of Alfred who was later, with Middleton and Turner Russell, to captain the first China steamers.

The West India enterprise did not last long. In 1864 competing interests, operating on a much larger scale than Alfred and Philip Holt could manage, bought out all the Holt steamers except the *Cleator*. Philip's name must now be added to Alfred's, for in January 1864 Philip had left Lamport and Holt 'and took his seat at 1, India Buildings, where his heart had been for years before', as Alfred wrote in his *Autobiography*.

The two brothers, with considerable capital from the sale of the West India vessels, now plotted to enter the China trade, a scheme determined upon by the autumn of 1864. Why China? More than 25 years later Alfred recalled a discussion with Philip which centred on finding a route they would 'have to ourselves', unlike the competition and opposition they had met with in the West Indies. By this time, of course, regular services were operating between England and all the major centres 'on this side of the Cape of Good Hope and Cape Horn', and the two brothers' thoughts turned naturally to the east and to the Pacific. China was settled upon partly because 'tea was a very nice thing to carry', and partly, Alfred recollected, because Samuel Rathbone, commenting on the prospects for sailing vessels, had once made the remark that the China trade, at any rate, remained safe for sailing vessels. 'I suppose the fiend made me say "Is it"?', wrote Alfred. But Alfred Holt's memory may also have played him a little false. We have seen already that the prospects for the China trade were burgeoning in the wake of the Treaty of Peking in 1860, and a tea-steamer had sailed from Hankow in 1863. With the American Civil War at its height the great American clippers had withdrawn from the China trade; freight rates rocketed and the doors were wide open for enterprising British shipowners. Alfred and Philip, and Liverpool generally, would have been only too well aware of this, and the decision to enter the tea trade can hardly have been so capricious as Alfred later recalled.

Having settled on the trade Alfred now used the *Cleator* to experiment with the engine which should carry the brothers' hopes to China, a compound single-crank high pressure engine (60 lbs/per inch) which proved itself admirably on a trial in the Channel. Further voyages, as far distant as Brazil, confirmed the greatly improved speeds and fuel consumption yielded by the new engine. The little *Cleator* had a notable history; she was later sent east and was possibly the first cargo vessel through the Suez Canal (and hence the first Holt, though not Blue Funnel, vessel); she then took the first cargo of new tea from Foochow to Melbourne, and was eventually sold to the reknowned 'king' of the Cocos Islands, George Clunies-Ross.

The Holts were now ready to order the first three ships for China, the *Agamemnon, Ajax*, and *Achilles*, 'the great venture of our lives', wrote Alfred, 'they started the Ocean Steam Ship Company, and the China trade, which have been for us "the tide in the affairs of men which, taken at the flood, leads onto fortune"'. The Ocean Steam Ship Company was registered on 11 January 1865, and the first ships were all ordered from Scotts of Greenock, at a total cost of £156 000.

This great venture was launched on the backs of what were, by later standards, very small ships, each a little over 2000 tons gross, and able to carry about 3000 tons of cargo. Yet so efficient were they with Alfred's engines, that they could non-stop journey the 8500 miles to Mauritius, via the Cape, carrying sufficient coal to drive them at 10 knots.

By 16 January 1866, the Holts were able to issue their first circular to the public. It read: 'I beg to inform you that I am about to establish a line of Screw Steamers from Liverpool to China', and went on to detail routes and times: 'Liverpool–Mauritius–Penang–Singapore–Hong Kong –Shanghae' [sic]. The estimated times taken at this stage were, outwards, Liverpool to Mauritius in 39 days, Penang in 54, Singapore 57, Hong Kong 66, Shanghai 76. The return journey from Shanghai was to be 77 days, from Foo-choo-foo (Foochow), 69 days, from Hong Kong 65, Singapore 57, Penang 54, Mauritius 38. The circular added 'The Captains have all been many years in my employ, and are well accustomed to the care and navigation of Steamers'; and 'My hope is to establish a reliable line of Steamers, which will carry Cargo, at moderate rates of freight, both safely and at tolerable speed.'

Several points about this circular are worth comment. We can see that from the outset Alfred Holt envisaged a regular line, working to a relatively fixed schedule of ports and times. Given the still relatively experimental state of compound-engined screw steamers this was a striking statement of confidence in his vessels. Still more remarkable was the decision to fix so absolutely the route and ports, for in the days before telegraphic communications, established agencies, and regular conference shippers, the prospects of loading cargoes at particular ports was by no means certain. This was in fact confirmed by the voyage accounts of several early Holt vessels which not infrequently visited additional ports in search of extra freight.

Yet even if the circular reflected a sound engineering judgment rather than a reliable commercial forecast, the fact remains that from this opening sally until the eclipse of the Blue Funnel line more than a century later, this Liverpool–Straits–China route remained the mainstay of the firm's business (though with Hong Kong becoming the focal point of operations, and with Japan becoming an important extension to the main trade). We may note also that from the outset the Holt brothers had their eyes set firmly on the China tea trade, as shown by the additional stop in Foochow, a major tea-loading port, on the return journey.

The estimated times proved uncannily accurate, although within two years the Blue Funnel steamers were regularly surpassing these speeds. In fact the *Agamemnon* reached Shanghai in 77 days, via the Cape, calling at Mauritius, Penang, Singapore, and Hong Kong, only one day more than announced in the first circular. And the return, with an additional stop at Foochow to load tea, was accomplished in 58 days steaming time. Notable was Alfred Holt's reference in the circular to the experience of his Masters and his lengthy acquaintance with them. All the Masters of his first vessels, Captains Middleton, Kidd, and Turner Russell, had served on Holt's West Indies steamers, and all were on terms of closest friendship and confidence. This personal touch, reflecting the value placed by Alfred Holt on the quality of his officers, remained a hallmark of the firm throughout its history. Long service and loyalty were evidently already a tradition even before the first vessel of the new line set sail.

Following the first circular of the Ocean Steam Ship Company on 16 January 1866, the first Annual Meeting was held on 8 February following. Four were present, with William Rowe (who soon left the Company) in the Chair, W. H. Swire, and the two Holt brothers. At this stage £98 000 of the original capital of £156 000 (in £1 shares) have been paid up, and of this sum the brothers Alfred and Philip held by far the largest share,

£48 500. Another brother, Robert Durning, contributed £10 000, and other family members were prominent among the 28 shareholders. Others in the list include W. H. Swire and John Scott of Greenock, the latter with a £5000 stake in the ships he was building. Each of the three Masters, Middleton, Kidd, and Turner Russell, held £1000 worth of shares, while several powerful Liverpool families were also represented.

Subsequent years were to see a steady expansion in the numbers of shareholders, but the family, friends and the local Liverpool connections remained dominant. In 1886, to mark the twentieth anniversary of the Company's operations, a silver tray and diamond necklace were presented to each brother and their wives by the other shareholders; there were (excluding the Managers) fifty 'co-owners'. This list, together with the dedication, is included as Appendix II, and the strong Liverpool presence is evident from such names as Eills, Forget, Melly, Rathbone, and many others.

For some reason, despite this dedication the diamond necklaces appear to have become the property of the husbands. In 1891 we find Philip Holt giving his necklace 'to the Ocean Steam Ship Co. as its own property absolutely'. He added, 'I hope the Managers may be able to arrange to keep it during the life time of my wife, Anna Holt, and that she may be allowed to wear it at any time without being responsible for its loss. After her death I hope it may be sold, and the proceeds added to the funds of the OSS Co using them (if thought desirable by the then Managers of the Co.) for pensions and allowances to old or disabled servants of the Co.' This mixture of individualism and concern is typical of Holt philosophy, and it is pleasing to record that Anna, too, signed her agreement with Philip's gesture, adding 'I concur in the above.'

III

In considering the early years of the Holt enterprise the question naturally arises as to whether the Holts had effectively established their new line on the 'old' route, before the opening of the Suez Canal at the close of 1869 transformed the situation. Or did the opening of the Canal give an essential transfusion to a stuttering and still insecure enterprise? The issue is not, perhaps quite so clear-cut as these questions imply. The four years prior to 1870 were years of general trade depression, with poor outward shipment, and low freights, whereas the period 1870–3 witnessed one of the most considerable booms in the entire nineteenth century with rising prices and rapidly expanding exports.

Although detailed accounts of the Ocean Steam Ship Company's performance prior to 1870 are fragmentary, there would, in fact, seem to be no doubt that by the eve of the Canal's opening a turning-point had been reached, and Blue Funnel can be considered a 'going concern'. After an initial period of some difficulty, though with adequate returns nonetheless, by 1869 the firm was securely established. New vigorous agents were in place in Singapore and Hong Kong, the first additions to the original fleet were already in service, profits were mounting, and businessmen were showing a willingness to invest in the concern of their own initiative. This was a striking achievement in a new trade and with trade conditions at a low ebb.

Alfred Holt in his diary summed up the first three critical years. At the close of 1866, with one voyage completed, he wrote 'My China scheme seems to promise success'; of 1867, 'My own affairs have progressed slowly but I hope surely. The China line will succeed I think'; 1868 'would have done very well from the beginning were it not for the number of totally unexpected accidents which have occurred'. By 1869 (unfortunately the diary stops before this) success was assured.

Understandably the earliest voyages were still experimental, with neither the commercial possibilities nor the technical performance of the ships tried and tested. Although Alfred Holt already had a wide range of business connections, the necessary contacts with shippers and agents had yet to be made. At first, too, Alfred was more concerned with the safe and trouble-free passage of his vessels than with their financial performance. As far as the performance of the ships was concerned the first voyages were a brilliant success, and well repaid the meticulous attention given to their design and construction, and to the extensive trials given to the engines prior to the first sailings. The *Agamemnon* had sailed from Liverpool on 19 April, 1866, the *Ajax* following on 1 July. By 15 August Alfred could record, 'Heard to my great satisfaction by telegraph from Calcutta of the safe arrival of the *Ajax* at Penang. This second successful voyage gives me great confidence in my China undertaking, in fact I now consider the "mechanical" part of the problem solved, the commercial one remains.' And when the *Achilles* sailed on 19 August, Alfred noted 'a very great improvement on the receipts of the previous ships and showing as I think that my line is beginning to be appreciated'.

None the less, success was not immediate, nor could it be with depressed economic conditions in Britain and with the fastest sailing clippers able to take their precious cargoes of new season tea at speeds equal to, or greater than, Holt steamers. The first *Ajax* voyage, for example, was nearly a financial disaster. The Master of this ship, Captain Kidd, has left a diary in which he sorrowfully recorded that 'we stayed a month in Shanghai waiting to see if times would mend, but found things getting no better. Started on the return voyage with very little cargo and very poor prospects . . . I was rather puzzled what to do with a fine ship going home one-third full. Not very promising for our new trade.' However, on the return trip the enterprising Captain decided to call at Port Elizabeth on Algoa Bay 'and try if I could get any cargo there'. Fortunately he was able to load all the spare capacity with wool, which 'enabled me to finish my first China voyage very favourable and got complimented by Mr Holt for the way I had managed'.

This extract shows vividly the difficulties of acquiring freight in distant lands (no wonder Alfred Holt was dissatisfied with his Shanghai agents), and it demonstrates how the enterprise and ingenuity of an individual Master could make the difference between a successful and an unsuccessful voyage. There was an echo of the latitude possessed by these Captains later when in 1868 Alfred Holt learned that the *Agamemnon* was loading teas in Hankow, on the Yangtze river. He wrote: 'It is not a trifle to take a steamer like the *Agamemnon* 700 miles up an inland river. If she has gone there Captain Middleton cannot be said to have erred on the side of timidity.' This emphasises yet again the part played by these early Masters in Holts' success, and no one was more aware of this than Alfred himself, whose diary frequently recorded his debt to his 'good Captains'.

The overall results of the earliest voyages were most satisfactory as Table 5.2 demonstrates, but the returns were soon overshadowed by subsequent voyages.

Thus the first four voyages earned an average of £6711 from passengers and freight over a period of a year. This may be compared with earnings from six voyages in the calendar year 1869 of £84 168, an average of a little over £14 000 a voyage, and around one third of the total cost of the five vessels then in service. The term 'earnings', also called 'profits' in the Holt shipping accounts, deserves some comment. Each voyage was considered a separate accounting unit, and total costs for each voyage, including fuel, wages, victualling, and so on, as well as port dues and handling charges, plus maintenance charges and any repairs were deducted from the total freight earnings to yield 'profits'. But these figures included no allowance for capital depreciation, nor for management costs, agents fees, or the like. The success of the early voyages

Table 5.2   Earnings on early voyages, 1866–7

Ship	Voyage	Dates	Earnings £	s	d
Agamemnon	(1st)	April 1866–November 1866	6793	17	3
Ajax	(1st)	June 1866–January 1867	2700	18	10
Ajax	(2nd)	March 1867–September 1867	12033	14	10
Achilles	(1st)	August 1866–April 1867	5315	11	1
			26844	2	0

Source;   Ocean Archives; Managers' Minute Books.

soon prompted Alfred Holt to order a new vessel from the Tyne shipbuilders, Andrew Leslie and Co. This was the 1850 ton *Diomed*, delivered in 1868. The cost of this steamer was £39 000 and of this sum Swires (in the name of W. H. Swire) took the largest single share, £5437.

Naturally enough freights fluctuated both in value and quantity. Some vessels sailed from Liverpool nearly empty, some with excellent receipts from cargo and passengers. Thus when the *Ajax* left on her third voyage in October 1867, the outward receipts from passengers and cargo were so good that Alfred noted 'the best yet, in fact the China trade is as right as can be if only I can make the vessels go regularly'. But a few months later, in February 1868, it was a different story. The *Achilles* sailed with a very poor freight, and 'unless she gets a good homeward freight she must lose money'.

The vicissitudes of the trade cycle and the difficulties of establishing a new service were not the only problems with which Holts had to contend. Soon Blue Funnel's very success was encouraging rivals. Of course, there had always been competition from sailing vessels and from steamers operating in eastern waters, but by 1869 there were various rival steamships journeying between England and China. When the *Agamemnon* set out on her fourth voyage in February 1869, the freight received was 'the worst yet', and Alfred mournfully related that 'the trade to China is excessively dull and a steamer [*Nile*] opposed her from London, we decided not to reduce our freight and therefore only obtained about 410 tons of goods'. The *Nile*, incidentally, was a Borneo Company ship; Holts were later to have long and cordial relations with this company who acted as their Bangkok agents from 1902, having earlier been represented by the German firm, Windsor & Co.

Another problem for the new line was a series of mainly minor but costly accidents (in terms of time lost and the expense of repairs). The most serious was the 'curious but immensely disgusting accident', in Alfred Holt's words, which befell the *Ajax* while docked at Shanghai. The vessel with all its cargo sank, and although the *Ajax* was successfully raised and repaired, the loss was in excess of £20 000. Typically, Alfred was more vexed by the design fault responsible than by the financial loss. Subsequent rectification (by fitting a water-tight division in the propeller shaft) made a lasting improvement to ship design.

Despite these various problems there were also some excellent results from voyages in 1868, the *Agamemnon* and the new *Diomed* in particular getting valuable cargoes of new tea from China. And some bad winter months gave way to a more hopeful spring in 1869. Already by April of that year trade prospects were brightening, and a number of existing shareholders and at least one outsider approached the Holts to take out shares in the company, 'making one think of building another vessel'. The *Diomed* arrived from China in May after a very profitable voyage,

10 *Nestor*, I, 1868–94

'her principal receipts on her second voyage were from passengers of whom she safely landed in the Thames 296. So many never went in one of my ships before.' The conditions under which these passengers must have travelled in a 2000 ton cargo ship can only be imagined. And on 10 June 1869, the *Diomed* departed from Liverpool with, for the first time, a full load of cargo and passengers (including, incidentally, the Bishop of Mauritius).

Success continued, so much so that in the autumn of 1869 Alfred proudly recorded the times of four of his steamers, the *Agamemnon*, *Nestor*, *Achilles*, *Ajax*, together with the times of two rival steamers. The Holt vessels were comfortable winners the fastest, the *Achilles*, making the passage from Foochow in only 62 days, and the Blue Funnel ships easily outpaced the clippers. Alfred was also delighted with the high freights and good condition of the cargoes, while the *Agamemnon* brought 'the largest cargo ever embarked in one ship, and earned the largest freight I ever heard of, viz. £28 087'. When the *Nestor* made her next outward trip, in October 1869, she departed with 'the largest outward earnings any of the fleet have made'.

Little wonder that the Holts were considering expanding their fleet, and already by June they were planning the *Priam* and contacting their old friends, Scotts of Greenock, to see if the once-bankrupted yard could again build for Ocean.

The opening of the Suez Canal, 'that dismal but profitable ditch' as Joseph Conrad called it, on 17 November 1869, thus found the Holt enterprise in a prosperous and confident position. Nor was the Canal itself a source of concern to Alfred, who had always supported the project and thought England's luke-warm attitude unworthy. In October he had sent Captain Alexander Kidd in the *Ajax* to visit the Canal and report on prospects for the Blue Funnel ships there. The Captain was even able to discuss prospects with Ferdinard de Lesseps, the visionary French diplomat who had pioneered the scheme, and asking 'for the information of Mr Holt about the Canal'. Alfred Holt rightly forsaw that the Canal would open a new era for England; 'We shall benefit by it more than all the nations of the earth put together.' And he realised too that the Canal represented a new era for the Ocean Steam Ship Company.

IV

In March 1870, the *Diomed* became the first Ocean ship to use the Suez Canal on an outward journey, and in August that year the *Achilles* made the first homeward passage. At around this date, Alfred Holt sent out one Captain William Stapledon to represent him at Port Said. Here Captain Stapledon founded the agency William Stapledon and Sons in a building originally belonging to a French engineer engaged in building the Canal, the firm's headquarters until a new building was completed in 1924.

Alfred Holt's speedy and wholehearted acceptance of the Canal contrasts with that of his P&O rivals. In June 1870 when Blue Funnel ships were all being routed through Suez, the P&O directors were telling shareholders, 'On the more important question of the time when it may be for the advantage of the Company to adopt the Canal as the route for the Company's weekly steamers the Directors can only say that they continue to watch its navigation with the utmost care.' And a year later they were still watching.

As we saw in a previous chapter, the opening of the Suez Canal enabled Blue Funnel ships to carry Moslem pilgrims to and from Jeddah after 1870. An incident involving the carriage of pilgrims in 1880 deserves not to be forgotten, and it formed the basis of an episode related by Joseph Conrad in *Lord Jim*. It was on 8 August that year that the *Antenor*, returning to London from Shanghai, and with 680 pilgrims on board bound for Jeddah, sighted the SS *Jeddah* (the SS *Patna* in *Lord*

Jim) in distress in heavy seas off Cape Guardafui. The *Antenor's* Chief Officer, Randolph Campbell, led a small boat party aboard the stricken ship to find a terrible situation. The *Jeddah* was drifting helplessly with engines disabled, boiler fires out, with 8 feet of water in the vessel, and with some 1000 panic-stricken pilgrims, adults and children aboard. The ship had struck a fierce gale, the engines started breaking up, and crew and passengers had worked all day and night to bail water with buckets. In the early hours of that morning the *Jeddah's* Captain, his wife, Chief Engineer, and a number of other officers, fearful of their safety as panic among the pilgrims mounted, decided to leave the ship in one of the lifeboats.

The dilemma facing the *Antenor's* master, Captain J. T. Bragg, was great. Should he put his own ship, cargo, and passengers at risk by going to the aid of the *Jeddah*, or could he abandon the *Jeddah* to inevitable foundering or shipwreck? Captain Bragg and his chief Officer decided to try to save the *Jeddah*. Since it was impossible to take the 1000 passengers on board the *Antenor* it was decided to take the *Jeddah* in tow, and, in the dark, and with much skill, the complicated and dangerous task of attaching a towrope was accomplished. Then, with Campbell aboard the *Jeddah*, and steering himself until he had taught two of the *Jeddah's* crew to steer, he encouraged the pilgrims once more to pump and bail, until the *Jeddah* was finally towed to safety to Aden. The heroic operation was the more striking for the contrast between the courage of the Blue Funnel Master and Chief Officer, and the shameful desertion of their ship by those in charge of the *Jeddah*.

The years between the opening of the Suez Canal and the mid-1870s were a time of considerable prosperity for Ocean Steam Ship. Five new ships were added between 1869 and 1871, which permitted regular sailings at shorter intervals. Total earnings from freights for the three years 1873–5 were almost double what they had been in 1869, and Alfred Holt recorded in his diary at the close of 1875 'our business during the five years has been uninterruptedly successful'. Although nine steamers had been added to the fleet in these years, business had been conducted with economy, so that reserves had been accumulated. By 1875 reserves stood at some £280 000 at a time when the average cost of new steamers was less than £50 000. For example, the 2000 ton *Sarpedon* and *Orestes*, built in 1877 by Leslie & Co. and Scott & Co. respectively, cost £40 579 6s 2d, and £44 770 1s 3d. This may be compared with the first three steamers, which cost £52 180 each. Ocean's reserves were invested in a wide variety of securities, including government bonds, railway shares, and various industrial holdings, and total investment income was some £70 000 in 1875. The policy of financial caution and the accumulation of reserves was to become a permanent feature of the Ocean outlook, though it was soon to bring criticism from no less an individual than John Swire.

New ships and growing business prompted Alfred and Philip to expand their staff. In 1872 Albert Crompton was taken on as a general assistant, Alfred noting in his diary 'Albert Crompton has come into our office and his brother Edward ... has come into Booths, and it is very satisfactory to me to see young men of the right stamp connecting themselves with us.' Nor did Alfred Holt's success go unnoticed. In 1874 he was offered the managership of that 'huge concern' the Pacific Mail Steamship Company, and recorded that 'I was tempted but principally owing to Philip's wise advice I declined.'

The pre-1876 era of prosperity was not destined to last, however, and the years from around this time until the beginning of the 1890s saw a period of some difficulty for the Liverpool firm. From the 1890s, however, there was a marked revival of fortunes. A long period of quite remarkable progress then ensued which lasted until, and in some ways throughout, the First World War, seeing the establishment of Blue Funnel as one of Britain's, and the world's, leading shipping lines.

The difficulties of the post-1876 period stemmed essentially from three interrelated problems. First, there was a growth of competition in the China trades, both for outward and homeward freight. This exercised a downward pressure on freight rates and sometimes meant voyages with little cargo. Second, the Blue Funnel vessels became increasingly un-competitive with those of their newer rivals such as the Glen, Shire, and Castle lines. By comparison with the best vessels of their competitors, Holts' ships by the late 1880s were small, underpowered, and slow to adopt such innovations as steel hulls and the triple-expansion engine. Third, there were some significant changes in the nature of the trade in both outward and homeward directions, which emphasised still further the unsuitability of some of the older Holt ships for the new conditions. To these general difficulties there were added others, such as an unfortunate and costly series of accidents, periodic trade depressions influencing both the outward and inward freights, and some unfavourable trends in costs; for example coal prices rose sharply between 1886 and 1891, while other costs (such as wages for crew members) could not be reduced.

The solutions to these problems were found in a number of directions which we have already noted. There was some diversification of trade, especially the development of local routes in the Straits in the 1880s. These years also saw the beginnings of the Far Eastern Conference under the leadership of John Swire, and by the 1890s the system was firmly established. This helped reduce the impact of cut-throat competition. Fundamental, perhaps, was a vigorous response from India Buildings which set the company on a new path from the beginning of the 1890s: new ships were built, old ones disposed of, new trades sought, and a new generation of Managers brought in to the firm.

Let us look a little more closely at mounting competition after the late 1870s. The most active competition came from British lines, especially the Glen Line, owned by MacGregor, Gow and Co., of Glasgow, the Shire Line, which had been founded by Captain David Jenkins, and the Castle Line owned by Thomas Skinner and Co. The Glen and Shire lines were later to become part of Alfred Holt and Co. but at this stage they posed a formidable threat to Blue Funnel. In addition to these lines there was growing competition from subsidised mail carriers such as P&O and Messageries Maritimes vessels, which had already provided rival steamship services in the Far East before the opening of the Suez Canal, and a few other concerns, among which may be mentioned the Ben Line and the British India Steam Navigation Company. The year 1875 also saw the first Japanese company, the Mitsubishi Mail Steamship Co., set up a competitive line with European steamships, from Yokohama to Shanghai. However Japanese competition did not become significant until 1885 with the formation of the NYK Line (Nippon Yusen Kaisha) of subsidised steamers from a fusion between Mitsubishi and another concern, the Mitsui-sponsored KUK (Kyodo Un'yu Kaisha).

Holts' vessels were increasingly outclassed in the late 1870s and by 1880 some of the ships of the Glen, Castle and Shire lines were able to make the passage from China to England in under 45 days, whereas the fastest Holt ship took 50 days. In 1882, the Castle Line vessel, *Stirling Castle*, made the voyage from London to Hankow in only 29 days. In 1875 the original three Holt vessels, the *Agamemnon*, *Ajax* and *Achilles*, were still the largest ships in the fleet of 16, at a registered tonnage of 2280 each, and not until 1892 did any Holt ship exceed 3000 registered tons. By this date the P&O had 34 steamers on the India and China service, with 8 of them over 3000 tons and the largest of 3742 tons capable of an average speed of 11 knots. The biggest vessels of the Messageries Maritimes in its China trade were then around 3600 tons, and the company also had 8 steamers over 3000 tons. The Glen Line was no less progressive, building its first 3000 ton steamer in 1882 (the 3749 ton *Glenogle*), and having added six more by 1892, four of them with steel hulls.

The first steamer of the Glen Line, the *Glengyle*, made its maiden sailing to the Far East in 1871 and the first Castle soon followed. That same year Jenkins ordered his first steamship for the Shire Line, the *Flintshire*, which was delivered in 1872 and journeyed to Colombo, Singapore, Hong Kong, Nagasaki, Kobe, and Yokohama. This vessel was later (in 1889) bought by Holts, and was subsequently part of the East India Ocean Steam Ship Company. By 1881 the Glen line had built 11 steamers, the Shire line 5, and the Castle line 7. In 1874 the new Glen steamer, the *Glenartney*, captured the record for the fastest homeward journey from the East, making the journey from Woosung to London in

41 days. The year following Alfred Holt recorded in his diary the impressive performance of the *Glenartney*, noting 'the OSS Co. cutting but a poor figure this year. We have, however, the solid satisfaction of knowing that if our voyages are slower our profits are greater.'

By 1876, however, Alfred Holt could no longer afford to look on the mounting competition with equanimity. In that year the average level of earnings per voyage was the lowest yet recorded, and the Managers, at the Annual General Meeting in February 1877, considered that 'it is of the utmost importance to the Company's interests not to let their competitors push them out of the trade'. In October the *Patroclus* sailed with 'the smallest cargo since we began the line' and by November of that year the outward trade was so poor that Holts were obliged to charter one ship to Lamport and Holt, and sent another in ballast to Jeddah. Alfred recorded in his diary that Ocean had 'very reluctantly and after everyone else has set us an example, reduced our outward freights'. Reviewing the year 1877 in his diary, Alfred nonetheless considered that 'our business during 1877 has continued successful, though smaller profits result each year, competition is so keen'.

The result of competition was to force freight rates downwards. When the Ocean Steam Ship Company was launched it had been able to charge outward freights of around £6 6s a ton. By the mid-1870s rates had fallen to around £4 10s a ton, and at the end of 1877 Holts were obliged to lower their rates to £1 10s. Oddly though, in the light of these figures, the China trade was not depressed. In fact, trade to and from China increased significantly during the years 1875–85. But the very success of the China trade encouraged otherwise unoccupied tonnage to enter this trade, thus forcing down rates. At the same time, Holts were meeting increasing competition in their traditional outward cargoes of Lancashire and Yorkshire textiles for which they had possessed a near-monopoly in the early years. Most serious was the formation of the China Shippers' Mutual Steam Navigation Company in 1882, which began operations with fast efficient ships. The China Mutual vessels commenced from Glasgow and loaded at Liverpool before sailing to the Far East, so directly threatening Holts in their home base and in their traditional trades.

The China Mutual ('Shippers'' was soon dropped from the title) had been formed with support from some prominent Manchester merchants, and with a prospectus promising shippers 'liberation' from conference ties. Their first vessel sailed in 1884 and in 1887 China Mutual and the Mogul Line joined forces to run a service outside conference rates. There followed a lengthy and often bitter period of freight warfare during which the conference system collapsed. By the beginning of 1891 freight rates on cotton yarns, which had stood at 52s 6d in January 1888 had fallen to only 20s per ton of 40 cu. ft. But this ruinous level forced the turning of the tide. China Mutual felt obliged to become a member of the Homeward Conference in 1891, and in 1894 a new Outward Conference was agreed from which time, until the two companies joined forces in 1902, Holts and China Mutual operated in close and friendly agreement.

Until 1875 the Holt vessels had sailed virtually without serious mishap. But from that time a series of misfortunes befell which happened to occur in the aftermath of the company's decision to undertake its own insurance in 1874. There were three total losses in the space of only eleven months, the *Hector* in China on 4 October 1875 (with the loss to the Company of about £35 000); the *Orestes*, with a cargo valued at £300 000 near Galle on 7 March, 1876; and the *Sarpedon* off Ushant on 4 September. The total loss to the company of these disasters was estimated at around £120 000. In addition to these total losses there was a disconcerting rash of minor accidents, Alfred Holt noting an 'epidemic of broken shafts' in 1877. There were several quite serious collisions, some of them involving costly delays, salvage operations, and expensive and tedious legal battles.

The decision to move to self-insurance after 1874 was a significant one. For one thing, such a policy dictated the highest standards of ship design and maintenance and an insistence on the best possible calibre of Masters and officers and qualities of seamanship. That Holt ships were built to higher standards than Lloyds A1, that Holt officers were hand-picked and generally the best that could be found in the merchant navy, and that the Managers were meticulous in ensuring that every detail in a ship's preparation for a voyage and rules for safety during a voyage were carried out, became an enduring tradition of the Line. Also, in addition to emphasising the quality of the fleet, self-insurance also necessitated the building-up of large reserves and a cautious policy in the distribution of dividends. High reserves underwrote the fleet, helped to provide extra funds for new ships on occasions, and facilitated the notable coup of 1902 when the China Mutual Steam Navigation Company was acquired. The Company's reserves stood at more than £350 000 throughout the period 1881–1914 (except during the year of the China Mutual purchase), not counting investments in the East India Ocean Steam Ship Company and the NSMO.

The spread of investments, too, showed caution, with security rather than yield the major consideration. In 1900, for example, around 40 per cent of investments with a total value of £396 000, was held in railways, British and foreign, and in American government stocks. A further 20 per cent was invested in English banks and insurance companies, and the remainder in a wide variety of 'miscellaneous' stocks. If caution and security was one aspect of investment policy, another was the direction of investments towards companies with Liverpool, family, or business connections. Thus the single largest investment (£80 000) was in Booths shipping interests, and other investments included the Taikoo Sugar Refining Co., the Eastern Telegraph Company, Mersey Docks and Harbour Board, South Lancashire Electric Traction and Power Company, the Bank of Liverpool, the Liverpool, London and Globe Insurance Company, Crosfield and Co., the English Sewing Cotton Company, Lever Brothers, Brunner Mond, and the United Alkali Company.

During the late 1880s the problems of competition and falling freight rates already experienced for several years developed into a major crisis. In the Liverpool–Far-East Trade Holts had to face the challenge of China Mutual, while in the Sumatra tobacco trade North German Lloyd was increasing its share. More generally, though, Blue Funnel, having pioneered so much, now seemed to lose its vigour. This was seen especially in the fleet, which became rapidly out of date at a time when competitors with greatly improved ships were appearing on the scene in growing numbers. The single experiment with a triple-expansion engine before 1890, the *Ulysses*, was apparently not successful, and Holts reverted to the compound engine. Moreover the Blue Funnel ships remained small. It is a striking fact that at the beginning of 1890 the average gross tonnage of the Blue Funnel fleet was lower than it had been at the end of 1866.

Until 1876 no ship had ever made a loss on a completed voyage. Thereafter, from time to time losses occurred, but this was inevitable when the accounting system deducted all expenses, including repairs, from the gross earnings of each trip. After 1885, however, losses mounted. In 1877 no fewer than 19 out of 55 voyages resulted in losses, and the Managers considered that the overall profit per voyage of under £900 was 'equivalent to a loss' when all overheads were taken into account. As Table 5.3 shows, the position rapidly became worse at the beginning of the 1890s, reaching a low point in 1892 when only £372 per voyage net profit was recorded. In this disastrous year no fewer than 27 voyages made losses, the total losses on these voyages being over £10 000.

The causes of these losses were various. For outward cargoes from Liverpool, Holts were forced to accept lower freight rates due to competition. As already discussed the decision of the China Mutual and Mogul lines in 1887 to offer a joint far eastern service outside conference rates led to a long and costly freight war. Not until 1891 was China Mutual forced to accept conference terms and a new Homeward Conference established, while a new Outward Conference was not formed until 1894. In the meantime Holts' freight rates, as we have seen, collapsed between 1888 and 1891. But this was not the only cause of Holts' difficulties. The Company felt it necessary to extend all their regular main line services to Japan in 1889 because, as shareholders were

Table 5.3   Blue Funnel fleet and financial results, 1881–1900

Year	Number of ships	Gross tonnage (000)	Average tonnage	Net profit (000)	Number of voyages	Profits per voyage
1881	23	48.4	2104	238.9	50	4778
1884	25	53.3	2132			
1885	28	58.4	2086			
1886	31	66.6	2148			
1887	30	64.7	2157	49.3	55	890
1888	31	66.8	2155	66.6	65	1025
1889	30	64.6	2135	95.4	66	1445
1890	34	78.2	2300	65.6	89	737
1891	33	76.1	2301	45.0	72	625
1892	37	90.3	2440	27.5	74	372
1893	35	86.2	2463	50.5	71	711
1894	34	86.6	2547	114.7	70	1629
1895	36	100.5	2792	123.6		
1896	38	117.6	3095	59.1		
1897	36	118.0	3278	106.1	67	1582
1898	35	115.8	3309	179.9		
1899	38	140.0	3684	274.5		
1900	41	165.6	4039	302.8	79	3832

Note: Gaps in the table indicate data not available.
Source: Ocean Archives: Managers' Minute Books and Voyage Accounts.

told, the company 'could not leave the Japan trade to its competitors'. But this extension added one month to the average voyage, and made the relative slow speeds of Holt vessels even more disadvantageous.

Most serious, though, were some permanent changes in the China trade which were threatening to undermine Holts' entire position in the Far East. In their Annual Report at the beginning of 1891 Alfred and Philip Holt, for the Managers, made a perceptive analysis of the situation. As far as outward trades from Liverpool were concerned, in former times cargoes had consisted 'almost entirely' of Manchester goods – that is, textiles – and a little iron and lead. But in the 1880s the picture altered substantially. Now cargoes consisted increasingly of a large variety of miscellaneous goods, which the Managers enumerated as old iron, chemicals, machinery and boilers for mills abroad, patent manures (that is, fertilisers), bricks, tiles, 'large marine boilers', and many others. These varied goods accompanied increased carryings of Manchester goods. The result was a sharp increase in the costs of handling, storing, and loading such goods, with additional pressure on space and time. In Alfred Holt's words, 'it will thus be seen that a good deal of the cargo consists of large, heavy, awkward pieces and packages – very different from a cargo of bale and case goods'.

Changes in the homeward trades had been no less revolutionary. As we have seen, the early cargoes consisted almost entirely of tea, and this continued to be the case into the 1880s. However, the Managers noted that in 1890 their ships had only been able to bring in one single cargo of tea, and, in fact, the year had seen only two other full cargoes of tea brought from China to England. Demand for Chinese tea in England was declining rapidly in the face of growing supplies of tea from India and Ceylon, and there was in any case a growing tonnage of shipping competing in the Far East. Thus shipowners were increasingly having to visit ports in the Far East outside China to find cargoes, and having to load with a greater variety of produce. As the Managers said, 'there seems to be no alternative but to fall back on rough heavy produce such as sugar, rice, sago, timber, gambier, wool, etc.'.

On this analysis it appeared that the current small ships possessed by Ocean were unsuitable. Larger ships were required, 'as large as our competitors', with an average speed of 10 to 10½ knots, and without passenger accommodation.

The solution proposed by the Managers was a courageous one. On the one hand they were determined to expand the fleet with appropriate ships, investing at once £300 000 on four new steel ships with triple-expansion engines (the money mostly coming from the accumulated reserves). On the other hand, they sought to transfer ten of the now obsolete China vessels to the newly-acquired Java trade. This latter was a master stroke, for it gave Blue Funnel an interest in an important new area in the Far East, and also established a new link with the European continent.

The proposal to build new ships was, the Managers admitted to the shareholders, 'risky', but the alternative was 'to bring the business to a stop, wind it up, and divide the proceeds among the shareholders'. Needless to say, the Annual Meeting of February 1892 took the risk. To provide cash for their new ventures the Company also sought help from their bankers. Prior to 1900 the Liverpool Union Bank acted as Ocean's bankers, and a harmonious relationship was built up between the two. In 1891 Alfred Holt made arrangements for the Ocean Steam Ship Company to have an overdraft up to £50 000 with interest at bank rate (with a minimum of 3½%), and in 1898 the overdraft was raised to £150 000, again at the bank rate, with a minimum interest of 2½%. In 1900 the Liverpool Bank was absorbed by the fast-expanding Lloyds Bank, and from this date began a strong and lasting connection between Ocean and Lloyds. Lloyds continued to allow Ocean to borrow – at bank rate – £150 000 without security and a further £100 000 on the security of stocks. However, such borrowing did not alter the overall picture of the Company's firm financial footing. The vast new building programme was financed almost entirely from accumulated reserves, and the

11 *Idomeneus*, I, 1899

achievement triumphantly vindicated the Holt brothers' earlier insistence on financial caution, in the teeth of John Swire's criticism, without
which they would have been in no position to embark on such an
ambitious programme of expansion.

The Liverpool Union Bank is not to be confused with the Bank of
Liverpool, another institution with which Ocean had close connections.
The Bank of Liverpool had been founded in 1831, with George Holt
(father of Alfred and Philip) an active promoter, an original shareholder
and director, and Chairman in 1847–9. Through amalgamation the Bank
of Liverpool became the Bank of Liverpool and Martins in 1918 (known
simply as Martins Bank from 1928) and Holt connections with this great
Liverpool institution were maintained. Among the Bank's chairman
were, in addition to George Holt, William Durning Holt (1884–5), Robert
Durning Holt (1897–8), Sir Richard Durning Holt (1937–8), and Sir John
Nicholson (1962–4).

The progress of the new policy was seen most vividly in fleet
construction and in the rapid growth of average tonnage. As Table 5.3
shows, the average tonnage per vessel nearly doubled between 1889 and
1900, while the total tonnage of the fleet not far short of trebled. The
building of twenty-three new ships of advanced design in less than a
decade was a heroic achievement, and in these few years Blue Funnel
became one of the world's great cargo lines. We should emphasise that
the years 1892–6 were generally depressed, and Holt expansion therefore took place at a time when many other lines were retrenching, and at
a time, moreover, when interest rates were low and shipbuilding costs

Table 5.4   New ships built, 1892–1900

Year	Number of ships in class	Average tonnage
1892	4	3627
1894–5	6	4650
1896	4	5570
1899–1900	5	6692
1900	4	7000

Source: Ocean Archives.

moderate. Perhaps the most striking picture of the progress achieved in these years comes through the average size of the five classes of new ships built in the 1890s (for the initial investment proposed in 1891 was just a start). These figures are shown in Table 5.4. As a result, by 1900 over half the fleet of 41 ships was less than eight years old.

With new ships went new blood. The year 1889 saw three young men brought into the Company, Richard Durning Holt, a nephew of the founders and destined to guide the firm from the early years of the twentieth century until the Second World War; George Holt (Alfred's son); and Maurice Llewellyn Davies, a nephew by marriage of Philip Holt. In 1895 the three were made Managers, thus joining Alfred and Philip Holt and Albert Crompton. Gradually the role of the younger men, especially Richard Holt, became more dominant as that of the founding brothers slackened. Philip Holt resigned from management in 1897, while Alfred, who had taken little active part in Company affairs after 1898, resigned formally in 1904.

If we return to Table 5.3 for a moment, we can see the apparent success of the new policy. Almost at once, from 1892, there was a rapid improvement in net earnings, and by 1894 earnings per voyage were more than four times their disaster levels of 1892. By the end of the decade the company was firmly launched on a long period of prosperity.

About this dramatic improvement in fortunes there is really no puzzle. The years after 1896 saw a world-wide boom in prices and production of primary products, exports of textiles from Britain to Asia rose rapidly, and many countries in South-east Asia in particular expanded their exports apace. The Blue Funnel fleet, with its more modern and better designed ships, was well placed to take advantage of the expansion of trade, and all branches of business, with the exception of the local Straits trades, benefited in the boom conditions.

In line with better conditions went better freight rates, especially after 1896. Moreover it was at this time, the early 1890s, that various internal economies were put in train by Albert Crompton. One such measure was a general reduction in salaries and wages of 15 per cent. Hitherto, the Company had paid their crews at a higher level than most other shipping lines. From 1893 the new lower rates were, for Chief Mate, £12 monthly for sea service, for Second Mate, £10, and for Third Mate, between £5 10s and £7 depending on experience and length of service. The Chief Engineers were paid £20, and Second Engineers £13 monthly. All these rates remained largely unchanged before the First World War, while during the First World War most rates were raised by around 50 per cent. Masters, too, were paid generously, especially in the early pioneer years. Thus, Captain Kidd's first voyage on the *Ajax* in 1866 earned him £102, and for the second half of the year he received no less than £255. This high rate (the same was paid to other Masters) remained until 1876, when remuneration was raised to £270 half yearly. Then, as part of Crompton's general economy measures in 1892 the rate was halved to £129 10s half yearly, and in 1914 was raised to £150 for six months. Even so, £300 was a good yearly salary at a time when Masters working for comparable lines received around £22 per month for sea service.

The economy measures of the early 1890s included the recruitment of Chinese crews, a system already adopted by other lines. Recruitment of Chinese crews was not undertaken directly, but was arranged through a Chinese compradore in Hong Kong. the compradore would be paid by Holts (through Butterfield and Swire) and the compredore would then make the necessary arrangements for recruiting and paying the Chinese crews, who came to form most of the ordinary deck ratings and all the engine room ratings on most Blue Funnel ships. On board ship the Chinese crews had their own quarters, and were under the general supervision of a Chinese who would act as interpreter, arrange foreign exchange for periods of shore leave, and so on. It became common practice to recruit engine and deck crews from among different regional areas of China, with the idea of limiting 'ganging-up' among the crews and making control and supervision easier. Especially on the earlier ships the Chinese quarters must have seemed strange and, to Europeans, almost unknown territory. The ship's Surgeon on the *Cyclops* in 1914 (a 9000 ton 'goal-poster', with a crew of 80, half of them Chinese) recorded in his diary, 'I considered the Chinese place a den of mystery – joss-burning and everything quiet. I was informed that no European ever went there but the Doctor and the Mate.' The Chinese quarters were of a lower standard than those of the European crews, and the Chinese were paid substantially less (how much less is unknown, since the compradore made arrangements and doubtless took an adequate share for himself). After the Second World War both conditions and pay for Chinese crews were improved significantly, yet even in 1961 the ordinary Chinese seaman was receiving take-home pay of £32 a month compared with his European counterpart who received £67.

12 *Telamon*, I, 1885

13 The launch of *Menelaus* (June, 1895, Alfred Holt (*third from left*), J. S. Swire
(*fifth from left*) (*photograph courtesy of John Swire & Sons*

Returning now to the early 1890s, we must ask if the 'new policy', and particularly the building of new ships, was responsible for the immediate revival from the low point of 1892? An examination of the voyage accounts shows that the four new ships delivered in 1892, the *Ixion*, *Tantalus, Ulysses*, and *Pyrrhus*, had net earnings in 1893 of £21 900. This was virtually the whole of the increase of net steamer earnings over 1892. In other words, four ships contributed between them some 44 per cent of total net earnings. The following year, 1894, these four ships and the first voyage of the new *Nestor* produced net earnings of £38 700, one third of the total. No wonder, therefore, that the Managers were keen to expand the fleet still faster, and introduce larger vessels. The critical success of these new ships on the China trade can also be seen in a different way. Although some ten of the older ships were re-routed to the Java trade after 1891, a few of the old vessels were kept on the China route for at least some years in the 1890s. An example was the original *Achilles*, while the *Telamon* (1885) also remained on the China run. Neither of these ships returned satisfactory profit figures in the 1890s. The *Achilles* returned an average of only £545 for 8 voyages made prior to 1896, while the *Telamon* returned £695. Such figures could be repeated for other vessels. The point is that without the new ships, the Blue Funnel vessels would still have been making inadequate net earnings on the main China run. The perception of the Managers in 1891 to launch their new policy, and their earlier careful husbandry of the reserves to enable them to do so, stood Holts in a formidably strong position when the new era of world trade expansion arrived in the decade or so before the First World War.

6

Combined Efforts: Holts and the Conference System

The Holt Line is stated to practically enjoy a monopoly of trade between this country and the East.

(J. S. Jeans)

The system of a Conference is by far the best you can possibly conceive.

(R. D. Holt)
(Evidence to Royal Commission on Shipping Rings, 1909)

Shipping conferences have loomed large in the history of Alfred Holt and Co. From the latter part of the nineteenth century conference affairs have been an enduring and integral part of Holts' prosperity and management responsibilities. The Holt story must be concerned with shipping conferences for several interconnected reasons. First, Alfred and Philip Holt were closely, though often reluctantly, involved with John Swire in the foundation of the first and ultimately one of the most important and long-lasting conferences in 1879, the China and Japan Conference (termed the Straits, China and Japan Conference in 1911 and the Far Eastern Freight Conference from 1941). Second, membership of various conferences has played a significant role determining the structure of freight rates, the routes operated, and the ports of loading and discharging. Thus both the prosperity and shape of Blue Funnel services have evolved for most of the company's history within the framework of the conference system. Moreover from the days of Richard Holt the Company's Managers have always been staunch upholders of the conference system and have devoted a great deal of time to its affairs. Richard Holt himself, Leonard Cripps, John Hobhouse, Herbert McDavid, Roland Thornton, John Nicholson, Lindsay Alexander, and others have made notable contributions to the management of the company's conference affairs. Third, since the conference system has had many critics, including those who have argued that the system operated to the long-term detriment of British shipowners, it is obviously relevant to raise the question of the overall impact of the system on Blue Funnel's commercial record.

The conference system which developed at the close of the nineteenth century was a product of special factors. One, certainly, was the inherent tendency for cargo-liner operators to face cut-throat competition, and the desire, therefore, to regulate such competition. Another factor was the dominant position of British lines on nearly all the long-distance trade routes and the fact that much of this trade centred on Britain itself, on parts of the British Empire, or areas where Britain had considerable commercial influence. Much trade, too, passed through British-controlled trading centres, and great entrepôt ports like Hong Kong and Singapore were of considerable significance in the structure of international trade. Thus, from the viewpoint of the liner operators, it was both rational and relatively easy to combine in some form of cartel arrangement which would restrict competition and raise freight rates. Later, in the inter-war years, the conference system had to adapt itself to growing competition from other mercantile nations. Still later, after the Second World War, further adaptation was necessary as the national lines of newly independent countries demanded shares in their own countries' trades. But the origins of the conference system were wholly British and provide an odd sidelight on a mid-Victorian period which prided itself on upholding the principles of unfettered competition and *laissez-faire*.

To understand the development and significance of shipping conferences it is helpful to remind ourselves of a few simple points about the economics of cargo liner shipping, which, like such other enterprises as railways, utility companies, and airlines, was a 'natural monopoly'. As such, it invited regulation. But it was a special feature of the shipping industry that regulation developed from within the industry, and from an early stage self-regulation involved not only national but international agreements. The very essence of cargo-liner shipping as it evolved at the end of the nineteenth century was the provision of regular services and fixed ports of call selling space in ships to shippers at specified freight rates. The typical cargo carried by a liner was composed of a great many, perhaps hundreds, of separate types of commodities emanating from a great number of small shippers. These shipments tended to be classes of goods which were relatively expensive in relation to bulk and where speed or regularity of service was important. There was, therefore, a rough division between the cheap bulky goods carried by slow tramp steamers (say coal, ores, or grain) plying from port to port in search of cargoes which would fill an entire ship, and the faster liners which would take goods in small 'parcels'. Obviously the liners' costs were greater, because of speed of services and the more complicated handling of many individual items, so that freight rates would naturally be higher. At certain times some cargoes, such as rice or wheat, might be taken by either tramps or liners, so that the two were to some extent in competition.

Since a part-empty ship could always carry additional freight at low marginal cost, and since some classes of freight could be carried by both liners and tramps, liner trades were vulnerable to rate-cutting. Moreover the seasonality of some trades and the need to provide both regular services and sufficient capacity to meet peak demands, meant the liners could by no means always sail with full loads. In the late 1870s two further factors precipitated the development of a Far Eastern Conference. One was the opening of the Suez Canal. On the one hand the Canal meant that existing steamships could make given journeys between Europe and Asia much more quickly than hitherto, which effectively increased carrying capacity. On the other hand the prospects held out by the Canal and new advances in marine technology encouraged such lines as Glen, Shire, and Castle, to build fast vessels in competition with established services like Blue Funnel.

The second factor we have discussed already: the imbalance of trade between outward and homeward trades. Excess tonnage on the outward journeys prior to the late 1880s encouraged competition for outward freight, while the changed pattern thereafter led to rate-cutting for homeward cargoes.

Conferences were develolped to solve these problems, shipping lines serving particular routes agreeing among themselves to regulate competition. This might simply mean rate-fixing, or there might be additional agreements covering the number of sailings, the ports served, the goods carried, or the sharing of freight and freight revenues ('pool' agreements). Additionally, conferences could be 'closed', refusing to admit outsiders to the arrangements. The means whereby conferences tried to prevent rate-cutting by outsiders, so that as much trade as possible was carried in conference vessels, were by measures taken to win the 'loyalty' of shippers. This might be the charging of a lower 'contract' rate to regular shippers, and a higher rate to those who sometimes used non-conference vessels. There was also the 'deferred rebate' system, whereby loyal shippers would receive a rebate on the charges they had paid over a certain period if, and only if, during the subsequent period they had not used non-conference lines. A typical rebate might be 10 per cent, and the deferred period six months, so that after a year of loyalty the shipper would get a six-months rebate. This was a powerful inducement for shippers to remain within the conference system.

The Far Eastern Conference, in which Holts were primarily concerned, was a 'closed' conference using the deferred rebate system from the outset (though under pressure from shippers during the great depression a contract rate was offered as an alternative after 1931). For many of the principal commodities carried and regions served the Far Eastern Conference operated pooling arrangements. A feature of the Conference was the existence of a main 'trunk' agreement between the United Kingdom and the Far East, outwards and homewards, and additional 'branch' agreements with different, though overlapping, membership. Thus before the First World War Blue Funnel participated in agreements covering the Straits–Europe trade, a 'Siberian' conference covering trade between Europe and North-China–Vladivostock and a number of others. After the war, when Ocean ships began loading at German ports, a North Continental Pooling Agreement operated within the Far Eastern Conference, covering outward trade between German and Dutch ports and the Far East.

As a rule, separate conferences were established for outward and homeward trades, and sometimes for individual legs of what might appear to be a single trade. Holts were full voting members of all the relevant conferences and subconferences covering their trades and the total number of conferences was considerable. A conservative estimate in 1909 put the number of formal conferences at 64, while by 1939 there were over 350 separate conferences in existence, while Holts (including Ocean Steam Ship, China Mutual, and NSMO) were members of more than 20 conferences at the latter date covering the Far Eastern, Australian, and American trades.

In theory, conferences were organisations representing shipping lines, not owners. Thus it was usual that when a shipping line was absorbed by another company, as when China Mutual was brought by Holts in 1902, the former line was maintained in name and as an accounting entity in order to keep an extra conference vote. When Holts acquired the Glen Line in 1935, Glen brought both its own conference votes and that of the Shire Line, the product of an earlier amalgamation. Thus from 1935 Blue Funnel had effectively four conference votes in the Far Eastern Conference. This practice explains the complicated charade whereby ships were assigned to China Mutual or Ocean, Ocean accounts presented China Mutual as an 'investment', and consolidated accounts incorporated returns for both China Mutual and Ocean. Yet it was a charade, for in every other way the fleets of the two companies operated as one and certainly the crews manning them were never aware of any distinction.

Shipping agents, of course, played a significant part in conference operations, since they were the ones who arranged cargoes, notified freight rates, and who tried to ensure that their principal's interests were not damaged by unofficial discounts offered by competitors. And this close relationship between agents, principals and conferences, brings us back to the very origins of the China Conference and the roles and attitudes of John Swire and Alfred and Philip Holt.

John Samuel Swire founded the first China Conference in London in 1879. He was the Conference's first Chairman until 1882, and when he died in 1898 the *Liverpool Journal of Commerce* described him with every justification as the 'Father of Shipping Conferences'. As Marriner and Hyde have written:

> The creation and continuance of conferences called for very great skill and perseverance and it was largely due to John Swire's efforts that certain conferences were conceived, came to fruition and survived the many hazards that they had to face... For nearly twenty years John Swire was *the* constant factor in Eastern liner conferences.[1]

The 1879 Conference, called officially 'The Agreement for the Working of the China and Japan Trade, Outward and Homewards', was signed on 29 August. It was not the first shipping conference. Already in 1875 seven British lines operating regular services between the United Kingdom and Calcutta agreed to regulate the sailings each would make, and to fix minimum freight rates in both directions of the trade. Later, in 1877, a rebate system was introduced. John Swire may or may not have been influenced by these arrangements, but he was certainly involved even earlier in various schemes to regulate competition on the Yangtze river following the foundation of his China Navigation Company in 1872. There was quickly an agreement with his erstwhile bitter opponent Russell and Co. to pool earnings and equalise sailings on the river. This was but one of a number of 'joint purse' and freight rate agreements. To some extent, therefore, the first China Conference grew logically from Swire's experience in the river trades.

But what appears to have precipitated the first China Conference was a combination of two factors discussed earlier which became apparent in the late 1870s. One was depression and competition in the China trades caused by post-Suez overtonnaging and the arrival of fast Glen, Shire, and Castle steamers. The second was the growing relative inefficiency and uncompetitiveness of the older Blue Funnel vessels. Here it must be stressed that from the moment he acquired the Holt agency John Swire always placed an absolute priority on maintaining the strength and prosperity of the Blue Funnel Line. The very foundation of Butterfield and Swire in 1869 had been largely as an adjunct to the Holt agency. Swire's solicitude could be put down to natural self-interest. Blue Funnel had extensive business in the Far East, John Swire perceived the great future in store for the line, and Swires handled considerable cargoes for the ships both through John Swire and Sons in London and through Butterfield and Swire in China and Japan. At the same time the Swire family were considerable shareholders in the Ocean Steam Ship Company, and the fortunes of both companies touched at various points. In short, the Blue Funnel agency was a valuable one for Swire. Yet as we saw in a previous chapter there was a strong non-commercial side to the relationship between the Holts and Swires. The Holts and Swires had deep ties of friendship, mutual esteem, and a common outlook on business affairs. Extraordinarily, it would appear that Swires made only a very modest profit from the Ocean agency during John Swire's lifetime. In 1880, for example, John Swire informed Philip Holt that the expenses of operating as London agents (over £2500 annually), which included one half of Swire's London office, considerably exceeded Swire's charges of £1740. In 1892 Ocean paid Butterfield and Swire a gross commission of £7000 for their representation in China and Japan, but total expenses were £5500 leaving a profit of only £1500. Again and again there are examples of John Swire foregoing his own immediate interests (for example, especially in the early years, by trading in textiles and other goods at a loss so as to provide cargoes for Holt ships), in order to protect Holts' position.

This close relationship between the two firms induced John Swire to advise, cajole, and criticise the Holts whenever he felt, which he did frequently, that they were neglecting their own best interests. By the late 1870s he was convinced that the Holts were being outclassed by their competitors, especially by the Glen Line. Swire wrote to his Butterfield and Swire manager in June 1879 'That *Line* is breaking my heart. Had my advice been followed those ships would never have been built. Now they are lapping up the cream of the trade and leaving the skim milk.'[2] Swire's worry was that with slower ships Holts would be forced to offer lower freight rates than the fast Glens or Shires, and that with rapidly descending levels due to overtonnaging such rates might fall to ruinous levels. One solution sought by Swire was to badger Holts to modernise and improve the Blue Funnel fleet, which he did without conspicuous success until a few years before his death. The other was to rationalise the existing China trade by the same sort of agreements among competing lines with which he was already familiar in the Yangtze trades.

Table 6.1, drawn from Richard Holt's evidence contained in the *Report and Minutes of Evidence of the Royal Commission on Shipping Rings* (1909), puts into some perspective the sharp deterioration in freight rates which afflicted the Far Eastern trades in this period.

The China Conference incorporated both a pooling arrangement, at first confined to British lines and to cargoes of Yorkshire and Lancashire woollen and cotton 'fine goods', as well as a deferred rebate system, as already mentioned. The original signatories were the five British lines and one French line already providing regular scheduled services, and two agency firms. The British lines were Ocean, P&O, Glen, Shire, and Castle. The French line was Messageries Maritimes and the agents were

Table 6.1   Far East trades, homeward freight rates, 1874 and 1879
(shillings per ton of 40 cubic feet)

	1874	1879 (lowest)
Japan		
Tea	105	60
Waste silk	108½	50
General merchandise	108½	62½
China (Shanghai)		
Tea	70–80	30
Waste silk	75–80	40
Bristles	90	45
China (Hong Kong)		
Tea	80	60
Waste silk	90	80
Straits (Singapore/Penang)		
Gutta percha	100[1]	55
General merchandise	100[1]	60

Note: [1] 1875.
Source: *Report of the Royal Commission on Shipping Rings, together with Minutes of Evidence* (London: HMSO, 1909) Appendix XVII, Evidence of Mr R. D. Holt, p. 102.

Norris and Joyner, and Gellatly, Hankey, Sewell and Co. Three of the ten departures a year given to Norris and Joyner (who, as agent for Shire, had to accommodate their services too) were allotted under a separate agreement to William Thomson's Ben Line, while the original Agreement noted that 'in the departures provided for, as well as in any further requirements which may hereafter become apparent, the steamers now owned by Messrs William Thomson and Company, Messrs John Warrack and Company and Mr Charles Williamson, all of Leith, shall follow the above-named eight vessels in the preference of the berth, provided their owners have not traded against the interests of those subscribing to this agreement'.[3] The first conference lasted only four months, ending in December 1879. John Swire, confident of his creation, thought the cessation 'will prove but a temporary suspension'. The Hong Kong paper, the *China Mail*, on the contrary, thought the end had come for 'One of the most ill-advised and arbitrary attempts at monopoly which has been seen for many a year', and added 'one is apt to get confused as to whom the ocean belongs'.[4]

The detailed history and vicissitudes of the conference system need not concern us. Suffice to say that John Swire's confidence was quickly justified and the Outward China Conference was, after the initial break, renewed for successive periods until 1887. Problems with the Homeward Conference, though, were bedevilled by overtonnaging and frequent rate-cutting wars. This conference broke down immediately and no new agreement was reached until 1881 which in turn lasted only until 1882, and no further conference was arranged until 1885. In 1887 both conferences were suspended during the important legal action against the system brought by the Mogul Line, an action which Mogul took right up to the House of Lords in 1891 when the legality of conferences was fully upheld. Not until 1893 were the conference agreements renewed, but then, apart from a short break in 1897, the China Conference became a permanent feature of international trade. Gradually membership was extended. In 1884 the Ben Line entered as a full member, and in 1885 the erstwhile anti-conference China Mutual Company, which only began operations the year before, was accepted. In 1893 two German lines, North German Lloyd and Kingsin, the latter to take pig iron to Japan only, were admitted. In 1894 Rickmers, another German line, gained brief entry, and the first Japanese line, NYK (Nippon Yusen Kaisha), became an associate member of the outward conference in 1896, a full member in 1899, and a member of the homeward conference in 1902. NYK's entry in 1899 brought them London loading rights which directly affected east-coast lines like P&O and Glen. In consequence P&O obliged Holts to give up their existing rights to three London loadings each year as the price for P&O not invading Liverpool. This was a typical conference arrangement, changes deleterious to one member usually leading to new terms with the burdens being shared.

Meanwhile a Straits Settlements Homeward Conference, with a membership largely overlapping that of the China Conference, had been formed in 1885. This Conference, too, suffered teething problems. The first Conference lasted only until 1887 and a further attempt in 1893 broke up in 1895. A constant problem was the power of a small group of some dozen British merchant houses who controlled much of Singapore's trade. From 1897, through the initiative of Holts, though later spearheaded by P&O, a permanent conference organisation came into being when these merchants were placated with a special rebate which continued until 1911 when arrangements similar to those in the Far Eastern Conference were established.

Despite Ocean's significant gains from the conference system the attitudes of both Alfred and Philip Holt were at best lukewarm and on occasions hostile. Philip Holt evidently played the key role in the Company's early conference arrangements, as made clear by the often acrimonious tones of John Swire's letters. Although agreeing to join the China Conference as founder-members, the Holt brothers' equivocation is perhaps understandable. Both Alfred and Philip were strong believers in competition. They had created their business through the competitiveness of their fleet and in the process they had built up what seemed an impregnable position in the outward carriage of Lancashire and Yorkshire textiles to the Far East. If their slower ships were less than competitive with the fastest Glens and Shires for the homeward carriage of tea to London, they could still make profitable round voyages, and the Holts wanted the flexibility to undercut their high cost competitors in order to maintain homeward cargoes. Rate-fixing was regarded as something of a threat to this flexibility, and in early conference negotiations Holts pressed to be able to charge lower rates for their slower ships. Swire's constant pressure upon the Holts to build faster and more efficient ships thus seemed, from the perspective of India Buildings, ill-judged.

Even before the first conference agreement had been signed John Swire was complaining that Holts 'would sooner miss earning an extra £100 000 for the OSS than allow the Glens, Castles, P&O, and MM to gain the same sum amongst them', and throughout the early operations of the conference Holts' actions were frequently a source of embarrassment to Swire. There was obviously ambivalence in Holts' attitude. The Managers were well aware that the scheme could be of considerable benefit. They told shareholders in February 1880, after the first few months of the agreement, that freight rates had risen and profits increased, and that 'Before leaving the subject it is only fair to add that the entire credit both of the conception and execution of the idea of a Conference is due to the energy of Mr John S. Swire who has so long been connected with the Company as its London agent and is a large shareholder.' Yet in the same year, 1880, Holts attempted to invade a P&O trade without any word to Swire. The break-up of the Homeward Conference in 1882 occurred when Holts refused to allow Shire and Glen the lower freights for their slower ships which Holts claimed for themselves. Disheartened by the attitude of his principals John Swire resigned as conference chairman in April 1882 although he subsequently took over the reins once more and remained the dominant force in conference affairs until his death in 1898. He wrote at the time of the break up, 'It is a serious affair – the old conference was made for him [Philip Holt] and he did all in his power to render it unworkable.' It appears too that Ocean was alone in opposition to Swire's plans, for Swire also wrote, 'In justice to the other members of the conference, I must say that no man ever drove a more reasonable and willing team. P.H.H. was the spare horse, and he amused himself by putting his head between his legs, and smashing the manger with his heels. When he looks round, he will see that he has lost his corn.'

Eventually, though, Alfred and Philip Holt became converted to conferences, and by the end of the 1880s were staunch upholders of the system. At the same time they were converted, too, to the need for larger and faster ships. We have seen in a previous chapter how the Ocean managers became convinced of the need for new ships as a result of changed trading conditions in the 1880s. Events in the decade, too, pushed them unwillingly towards the conference. A significant factor was the organisation in 1882 of a new line formed specifically to fight the conference, the China Shippers' Mutual Steam Navigation Company, a company destined to play a big part in Ocean Steam Ship affairs, which sailed its first ships in 1884. China Mutual posed a major threat for Ocean, loading in Glasgow and Liverpool and thus competing directly with Blue Funnel cargoes of Lancashire and Yorkshire fine goods. This transformed Holts' competitive position. Holts responded partly by loading their own ships in Glasgow in 1887, but partly by becoming more supportive of the conference. By 1888 the Managers were informing the Annual Meeting of Shareholders that, when legal action against the conference was threatened, 'they do not think it possible to carry on their business without combinations of a character similar to that of which the legality is questioned'. China Mutual meanwhile, despite its anti-conference origins, soon joined the China Conference and, except for a period between 1887 and 1891 during the conference fight with the Mogul Line, became a staunch member. In 1902, as we know, China Mutual entered the Ocean stable, bringing with it the conference vote which it then possessed. Already in the early 1890s Ocean and China Mutual were operating a secret 'joint purse' agreement covering Lancashire and Yorkshire textile goods exported to China and Japan in which earnings were pooled and then allocated between the companies in proportion to the gross tonnage of ships in the trade. Such 'pools' were usually confined to certain classes of cargo, and over-carriers would compensate under-carriers at an agreed rate.

The growing importance of Holts' conference business warranted the formation of a new 'Pool Department' in 1902, headed initially by the newly arrived Secretary from the China Mutual. Before this, work on the various freight pools had been distributed rather widely, the Lancashire and Yorkshire pool by the Outward Department, the Straits and Java homeward by the Inward, and the China homeward by John Swire and Sons in London. The rapid growth of paperwork led the Company to purchase in 1906 a specially designed electric adding machine from the Burroughs Company – the largest and most complex such machine yet made by Burroughs and it remained in use in the Pool Department until May 1941 when Hitler's bombs brought its eventual demise.

From the outset the conference system proved able to raise freight rates and also, by providing an organisation within which various types of pooling and other arrangements could be made, produced a system capable of future development. John Swire himself thought that the first six years of the China Conference had resulted in higher rates there than in any other trade. Since we are here concerned only with those aspects of the conference system of direct relevance to Holts the detailed development of conferences cannot be considered. We should stress, though, that once the Ocean managers had embraced the system they played a leading and usually dominant role in conference policy. They were active too in the further extension of the system, as in the arrangements for the Straits Conference in 1897. Richard Holt was one of the leading witnesses before the Royal Commission on Shipping Rings which reported in 1909. His closely reasoned testimony in favour of the closed conference system and deferred rebates was vindicated by the Commission's majority report.

Mention of the Royal Commission on Shipping Rings should remind us of the shippers, or 'merchants', who were the ultimate providers of cargo for shipping companies like Ocean, and whose existence was the fundamental purpose of liner services in the first place. To go beyond the broadest of generalisations is difficult, for shippers came in all shapes and sizes, from the regular provider of seven or eight hundred tons of Manchester textiles to the Far East to the occasional provider of small parcels; from the manufacturer dealing directly with the shipping company to the merchant dealing separately with independent manufacturers and then acting through a cargo-broker. In a chapter dealing with the conference system, however, it is convenient to say something about shippers since it was often held that conferences acted against the interests of shippers and also because the Royal Commission already mentioned took evidence from shippers and hence we have an opportunity to hear from them directly on the quality of services with which they were provided.

The Report and Evidence of the Commission, must have been very pleasing to Alfred Holt and Co. Not only were there virtually no criticisms from shippers either on freight rates or the quality of service on the main outward services in which the Company was engaged, but some flattering things were said about Holts themselves. This was perhaps the more surprising in that a characteristic of the China trade was the dominance of large shippers, for it was sometimes said that conferences operated in favour of small shippers and against the interests of the large (who deserved special rates for large shipments). Thus Alfred Zimmern, partner in the firm of Reiss Brothers, one of the largest shippers of cotton goods to China, was asked by a Commission member: 'I understand that during your experience you have always found these qualities of regularity of departures, excellence of service, and steadiness of rates of freight, and that you have nothing to complain of in your experience with regard to these matters?' And Zimmern replied, 'that is so, especially in the case of Messrs Holt, who have done more for the China trade than any other firm'.

Holts seem to have been responsive to representations from shippers. In 1902, for example, a deputation of merchants approached the Company during a trade recession, asking for a reduction of freight rates. The then President of the Manchester Chamber of Commerce, Mr Thompson, who was partner in the firm of Stewart and Thompson, 'one of the largest China houses', was part of the deputation. The Commission were told of Holts' readiness to meet reasonable requests, although, as we will see shortly, the readiness may have been connected with the coincident purchase of the China Mutual Company. Richard Holt himself received the deputation, 'they put their case before us, and we considered the whole matter, and thought it was a proper occasion for meeting them'. Accordingly freight rates on cotton goods were lowered substantially, on heavy piece goods, for example, from 65s to 45s per ton.

The first Far Eastern Conference chairman was John Swire, and from 1879 until 1976 British lines provided every successive chairman. Prior to 1963 John Swire's successors always came from P&O, but subsequently there were three Ocean chairmen, Sir Herbert McDavid (1963–5), W. H. McNeill (1969), and Sir John Nicholson (1969–72), while Nicholson's successor was H. O. Karsten from Overseas Containers Ltd, former chairman of Glen, and chairman of the Far Eastern Freight Conference until 1976. The Conference secretariat was at first in the London offices of John Swire and Sons, initially consisting of a part-time member of Swire's staff and a desk. By 1932 four rooms were rented in the same building as Swires, and the staff had risen to three: the Secretary, a clerical assistant, and a typist. In 1934 C. R. Hawkins was seconded from Blue Funnel; in 1935 he became Assistant Secretary and, in 1939, Secretary, in which post he remained (except for the war years) until his retirement in 1964.

The key feature of the conference system upon which much of Blue Funnel's prosperity rested was the so-called Lancashire and Yorkshire Agreement of 1911. This Agreement was made between the British members of the Outward China Conference, together with NYK of Japan. The Agreement formalised and refined various arrangements which had been made from time to time since 1887 and had resulted in a clear division between 'west coast' and 'east coast' lines; that is, those lines like Ocean agreeing to load outwards only at the United Kingdom's western ports, and the other lines which would restrict outward loadings to London and other east coast ports. These groups then further agreed on a 'pool' of Lancashire and Yorkshire fine goods. In 1894 there had been three west-coast lines, Ocean, China Mutual and Mogul, but by 1911 the Mogul Line had ceased activity and China Mutual was now joined with Blue Funnel.

The Lancashire and Yorkshire Agreement was destined to remain the basis of the Far Eastern conference system for many decades, only disappearing in the age of containerisation in the 1970s. A P&O memorandum in 1951 recorded that 'In our view the Lancs/Yorks

Agreement remains the basic agreement of the FEFC.' It is worth recording the opening paragraphs of this important document in full:

AGREEMENT between the Peninsular and Oriental Steam Navigation Company, Nippon Yusen Kaisha, Glen, Shire and Ben Lines, hereinafter called the East Coast Lines, and the Ocean Steam Ship Company, Limited, and China Mutual Steam Navigation Company, Limited, hereinafter called the West Coast Lines, for working certain portions of the outward Straits and China and Japan trades.

1. Yarn and all fabrics of wool, cotton, silk or mixtures thereof, manufactured and packed in Lancashire and Yorkshire, carried by the above seven Lines for the Straits, China and Japan, shall be considered as cargo in common, in proportions as follows:

Straits

P&O, 12 points; NYK, 8 points; Glens, 5½ points; Shires, 5½ points; Bens, 7 points; OSS and Mutuals 70 points.

China and Japan

P&O, 17½ points; NYK, 16 points; Glens 6¼ points; Shires, 2¼ points; Bens 3 points; OSS and Mutuals 71 points.

2. Separate statements shall be made for cargo to the Straits and for cargo to China and Japan. Over Carriers shall return to Under Carriers freight on the excess tonnage carried, calculated at the average freight and primage earned by the former during the period in question on the cargo above enumerated, less twenty-two shillings per ton, out of which deduction they are to pay any 'returns' arranged with Shippers, but when East Coast Lines overcarry West Coast cargo they pay on that portion at Ocean and Mutual average rate less twenty-seven shillings per ton.

3. None of the Lines shall carry bale or case goods from Singapore northwards under a rate of twenty shillings per ton measurement.

The significant points are two. First, the formalised division among British lines of loading rights between west and east coasts. Apart from limited rights for Japan's NYK between 1919 and 1941, restricted to 13 sailings a year, the west coast ports, which included Liverpool and Glasgow, were allotted solely to Ocean and China Mutual. Thus Blue Funnel ships had a virtual monopoly of west coast sailings to the Straits, China, and Japan. Secondly, the agreement specified a pooling arrangement for Lancashire and Yorkshire fine goods, with other manufactured goods added later. Not all cargoes were pooled. There were classes of 'general cargo' which were subject to fixed freight rates but not pooled (though general cargo from most continental ports was pooled in the 1920s, and a separate North Continental Far East Conference established in the 1930s), and a third category of cargo was 'open' and shippers

Table 6.2   Far Eastern Conference: west-coast loadings (%)

	1927–9	1932–4
A. Holt and Co.	84.5	90.0
NYK	12.3	9.5
Blue Star	3.2	0.5
	100.0	100.0

Source: Ocean Archives: Conference documents.

could use non-conference ships without penalty. But the pooled cargo and monopoly of west coast berths gave Blue Funnel a powerful position, assuring Holt ships of around two-thirds of the pooled cargoes for the outward trade to both the Straits and Japan. The combination of large shares of trade and conference-maintained freight rates was a propitious one.

The extent of Holts' monopoly on the Mersey can be seen in Table 6.2 in the figures for the shares of total carryings from west coast ports to the Far East in the inter-war years.

Table 6.3 shows the pool shares allotted to Blue Funnel, shares which remained unchanged until the Second World War, though there were revisions for other lines in October 1936. Under the Agreement an overcarrier could keep a certain proportion of the excess and pool the remainder, in order to encourage competition within the pool. In 1922 the proportion was set at 60 per cent, so that members pooled only 40 per cent of earnings above the allotted proportions. In 1934 Holts carried 64.3 per cent of the goods covered by the Agreement to the Straits, and 59.5 per cent to China and Japan. By 1938 the Holt shares had risen to 98.2 per cent and 89.6 per cent respectively.

The separate North Continental Pool gave Blue Funnel ships around 9 per cent of all loadings from the specified ports to the Far East, in the 1930s equivalent to a total by 1936 of around 6500 tons a month, or one

Table 6.3   1911 Lancashire and Yorkshire Agreement, pool shares, 1911 and 1936

Company	Straits trades		China and Japan trades	
	1911	1936	1911	1936
A. Holt and Co.	64.8	64.8	61.2	61.2
P&O	11.1	20.4	15.1	12.9
Glen/Shire	10.2	6.9	7.3	5.2
Ben	6.5	4.2	2.6	1.7
NYK	7.4	3.7	13.8	19.0
	100.0	100.0	100.0	100.0

Source: Ocean Archives: Conference documents.

monthly sailing. At this time Blue Funnel ships were carrying roughly one half of all cargoes carried from United Kingdom ports in British ships to the Straits, China and Japan.

Most homeward cargo was also pooled, the shares in the pool being adjusted frequently after the First World War to take account of new entrants (there were 14 full members in 1935). Blue Funnel carryings in the mid-1930s were some 17 per cent of total homeward loadings from Hong Kong and 15 per cent from Shanghai, and no other line, British or foreign, carried a greater share.

What the homeward pools meant for Blue Funnel in terms of volume of shipping can be seen from the figures for 1938 shown in Table 6.4 and already discussed in another context in Chapter 3. What is striking is how by the eve of the Second World War three-quarters of the pooled cargo space from China, the Philippines, and the Straits was allocated to freight from the Straits alone.

As the conference system itself developed, and as Ocean became involved in more trades, so the Company became a full or associated member of a growing number of conferences. The Far Eastern Conference remained the cornerstone of Holt conference policy, but by the eve of the First World War the Company was a party to agreements covering nearly all the trades in which it was engaged. Soon after commencing the direct Java–Europe service in 1891, Holts came to an arrangement with the two Southampton-bound Dutch Mail lines (Netherland Steamship and Rotterdam Lloyd) on freight rates and sailings. This inaugurated a long period of cooperation between Blue Funnel and the Dutch lines. In 1900 the Dutch Mails and NSMO formed the Batavia Freight Conference, an agreement to regulate the Deli tobacco trade was drawn up between KPM (Koninklikjke Paketvaart Maatschappij) and Ocean in 1908, and a more general Deli Freight Conference formed in 1921. The latter year also saw the start of a joint New-York–Java service between Blue Funnel and the Dutch Mail lines, covered by conference arrangements.

Table 6.4 Blue Funnel Far Eastern Homeward pool allocation, 1938

From:	Space (000 tons)
Shanghai	19.1
Hong Kong	38.4
Manila	38.5
Straits	360.0
	456.0

Source: Ocean Archives: Conference documents

The Australian services were also regulated by conference agreements. When Blue Funnel ships first commenced regular Australian services in 1900 there existed an outward United-Kingdom–Australia conference only. With the growth of refrigerated homeward cargoes thereafter (a trade in which Blue Funnel ships played a full part) pressure grew among shipowners for a homeward conference, and this came into effect in 1913. Lord Inchcape, of P&O, took the initiative in forming the conference, and the Oversea Shipping Representatives' Association was set up in Sydney to represent the conference lines. Blue Funnel was an original member of the group which, in addition to P&O, included White Star, North German Lloyd, Messageries Maritimes, and a few others.

Australian conference history was dominated in the early years by hostility towards the 'Great Combine' from government and shippers alike. An Australian Prime Minister of the period, W. M. Hughes, argued that 'Except for the Commonwealth Line, there is no way to the markets of the world, save at the price that the Great Combine's Lines determine.'[5] From 1909 until 1929 deferred rebates were illegal for outward trades (from Australia), while the existence of the state-run Commonwealth line, operating outside the conference, caused considerable friction. The Commonwealth line was set up in 1916, and after some intense post-war struggles was forced to come to terms with conference agreements in 1923. The line was eventually sold in 1928. An added problem was endemic labour problems on the Australian waterfront, leading to rising costs for shipowners and constant pressure to raise freight rates. Eventually, in 1929, new arrangements led to the formation in Australia of an official body, the Australian Oversea Transport Association, which could ratify conference agreements, including deferred rebates. Effectively, therefore, the conference monopoly was recognised in return for some safeguards for Australian shippers.

Conference arrangements were always weakest in the American trades where, as a result of the Merchant Shipping Act of 1916, both the closed conference and deferred rebate were declared illegal. The result was that conference terms were hard to enforce and both Blue Funnel's Trans-Pacific service and the 'round-the-world' New York service (regulated by the Straits–United-States Conference) were plagued throughout the inter-war years by low freight rates, intense competition, and low profitability.

By 1934 Holts were full members of some dozen conferences, including the Far Eastern Conference and its various branches (the main Conference having 23 members at this date), the Outward Continent–Far-East, Outward and Homeward Europe–Philippines, the Colombo Homewards and Colombo–Continent Conferences, the Batavia Freight

Conference (through NSMO), the Deli Freight Conference, the UK–Australia Conference, the Continent–Australia and Australia–Europe Conference, and various conferences dealing with the American and Pacific trades. Holts was also an associate, non-voting, member of a great many conferences and agreements. An example here was Blue Funnel's associate membership in 1938 of the Indo-China General Freight Conference of Saigon. Under this agreement Blue Funnel ships could load in Saigon on a limited number of occasions (up to 13 a year) for a limited number of ports and destinations – the United Kingdom, Rotterdam, Amsterdam, Bremen, and Hamburg.

II

Particular aspects of Ocean's conference dealings, such as the company's policy towards the re-admission of ex-enemy shipping lines after the two world wars and various problems with competition inside and outside the conference will appear in other chapters. However, before leaving the subject of conferences we should at least touch on the criticisms which have been levelled against the system. Apart from the opening few years of the conference system, Holts have been firm in their support of conferences, have done much to initiate and sustain various agreements, and remained dominant in the Far Eastern Conference. The Managers were, in consequence, highly sensitive to the many attacks made upon the conference system and always defended the system with vigour.

Shipping conferences have never been without their critics. Against conferences it has been argued that rate-fixing and the practice of deferred rebates penalised shippers at the expense of shipowners, that they promoted inefficiency and complacency, that they discriminated against the national-flag-carriers of newly developing nations who were denied access to 'their' trades, and a host of other charges. Defenders of the system pointed out the advantages to shippers of regular, high quality services and stable freight rates paid by large and small shippers alike. Cut-throat competition and the disappearance of such services would be in no-one's interest. Defenders also argued that conference members had to provide a competitive service. Within the conference there would be competition between members in services while the freedom of shippers to use tramps, non-conference liners, or chartered vessels, would provide an effective spur to the performance of conference lines.

The long and complex history of attacks upon and defence of the conference system should warn us against snap judgements. Certainly, though, we should note that running through all the official and semi-official enquiries which have been held into the system is general agreement that some sort of rate fixing agreements among liner companies has been necessary. As early as 1902 an enquiry into the Straits Conference vindicated the principles upon which the Conference rested, while a British government inquiry commenced in 1906 and produced a Report, mentioned earlier, in 1909. In the United States a similar investigation was started in 1912 and reported in 1914. Both these latter investigations concluded that unrestricted competition in liner trades could not be sustained, and that shipping conferences were a necessary device to give shippers freight rate stability, regular services, protect the small shipper against the large, and allow the shipowner to operate at a reasonable profit. The deferred rebate system was accepted in Britain (though a Minority Report was more critical) but in the United States, where it was felt to conflict with anti-Trust legislation, was outlawed. The Merchant Shipping Act of 1916 therefore made deferred rebates and some other methods used by conferences to induce 'loyalty' illegal. This legislation, incidentally, was largely responsible for the fiercely competitive conditions which afflicted all the American trades after the First World War. Between the wars, too, a number of further inquires outside America, by the Imperial Shipping Committee and other bodies, were also generally favourable towards both the principles and methods of operation of the conferences.

The matter of the impact of the conference system upon efficiency, though, is elusive. Really there are two separate considerations. A simple rate-fixing agreement among lines otherwise in competition should not prevent improved services (for example, the introduction of faster ships), for each line would try to maximise its share of the trade. However, pooling and other measures to preserve shares of the trade for particular members were different. Where arrangements were such that an over-carrier paid an under-carrier in full the excess freight earned, or where one line had a protected monopoly to or from a particular port or in a particular commodity, an inefficient line might well be protected under such a system.

Does Blue Funnel's history shed light on the issue? At any rate from the experience of Alfred Holt and Co. there is little direct evidence that conferences promoted inefficiency. On the contrary, it was Blue Funnel inefficiency *before* the conference system which had to a large extent brought the China Conference into being. Holts' revival after the 1890s was achieved within the framework of the conference system. Between the wars the company was quick to introduce diesel ships, pioneered the shipments of bulk latex, and sailed the first ship to carry successfully gas-chilled beef from Australia. Moreover the new Glen ships of the 1930s saw Glens set to capture an increasing share of the east-coast trade before war intervened. After the war, adherence to the Far Eastern

Conference did not prevent the company leading the way in the development of containerisation, the single largest improvement in shipping efficiency since steam replaced sail. Moreover if it is possible to see some decline in Holt competitiveness after 1950 by comparison with, say, the expansive Ben Line, it should be remembered that Ben, like some of the dynamic Japanese and Scandinavian Lines, was a conference member. There is, in short, no obvious relationship between conference membership and inefficiency.

This reference to Ben competition after 1945 does, however, raise a significant perspective on Holts and the conference system. For Blue Funnel, more than for any other British shipping line, the conference system bestowed a monopoly the preservation of which became a principal object of policy. The monopoly was the Liverpool berth to the Far East established in the 1911 Lancashire and Yorkshire Agreement, as we have discussed earlier. It is at least arguable that Holts' long-term interests were damaged in two respects. First, they were fearful of expanding in other trades lest disgruntled lines should retaliate by challenging Blue Funnel in the west-coast trades. An example of what might happen occurred in 1910 when Blue Funnel's new Australian passenger service via the Cape offended the Ellerman Line, which promptly tried to enter the Liverpool berth. Moreover Holts felt they needed the full support of other British lines to preserve the west coast from foreign invasions as far as possible.

The second problem was that Holts felt obliged to run as many services as possible to and from Liverpool and Glasgow, even in depressed years with half-full sailings. Such a policy, which was deliberately pursued by Richard Holt between the wars, may have given an excellent service and staved off competition from outsiders. But it cannot have done much to promote profits or efficiency. As we will see, by the outbreak of the Second World War this saturated service was being conducted in increasingly ageing and relatively small ships. Arguably the conference system induced a somewhat defensive attitude which both encouraged over-provision of services in bad times, and a reluctance to expand in new trades in good. Arguably, too, it produced a management mentality drawn towards cooperation and combination with like-minded associates. The extent to which Holts pursued post-1965 diversification in concert with 'friends' is striking.

7

The New Century:
Profits and Perils

The shareholders were so pleased.
(R. D. Holt, *Diary*, 1916)

The twentieth century dawned with Blue Funnel in the midst of an expansive and prosperous phase. New ships and tighter conference arrangements had given Holt vessels an edge over their competitors, while large reserves and conservative finance enabled the Company to ride with some ease the ups and downs of the trade cycle which were part and parcel of commercial activity. Despite such cyclical movements, the years between the late 1890s and 1914 saw a sustained growth of Britain's imports and exports, and all the main Blue Funnel trades participated in the expansion. For example, the volume of imports from Malaya and Singapore to the United Kingdom approximately trebled between 1900 and 1914, imports from Japan doubled, and from the Dutch East Indies nearly trebled (far more significant here was a doubling to the Netherlands, since the absolute amounts carried were so much greater). All these regions provided sources of a growing variety of primary products which formed the bulk of Blue Funnel cargoes. As far as Britain's exports were concerned – still dominated by cotton textiles and other basic manufactured goods – the great Blue Funnel markets of China, Hong Kong, and Japan were all buoyant in these years. China's direct purchases of British goods roughly trebled, while China's imports from Hong Kong (which included transhipped British goods) also expanded rapidly. China's imports of British cotton goods were valued at 75 million Haikwan taels in 1900, and 183 million in 1913 (the tael then worth about 2s 6d). The same story of thriving imports and exports could be repeated for other trades.

Recalling the period around 1900, George Pridgeon, a stevedore foreman with the Company noted the tremendous advances made during the preceding decade.[1] He wrote:

Ten years before, the China Coy's [i.e. Holts'] trade to the Far East comprised principally fine goods only. Consider, now, the great advancement which had been effected since that time, from a commercial point of view.

Fine goods were still being shipped, but in addition we loaded miscellaneous machinery of all descriptions – railway materials, bicycles, soap, sewing machines, whiskey, brandy, beer, tobacco, and large quantities of chemicals from Brunner-Monds, and numerous other items. The Far East trade had developed: the Far East itself had developed!

Ships were bigger still. Cargo-carrying capacity had advanced to about 7,000 tons deadweight. The ports of call were as follows:

China . . . Penang, Singapore, Hong Kong, Shanghai, Kobe and Yokohama
Java . . . Batavia, Semarang and Sourabaya.

The foreman also wrote of the famous 'goal-poster' ships, introduced a few years later:

They carried cargo for Hong Kong, Vancouver, Victoria, Seattle, Tacoma and Los Angeles. Generally, like all the other ships of the Company, they commenced loading at Glasgow and usually arrived at Birkenhead requiring about 7000 tons to complete the cargo. Having so many ports of call the cargo they carried was very miscellaneous. To work on them was to be carrying out a real tough job, and almost always resulted in a heavy run of overtime – very often all night through . . . At about this period more ports of call in China came on the Company's list, and each fortnight a ship was loaded on the berth for Shanghai, Tsintau, Tientsin and Dalny. The Company, on the opening of the Vittoria Dock, took over all of one side of it and relinquished their berths at the 'Tin Shed' for the Vittoria Wharf. They now had four berths close together and, divided out amongst different ships, they loaded cargoes for no less than approximately 20 different ports of call. Thus then they had advanced from two ports to twenty in about twenty years! Progress with a capital 'P'!

And:

advancing now to the year 1913, I can write of it as the peak year for the Company. Never before in the history of the firm had there been the amount of cargo offered for shipment as in that year. In 1913, from Birkenhead, we shipped no less than 550 000 tons deadweight of cargo for the ports in Java, the Malaya Straits, China and Japan. A record! And also in that year another record was made. In one week, we loaded in Birkenhead, 22 000 tons of cargo for the ports mentioned above.

And what a change there was in the cargo! From cotton goods only,

25 years ago, we now shipped every possible commodity one could think of! I have looked at the cargoes piled up on the quayside and in the sheds and tried to wonder at the progress made by the peoples of the Far East in so short a time.

Not only was there an underlying prosperity in the years 1900–14, but even events which might be expected to be deleterious proved otherwise. The anti-foreign uprising in China in 1900, called the Boxer Rebellion, and which provoked Western retaliation, was one such. Reviewing the trade of that year, the Managers reported to shareholders that 'The political disturbances have had surprisingly little immediate influence upon the Company's business, shipments of coal for the warships having taken the place of ordinary cargo during the short time when there was a substantial falling off.' The war between Russia and Japan in 1904 provided a timely boost to Blue Funnel trade at a time of depression, with the temporary withdrawal of Japan's merchant fleet and a booming demand in Japan for war goods.

As we have seen, Blue Funnel's expansion was a product not only of buoyant traditional trades, but of new routes and new ventures in property overseas. In 1901 Ocean had started a service to Australia from Glasgow and Liverpool via the Cape, and in 1910 entered the Australian passenger trade. In 1902 came an event of great significance for the Company, the acquisition of its main rival, China Mutual, which had an excellent Far Eastern service and a fine fleet of modern ships. This purchase, which was something of a coup, brought at one swoop a virtual Holt monopoly of western berths to the Far East, an additional seat on the Far Eastern Conference, and a new trade between China and the western seaboard of Canada and America across the Pacific Ocean. This Pacific trade proved very profitable in the pre-war years, and, together with the Australian and Java trades, was producing total earnings for the Company by 1913 about equal to those of the main line China and Japan service.

Although China Mutual had been born in anger, committed to fighting Holts and the conference systems, relations between the two lines had become during the 1890s close and cooperative. From 1894, when the Outward Conference was re-established, China Mutual and Ocean Steam Ship had operated a secret 'pool' for Lancashire and Yorkshire fine goods. China Mutual had an efficient and modern fleet, but the timing of their new acquisitions was less fortunate than that of their Liverpool rivals. In 1901 China Mutual ordered three 9000 ton twin-screw steamers (one from Workman Clark, two from D&W Henderson). This was at a time when the largest Blue Funnel ship was 7000 tons, and none of the fleet had twin propellers. However, in 1901–2 a sharp trade recession found China Mutual over-extended. Indeed, it is said that in

1902 some 80 per cent of Britain's merchant fleet sailed at a loss. On Richard Holt's initiative, Ocean Steam Ship was able to acquire nearly all the shares of China Mutual. Apparently the newly-appointed Chairman of China Mutual, G. B. Dodwell (Dodwell and Co. were China Mutual's agents in the Far East and on the Pacific Coast of North America) was at that time travelling by train in the United States and was uncontactable. Learning eventually by cable that the fleet had been sold to Holts he hurried back to block the deal, but too late.[2] Thus began a long and fruitful association between Holts and Dodwells, Dodwell and Co. becoming agents for Blue Funnel on the Pacific Coast of America from 1902 until the trades were disrupted by the Second World War.

There is just a hint of collusion between Holts and China Mutual shareholders in what is really a very mysterious transaction. While the China Mutual chairman was in the United States, one of the principal shareholders in the company (most of whom were Manchester merchants), Donald Stewart, 'induced the other Directors, mostly old men', to join him in selling their shares to Holts.[3] Stewart was also a partner in the large merchant firm of Stewart and Thompson, and it was Thompson, then President of the Manchester Chamber of Commerce, who was a member of the deputation which saw Richard Holt in 1902 prior to the acquisition and obtained large reductions in the freight rates on cotton goods shipped to China from Liverpool (reductions of about one-third). It is not outside the bounds of possibility that the two matters were connected, Holts agreeing to substantially lower freight rates – the chief concern of the Manchester merchants in setting up China Mutual in the first place – in return for the sale of the shares. The episode is curious, since the lower freight rates seem to have been agreed by Holts without reference to the Far Eastern Conference, as Richard Holt's evidence to the Royal Commission on Shipping Rings quoted earlier indicates.

The cost of the acquisition to Holts was around £300 000, of which two-thirds came through the sale of securities, and the remainder from bank loans and the immediate sale of China Mutual preference shares. To the Holt fleet were added 13 vessels, all under ten years old, with an average tonnage of nearly 6000. Rather uncharacteristically the China Mutual ships maintained their former names and, for four years, their traditional colours (white funnels with black tops), but in other respects there was speedy integration of the two fleets. Almost at once insurance of the China Mutual vessels was dropped, agencies rationalised, and pay and conditions for crews of the two fleets brought into line with Blue Funnel practice.

The expansion of Company business, both before and after the China Mutual takeover, can be seen in figures given by Richard Holt to the Royal Commission on Shipping Rings for the outward carryings in Blue Funnel ships to China and Japan. These cargoes had increased from

Table 7.1   Voyages and earnings, 1901–13

Year	Number of voyages	Net earnings (£000)	Earnings per voyage (£000)
1901	79	218.0	2.8
1905	82	352.2	4.3
1913	95	670.8	7.1

Source: Ocean Archives; Voyage Accounts.

85 000 tons in 1884 to 160 000 tons in 1894, and to 294 000 tons in 1902, before the arrival of the China Mutual fleet. The fleet's gross tonnage, meanwhile, had increased from 53 000 tons to 87 000 tons and to 184 000 tons on the eve of the amalgamation. In 1903 outward carryings (with the new ships) were 430 000 tons and in 1907 reached 670 000 tons.

When the Great War, expected by nearly everyone to be a short one, began in August 1914, Blue Funnel was on the crest of a wave of unparalleled prosperity. Table 7.1 gives a snapshot impression of Holt's very great advance in these pre-war years. A trebling of net earnings in a dozen years and a two-and-a-half-fold increase in profits per voyage was an outstanding performance.

Moreover the growth of Holts in these prosperous years was not based solely on ships. In the early years of the new century the Company also took the major step of investing in its own wharves and warehouse facilities abroad. A minor project was started in 1904 when Holts formed, jointly with the China Navigation Company, the Tientsin Lighter Company with an initial investment of £5000. On a much greater scale was an undertaking started in 1905 when Ocean purchased some land at Kowloon in Hong Kong for the construction of wharves, warehouses, sheds and other facilities. Similarly in the following year land was purchased at Pootung in Shanghai, where wharf facilities were built opposite the international settlement on the Whangpo River. By the end of 1913 around £600 000 had been invested in the Hong Kong and Shanghai property developments. More followed in 1911 with the purchase of land in Hankow for wharves and warehouses. These properties were beginning to show a significant return on the substantial investments by 1913, and at that time they were all being extended. In 1913 plans were drawn up to extend the warehouse and wharf at Hong Kong, and a pontoon hulk, the *Laestrygon*, was moored near the Hankow property. A new property in Sumatra was also purchased just before war broke out, Holts building a wharf and warehouse for tobacco and rubber at Belawan.

Notwithstanding these property investments the purchase of China Mutual and new ships, Ocean's financial strength, backed by large reserves, remained intact. The Company continued its traditional policy

Table 7.2   Net voyage earnings, 1909–13

Year	Net earnings (£)	Gross tonnage	Net earnings per voyage (£000)
1909	247 000	350 853	2.8
1910	358 580	373 521	3.9
1911	301 536	389 391	5.2
1912	311 309	409 335	5.5
1913	670 847	467 815	7.1

Source: Ocean Archives; Voyage Accounts.

of paying low dividends, of allowing generous depreciation for the fleet, and adding to the reserves and investments where possible. Ocean's prosperity was solidly rooted in the growth of voyage profits, and many of the company's trades experienced a boom in the five or six years before the war. If we compare Table 7.2 with Table 7.1 we can see that progress was by no means unchecked, as the vicissitudes of the trade cycle and other circumstances influenced trading returns. 1913 was an exceptional year, but there is no disguising the very clear upward trend of profits after 1909 shown by 'net earnings per voyage'.

If cold statistics culled from balance sheets and presented in a table give no real flavour of success or otherwise, the comments of Richard Holt in his diary certainly do. On New Year's Eve, 1911, he reflected that Ocean had had 'a remarkable year – Trade, probably for everybody, has been very good and certainly our firm has done quite extraordinarily well'. A month later he was writing 'Annual meeting of OSSC – Profit over £500 000, an immense record.' The following year, 1912, was also excellent. In August Holt wrote, 'Business successful, but for incessant strikes, brilliantly so.' And looking back at the end of December 1913, 'again it has been a most prosperous year for Alfred Holt & Co. ... unexampled wave of prosperity which we have enjoyed for 2 or 3 years'. Overall the financial returns showed a 'splendid result, the best yet recorded', and the company was able to transfer £500 000 to its growing reserves.

The outbreak of war thus found the Ocean Steam Ship Company in a strong financial position. The company was also in the process of modernising its fleet. In 1913 no fewer than 6 ships were delivered with 8 more on order for delivery the following year – though due to the war only 5 were forthcoming in 1914. Two of the 1913 orders were placed with Workman Clark for very large 19 000 ton passenger liners for the Australian trade, and at this stage the company looked set for a major entry into this service. However, because of the war the ships were never built, and in the stagnant post-war years the company continued with the modest service established before 1914.

The new vessels built and planned on the eve of the war formed a large proportion of a fleet which, at the beginning of 1913 numbered some 64 vessels. The designs and capabilities of the new ships, inspired by Wortley, were fully competitive with the vessels of any other fleet in the world. Professor Hyde, summarising with only a little exaggeration Holts' performance in ship design over the entire pre-war period, put it thus:

> The technical progress made in the design and construction of the company's ships between the years 1865 and 1914 can be readily measured by comparing the leading particulars of *Agamemnon* (1865) and *Lycaon* (1913). Both ships were designed for the service between the United Kingdom and the Far East, and each was in her day the fastest and most efficient cargo liner in the trade.[4]

In other trades, though, some lines were matching Blue Funnel in terms of speed and efficiency. Very large cargo-carrying passenger ships and motor-driven cargo vessels were beginning to make their appearance. *Lycaon*, 7500 tons, with a speed of 14 knots, was a superb ship, but there could be no room for complacency in an increasingly competitive world.

Not that the commercial horizons of 1914 were unclouded. There were signs that the long trading boom was coming to an end while industrial troubles, which had been simmering and sometimes boiling over from 1910, continued. Already in 1911 a seamen's strike for union recognition resulted in some of the big companies, like Booths, Cunard, and Holts, making separate agreements and giving a general increase of 10s a month. In Liverpool there followed a dock strike and general transport strike, with troops called in to confront strikers. Holts again led the way in making a separate settlement with dock workers (they had already employed some permanent dock workers for over a decade, although most were casual). A happy feature of this bleak period in Liverpool's industrial history was the work of Lawrence Holt is setting up a clearing house for dock workers in Liverpool in July 1913, Holt working tirelessly to help inaugurate the scheme (although the principal role was played by the dockers' labour leader, James Sexton). There were problems abroad, too, especially in China where the ancient Manchu dynasty was overthrown by Sun Yat-sen in 1911. True, the upheavals had little immediate impact upon Blue Funnel trade, and at the Annual Meeting early in 1914 Richard Holt thought it 'remarkable that all the political disturbance in China should have produced so small an effect upon trade: indeed, during the last few months larger cargoes of ordinary merchandise have been shipped to Shanghai than in any previous period'. But the situation was fluid, and the China Navigation Company certainly suffered considerably as a result of the political unrest in China.

Moreover by 1914 the China Conference was in serious difficulties with competition from a German outsider. The outsider was the Rickmers Line, a company destined to have a long and tempestuous relationship with the conference. Rickmers had begun in 1911 to trade between the continent and 'Siberia' (Vladivostock and Nicolaesk), a trade in which Holts and Swires were involved. A short fierce freight war resulted in the breakup of the Siberian conference, as Rickmers' vessels became established. Then in 1913 Rickmers entered the Antwerp–Japan trade and in February the following year started loading ships on the west coast (Glasgow, Liverpool and Swansea) for China and Japan. This was indeed a dangerous situation, and the German line had planned to put some large new steamers (around 9000 tons) into the trade. To some extent the problem was caused by great bitterness between Rickmers and the German conference lines, Hamburg–Amerika and North German Lloyd. In any event the war put an abrupt end to a complicated situation, but conference trouble with Rickmers was to re-emerge for much of the inter-war period, being settled only in 1936 by an agreement between the three German lines. In the early 1960s Rickmers returned once more to attack conference arrangements and, after another struggle, was admitted as an associate member in 1963.

There had been some significant managerial changes in the years before the Great War. George Holt and Maurice Llewellyn Davies, who had become Managers with Richard Holt in 1895, retired in 1912 and 1913 respectively. George Holt died prematurely in 1916 at the age of 44. His health had never been strong and consequently he had never been able to play a dominant role in company affairs. His main responsibilities had lain in the fields of accounts, pooling arrangements in the Far East trade, and with Mansfields. That he was not without some of his father's courage is shown by George's proposal in 1895 at a shareholder's meeting that the company should adopt limited liability. This was against the wishes of both Alfred and Philip Holt, who were able to stave off the move until 1902. The retirements of George Holt and Davies came in the midst of the deaths of the founding brothers, Alfred dying in 1911 and Philip in 1915. There was, as it were, a shift within the Holt clan towards the Durning–Potter side which was strengthened by the subsequent appointments of Hobhouse, Cripps, and Thornton (these relationships are shown in the family tree below). Lawrence Holt, Richard's younger brother, had become a Manager in February 1908 at the same time as Henry Bell Wortley, the naval architect. To replace George Holt the company brought in Charles Sydney Jones in 1912. In this year also John Hobhouse entered as a young assistant, while Charles Wurtzburg (like Jones, outside the family nexus) came the following year. The Managers felt unable to replace Davies, whose

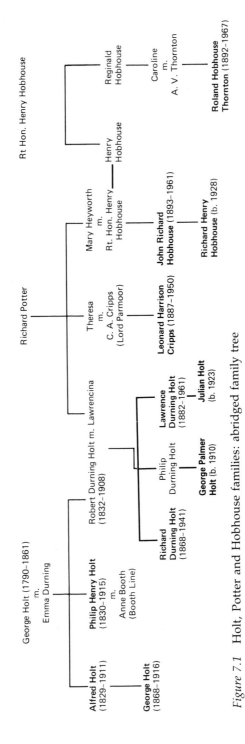

Figure 7.1 Holt, Potter and Hobhouse families: abridged family tree

Note: Ocean managers and directors in **bold** type.

retirement seems to have come as a surprise. Davies had been ships' husband, and, as the Annual Report for 1913 expressed it, Davies 'has controlled masters and mates and had general supervision in all matters relating to the navigation'. Davies's tasks fell largely on the shoulders of Stapledon and Lawrence Holt, and the Company thus entered the war with five partners: the Holt brothers, Stapledon, Wortley, and Jones.

It is appropriate here to say something of these new partners, especially Lawrence Holt, whose period of managership, extending until 1953, was only a few months short of Richard's. Lawrence Holt became a legendary figure among generations of Blue Funnel masters. He found in Alfred Holt and Co.'s fluid management structure and tradition of public service a perfect vehicle through which to pursue his own interests and beliefs. The indelible impact he made on the company over half a century can best be understood by stressing various aspects of his outlook and complex personality. Above all he was enamoured of the sea and seafaring, and of the training of youth, his interest in the two often going hand in hand. He was proud of Liverpool and its history, and was devoted to public service within Merseyside; with less enthusiasm he participated in various national committees and organisations. Also, he had a genuine and deep attachment to individuals, being understanding of human weaknesses and sympathetic towards the less fortunate. Finally, he saw himself as the upholder of the principles and virtues of the company's founders, and as time went by became almost mystical in his references to 'duty', 'discipline', and 'the flag'. He was fond of quoting the family and company motto – *certum pete finem* – which he freely translated as 'see the job through'. The motto, incidentally, had been used by Alfred and Philip's father, George Holt, and seems to have derived from a passage in Horace's Epistles. When Philip Holt died in 1915 Lawrence moved into 52 Ullet Rd, and this doubtless strengthened his sense of continuity and family tradition.

Lawrence Holt's interest in the sea was essentially romantic and humanist. He was not without vision, but it was the one-eyed vision of his hero Admiral Nelson. He had a blind spot for financial and commercial affairs for which he showed neither aptitude nor interest. Although he took responsibility for the fleet's technical development in the 1920s, this was not his forte either.

Against this background we can readily understand the particular spheres in which Lawrence made his mark. As far as Ocean was concerned Lawrence Holt's legacy was principally twofold. One was his work as ships' husband and his tireless efforts to look after all aspects of the welfare of seamen on sea and shore. He knew every nook and cranny of the ships, made frequent voyages, and in this way came to know the masters personally, knew their problems, their backgrounds and home life, and was often prepared to help personally in cases of,

14 Lawrence Durning Holt, Manager, 1908–53

say, bereavement or illness. He liked nothing better than prowling round the ships in the Liverpool docks, and was well known to Holts' stevedores and other dock workers. Not that he mixed on terms of equality with masters and men. His relations were paternal and kept at a somewhat austere distance, but the masters returned his interest with genuine and enduring loyalty and respect. For many of them 'India Buildings' meant Lawrence Holt. Holt became ships' husband in 1931 on the retirement of Stapledon, and, working closely with marine super-intendents, he remained in charge of most matters relating to the welfare of seafarers until his retirement in 1953. And this introduced the second major legacy. Lawrence Holt, who became senior partner in 1941, guided the seafarers through the horrific years of the Second World War. In Sir John Nicholson's words, he was 'wholly absorbed in maintaining the strength and morale of our seafarers' and the Foreword to Captain Roskill's *A Merchant Fleet at War, Alfred Holt & Co. 1939–45*, records that 'By character and training he was uniquely fitted to sustain our fleets under the desperate stress of war, and their achievement owed much to his inspiration.'

On wider fronts Lawrence Holt's achievements were considerable. Rather than give a lengthy and not very informative catalogue let us in-stead look briefly at three areas where his influence was both significant and characteristic. One was in the training and education of boys. Before the First World War he had worked in the Sunday School of the Ullet Rd Unitarian Church and in the Unitarian Domestic Mission, and organised and sponsored camping trips in the Isle of Man for deprived youngsters. He also followed the century-old family tradition of being governor (and Chairman) of the Liverpool Institute High Schools for Boys and Girls. In 1934 he became Chairman of HMS *Conway* training ship at a time when the fortunes of the old-established cadet ship were at a low ebb, and he soon transformed the ship by the application of his ideas on sea training. Sadly the old ship was wrecked in 1953, the year Lawrence Holt retired from Ocean. Lawrence Holt was also the founder, together with Dr Kurt Hahn, of the first Outward Bound Sea School at Aberdovey in 1941 (the term 'Outward Bound' was Lawrence Holt's), and for the first five years of its existence Blue Funnel underwrote the losses made by the school.

A second notable area of Lawrence Holt's interest lay in the city of Liverpool. He became a City Councillor in 1913 and remained one for 20 years; he was a magistrate for even longer; and became Lord Mayor in 1929. Throughout his long association with the city he was prominent in a multitude of educational, public and charitable organisations. He was also Chairman of the Liverpool Local Employment Committee between 1918 and 1929 in the developing years of National Health and National Insurance schemes.

A third involvement was his long-term effort to improve the conditions of casual labourers, especially in the Liverpool docks. On becoming Lord Mayor of Liverpool, one of his first acts was to commission an Enquiry into Casual Labour in the Merseyside Area. The Report was published in 1930, and in his Foreword, Lawrence Holt wrote 'It is manifest that the practice of carrying on the industry of Merseyside so largely by Casual Labour has brought in its train economic waste, besetting poverty, much physical and moral degredation, as well as almost insoluble difficulties of social ordering and civic government.'[5] Lawrence Holt's concern was of long standing. Before 1914 he was Secretary of the Merseyside Joint Dock Labour Committee, and through this he was largely instrumental in creating the Dock Labour Clearing House in 1912. This initiative followed in the wake of the paralysing Merseyside Dock Strike of 1911, and early the following year Ocean shareholders were informed, 'the Managers have directed their efforts, in common with other shipowners, towards securing better and less casual arrangements for the labour employed at the Liverpool docks'. Lawrence Holt's enlightened views were noted in typical caustic fashion by Aunt Beatrice Webb in 1913. 'The younger brother Lawrence', she noted in her diary, 'has I think slightly more originality (than Richard Holt) and decidedly more social compunction; he worries over casual labour and feels to some extent responsible for its existence. But he is a prig and almost pathologically self-centred.'

The last remark is worth dwelling upon a little. For all his many contributions to the firm, Lawrence Holt did not endear himself to his fellow Managers and made in many respects an uneasy colleague. Richard Holt described him as 'always looking in the glass',[6] and personal friction between the various managers, Lawrence Holt, Cripps, Hobhouse and Thornton, kept in check only by Richard Holt's unquestioned authority, seems to have been a feature of the inter-war quarter-deck.

Lawrence Holt also sat on a number of national committees and commissions, including the Coal Mines Reorganisation Commission in 1931. He was offered, but declined, a knighthood in 1925, perhaps in deference to his father's decision in 1895. But he certainly did not seek recognition, being essentially a shy man, and only fully in his element in the company of seafarers or youngsters with whom he established an evident rapport.[7]

Lawrence Holt's career with Ocean spanned the years 1904 to 1953, when he retired at the age of 71 (henceforth the Managers decided on a retirement age of 60). Educated at Winchester like his brother Richard, he was not considered to have the ability for a university education and joined the shipowners Alfred Booth and Co. as an office boy in 1900. The connection between the Booths and Holts was a close one, the Booth Line starting in 1865 with two small steamers built to Alfred Holt's specifications, and the Holts were substantial investors in the Line. Philip Holt had married into the Booth family and for a time the two shipping firms shared the second floor of India Buildings. It was Philip Holt who arranged for Lawrence to enter Alfred Booth and Co. From Booths, Lawrence Holt graduated to Ocean at the beginning of 1904, at Richard's invitation. In true Holt tradition Lawrence soon made an extensive tour of the Far East during 1906–7, journeying to the Straits, Java, China, Japan and Australia; and in February 1908, still only 25 years of age, he became the youngest-ever Manager of the Ocean Steam Ship Company.

We have mentioned already that Lawrence Holt's interests lay in individuals rather than in the commercial or purely technical side of operations. His interest in ship design was principally one of concern for safety and standards, and he was always fully supportive of Flett, the ultra-conservative Chief Naval Architect for most of the inter-war years. Not that Lawrence Holt was unprepared for change. He founded the company's laboratory during the First World War, and engaged Brian Heathcote, later an Assistant Manager, to oversee the laboratory and undertake the training of deck apprentices. Lawrence Holt also initiated the fleet's transition to diesel engines after the First World War, with the purchase of the first Burmeister diesel engine in 1921. Undeniably, though, his role was not always positive. He became senior partner late in life (he was 58 when Richard Holt died) and after the war appeared to retreat into a mystical welter of slogans and attitudes hardly appropriate to the real tasks of rebuilding the enterprise in a competitive and changing world. 'Remember Britannia, and look after my flock' is how, on his retirement, Lawrence Holt bid farewell to Lindsay Alexander.[8]

We have noted Lawrence Holt's interest in the conditions under which dockers worked and a fascinating glimpse of Birkenhead dock labour is given in an unpublished manuscript written by a George Albert Pridgeon, who worked at the docks for 54 years, mostly as a Ship's Foreman with Alfred Holt and Co. at Birkenhead.[9] He started work there in 1886 as a boy of 12 (though lying that he was 14), and retired eventually in 1940. His first job was as a steam boy ('they did nought else but turn the steam on to drive the winches'), and he took his place on the 'stand' in the old China Shed at the Morpeth Triangle Dock, earning 2s a day plus 1d for each hour of overtime. At that period Morpeth Dock was the principal loading dock, and Pridgeon recalled how:

> after discharging her in-coming cargo, a ship would normally proceed to the Cross-berth in the Morpeth Dock to take on Horse-shoes and old Cart Tyres. This would take maybe two days. At one o'clock on Friday's she would – all going without incident – be due at her loading-berth proper to take on main cargoes for Hong Kong and Shanghai – note: two ports only. This would generally take until the following Friday noon, the principal cargo loaded being fine goods, bales and cases, from the Lancashire and Yorkshire Mills who were enjoying prosperity at that time. The loading facilities were pretty bad. The fact that we, working on the China boats, had a full week in which to load was often our saving point. Later the Alfred Holt Co., gave a lot of consideration to the gear for loading and discharging, and in later years when I was foreman in the firm, it was common practice for us to load 2000 tons deadweight in one day, working 20 gangs of men in a 6 hatch ship, for transit to 9 foreign Ports.

Pridgeon's memoirs give a vivid picture of harsh conditions, arduous work, and endemic hostility between men and shipowners (though Holts emerge as one of the more humane and conscientious employers). Certainly the realities of casual labour must have been severe and often brutal, with hundreds of men packed into a 'stand' waiting for work which might not materialise, at the mercy of foremen who might or might not favour them. We learn that in the late 1880s Holt ships at Birkenhead were loaded by dockers picked from a stand under the control of a T. J. Dodds, who was Master Stevedore not only for the 'China Company', but also for the Anchor and Rathbone Lines. A few years later Dodds retired, and employment for each Line was then separated, Holts managing their own recruiting under a Stevedore Superintendent (initially a W. Hagen, succeeded in turn by Captain Dixon, and, after the First World War, by Captain Murphy). According to Pridgeon, the 'Master-Stevedores each had a following of men who depended on them for their living and seldom worked at any other firm', and when the recruiting was separated lines like Blue Funnel, too, had 'their' pool of labour at the stand. Pridgeon recalled that at this period, in the late 1880s,

> working conditions were hard and unjust. Men were compelled to work under whatever conditions governed local labour, and had no redress whatsoever to any injustice which came their way. The working day was split into two parts – morning labour and afternoon labour. In the former, the hours were from 7 a.m. to mid-day, and in the afternoon 1 p.m. to 5 p.m. – 5 hours and 4 hours . . . This splitting of the working day caused trouble and hardship to the men, inasmuch that a man who had been hired to work on the morning stand was compelled to go on the stand again at twenty minutes to one if he wanted to complete a day's work. Although, then, a man officially got one hour break for his dinner, it was reduced to 40 minutes by his having to get back on the stand for 20 to one – or even earlier, to be re-hired for the balance of the day. Another point about the split-day system: the employers, whenever it suited them, made a point of employing more men in the afternoon (4 hour period) than in the morning, (5 hour period) thus saving themselves money at the workers' inconvenience. This, together with the fact that the men hired for work from 1 to 5 p.m. often commenced work at 10 minutes to one, become known as time-stealing and caused friction and bitterness amongst the men.

The villain, of course, was casual labour, and Holts were one of the few shipowners to try to improve conditions. A revealing episode occurred in 1888, the year the Dock Labourers' Union was formed. Resistance by Liverpool shipowners to Union recognition and, particularly, the demand that only Union labourers be employed at the dock, brought matters to a head. George Pridgeon remembered that

> the firm of Alfred Holt & Co. then came to the fore. Representatives of this Company called on their men to attend a meeting of all shipping firms. This meeting was addressed by Mr T. J. Dodds, and supporting him were Mr Phillip [*sic*] Holt, Mr Crompton and Captain Russell.

They offered to make a permanent employment of five gangs of men (approx. 80) and to accept the Union Rules as a basis for working conditions.

One of the Dockers was elected as spokesman for all present at the meeting. His reply was 'We thank you for your generous offer, but as Union members we sink or swim together.' Those men were fine fellows, refusing as they did that generous offer, they chose to stand by their comrades, and in doing so, probably saved the collapse of the Dockers' Union.

Things then came rapidly to a head. An ultimatum was sent by the Union to the Employers demanding recognition of their rules, and informing them that after a certain date, Union members would refuse to work alongside non-Union members. This ultimatum was ignored by the Employers. The date set arrived and the long-anticipated strike came.

The employers decided to fight. On the quays and at the shed entrances they erected barricades, and then they called in Police protection for the men they employed. I don't know where they got these men from but they got plenty of them, and before long all ships were working at top speed except those of the Alfred Holt Co. Throughout the strike this latter firm employed no 'imported' labour and their ships remained idle.

Eventually, though, the shipowners yielded on most of the men's demands.

The problem of casual labour was destined to remain unsolved until the Second World War, although it was an issue in which Alfred Holt and Co., especially Lawrence Holt and John Hobhouse, had a deep concern. An interesting episode occurred in 1925 when the Managers asked Pridgeon (who had impressed Richard Holt with his handling of the vexed question of pilfering at the docks, demonstrating that the main problem lay not in Birkenhead but in the east) to draw up a scheme for permanent employment at the docks. Unhappily the scheme, which was accepted by the Managers and was, according to Pridgeon, 'a very fine and genuine offer, and one that would have cost the firm a lot of money' was rejected by the dockers' union, despite strong support from the Holt workers.

To return to managerial appointments before the First World War, these included, in addition to Lawrence Holt, Henry Bell Wortley in February 1908, and Charles Sydney Jones in 1912. Wortley's role in particular is worth stressing if only because the rather family-centred traditions of Alfred Holt and Co. have led to his unjustified neglect.

Wortley came to the company in 1892, only 24 years of age, as naval architect. This was the same year that the last ships designed by Alfred

15 Sir Charles Sydney Jones, Manager, 1901–30 (*reproduced by kind permission of the University of Liverpool*)

Holt, the 3500 ton *Ulysses*, *Pyrrhus*, *Tantalus* and *Ixion* appeared. Wortley's first ships (designed in cooperation with Alfred Holt, of course), were the *Orestes*, *Dardanus*, *Sarpedon*, *Diomed*, *Menelaus*, and *Hector*. These ships were of around 4600 tons and had powerful cargo gear which greatly impressed the shipping world. They were delivered in 1894 and

16 William Clibbett Stapledon, Manager, 1901–30

1895 and had an immediate impact upon the quality of service Blue Funnel could provide. Later came the larger and faster ships, mentioned in an earlier chapter, including the renowned 'goal-posters' inaugurated with the 9000 ton, 14 knot, *Bellerophon* in 1906. By this stage the Blue Funnel fleet was rapidly becoming the most efficient fleet on the seas, and much of the credit was due to the brilliance of Wortley's designs.

In surveying the company's fortunes before 1900 it became clear that, as we saw in a previous chapter, the mid-1890s marked a watershed between the earlier difficulties and subsequent prosperity. But the easy association of this change with the year 1895, the year of the appointment of three young managers, Richard and George Holt and Llewellyn Davies, is misleading. Already by 1895 the new ships had arrived, many of the new trades had been opened, and the conference system improved and regularised. From around 1896 there was a sustained boom in international trade which the Holt fleet was well prepared to meet. Thus the Holt revival in the 1890s was many-sided, and involved far more than the arrival of the new managers. Professor Hyde, mistakenly dating Wortley's arrival as 1895, has unfortunately missed the contribution of the first Wortley-designed ships.[10] Wortley's contributions as naval architect were fully recognised by the partners when he became a Manager in 1908, the shareholders being informed he was 'a man of exceptional ability and familiar with many aspects of the

17 *Bellerophon*, c.1910 (built 1906) one of the famous 'goal-posters' (*photograph courtesy of Liverpool Museum*)

(Company's) business'. Tragically Wortley died in February 1919, a victim of the virulent post-war influenza epidemic which ravaged Europe. Wortley was only 51 years of age, and the shareholders were told:

> No greater loss could have been sustained by the Company. In his profession of Naval Architect Mr Wortley had no superior and it is to his genius that the Company owes a fleet acknowledged by all to be unsurpassed as cargo carrying vessels. At a time like the present when great changes in the method of propulsion are being introduced the sudden loss of a genius so rare and so experienced is a catastrophe of the first magnitude; Mr Wortley was not only a great Naval Architect – he was a sound and shrewd businessman whose good judgment in all branches of the Company's affairs was of the greatest value to his colleagues; his high personal character and kindness of heart gave him great influence of the best sort with the Company's staff and secured through his own branches of the work a most competent and loyal band of subordinates.

In the context of the company's traditional reticence in recording the deaths and departures of its servants the language sounds effusive, and even distraught. In Wortley's memory the company established a Chair of Metallurgy at the University of Liverpool, and provided money not only for the Chair but for a 'suitable laboratory'.

The final managerial appointment before the First World War was that of Charles Sydney Jones (he was knighted in 1937) who was brought into the management when George Holt retired in 1912. Sydney Jones ('Charles' was never used) was not a member of the Holt clan, nor had he been employed formerly with Alfred Holt and Co., but in other respects he was well suited to the post. He had been educated at Charterhouse and Magdalen College, Oxford, and came from a wealthy Liverpool family. He was both a Liberal and a devout Unitarian, and only agreed to join the Ocean Steam Ship Company after assurances that he could cultivate his interests in various aspects of public life. Jones was 40 years of age when he joined the company, already an established figure in Liverpool local government and a partner in the shipping firm of Lamport and Holt. His arrival at Ocean may well have been a consequence of Lamport and Holt's absorption in 1911 within the growing Royal Mail empire. At any event he was invited in 1912 by Richard Holt to join Ocean, and he remained with the company until his death in February, 1947, at the age of 74. Although shareholders were told that 'his business experience and personal qualities should enable him to render very valuable services to the Company', Sydney Jones's effectiveness certainly lay more in those personal qualities than in his business capabilities. His contributions to commercial and technical matters were always slight, but he played a major role in staff recruitment and welfare and did much to foster the spirit of paternalistic benevolence which was a characteristic of India Buildings. In particular, through his connections with local schools, he built up fruitful relationships through which the company, with its established local standing, was able to attract a stream of bright recruits, doing much to maintain the general quality of staff in the company. Jones also helped to maintain staff morale during the bleak days of wage reductions and staff cuts in the early 1930s, and his personal sympathies did not stand in the way of his enforcing a number of redundancies in 1932. On a wider front Jones made many and notable contributions to Liverpool University, of which he was successively Treasurer, President, and Pro-Chancellor; and he paid anonymously for the higher education of local boys, some of whom – like Tom Williamson who became General Secretary of the Municipal and General Workers' Union – themselves came to play a part in public life. He was briefly Member of Parliament for the West Derby Division of Liverpool in 1923–4. In Liverpool he long served as a Councillor and as member of the Education Committee, and he was Lord Mayor in the critical years between 1938 and 1942.

I

The First World War proved a short-term bonanza and a long-term disaster for the British shipping industry. While rocketing freight rates and profitable government requisitioning brought large short-term gains for shipowners, the post-war years by contrast saw general stagnation and depressed trading conditions. Wartime profits were then often quickly dissipated in a short-lived post-war boom which soon collapsed, leaving a greater number of firms with expensive purchases and capital extensions at inflated values. The Blue Funnel Line, as we will see, shared in the wartime prosperity and did far better than most thereafter, but even here the post-war picture was largely one of trying to hold on to previous gains. The pre-war days of expansion and large new initiatives had gone forever. Indeed, before the decade of the 1920s even commenced Blue Funnel had more or less reached its maximum position in terms of ships and tonnage.

When war broke out in 1914 the Company's operations were for a time very much 'business as usual'. Some changes were immediate and unavoidable. War insurance was taken out, while, as we saw in an earlier chapter, the withdrawal of German feeder services based on Singapore led to Ocean's takeover of the Straits Steamship Company, with Mansfields becoming managers. At an early stage the government, fearful of sugar shortages and inflated prices as distant cane sugar replaced European beet sugar, imported sugar by centralised chartering through Alfred Holt and Co. So successful was the scheme that in the four months ending January 1915 imports were more than one-third of the usual peacetime levels, and accumulation of sugar at the docks was leading to congestion. War also soon made an impact on the Australian service, where Admiralty requisition of the passenger vessels meant the abandonment of the service in 1915 for the duration of hostilities. The earliest Holt venture into the carriage of bulk liquids also came in response to war needs, in 1915. Supplies of fuel oil were at that time urgently needed for the navy in European waters, and large quantities of oil from Borneo and Sumatra were carried from Singapore to the United Kingdom in the ballast tanks of Holt, Shire, and P&O vessels. To Blue Funnel went the honour of the first such delivery, the *Keemun* being loaded in Singapore and Pulo Bukom (where Shell had installed facilities) in October 1915, 1089 tons of oil being pumped into the ballast tanks in less than seven hours.

However, the most immediate and visible impact of the war upon the Company's fortunes was upon profits, especially as the war erupted at a time of slackening trade and falling freight rates. Shipowners typified more than most the 'hard-faced men who had done well out of the war'. They were frequently vilified in the popular press, more especially since it was hard to avoid the contrast between the sedentary tycoon reaping profits and the brave merchant seamen who risked their lives in the making of such proftis.

When we put Ocean's large maritime earnings in perspective a number of background points should be remembered. Britain was a small industrial island, able to produce only a fraction of the foodstuffs her population consumed, and wholly dependent on imports of supplies of many vital raw materials. And to pay for imports exports were necessary. The country could not possibly have survived in wartime without the commercial services of the merchant marine. For the first two and a half years of war the government's policy was mainly to allow companies to operate such ships as were not requisitioned by the Admiralty to sail under the White Ensign, although there was increasing direction of routes. Under these circumstances, given the need for shipping space and rising freight rates, large profits were inevitable. Whether this was a sensible policy is another matter. Certainly imports of American and Canadian wheat and war supplies were crucial for ultimate victory, but it is not quite so obvious that it was in the national interest for Holts to make money by carrying Moslem pilgrims to Jeddah, lifting horsebeans from Hankow, or by taking Australian cattle to Singapore. But this brings out the incoherence of government policy in the early stages of the war. All merchant ships were subject to requisition, but this meant at first that only a small number of ships were commandeered as troop-carriers, hospital ships, merchant cruisers, stores transports, and the like. Gradually, as war dragged on, the arm of requisition lengthened and its grip tightened. The government increasingly ordered supplies on its own account, which non-requisitioned ships were obliged to carry at Blue Book rates (rates agreed between the government and a small committee of shipowners which were much lower than free market rates). But space not needed for these government account goods could be used for normal trading at market rates. In 1915 all refrigerated space on cargo ships was requisitioned at Blue Book rates, but non-refrigerated capacity on the same ships could be filled at market rates. This haphazard system continued until the beginning of 1917 when Lloyd George's government introduced general requisitioning and a Shipping Controller was appointed, and until then considerable anomalies existed in the extent to which ships of various lines were sailing with cargoes at government or non-government rates.

Another consideration when reviewing Ocean's profits is that although we refer to the 'world war', the term 'European war' would be far more accurate, at any rate until America entered hostilities in April 1917. The areas normally served by Blue Funnel ships were to a great extent unaffected directly by the war. In contrast to the Second World War both Italy and Japan were allies, and the Mediterranean and eastern seas (once the German cruiser *Emden* had been put out of action) were relatively safe for British shipping. Throughout the war it was possible for Blue Funnel to maintain its China, Java, and West Australian services, though at a reduced level. The indirect effects of war in the

Far East, such as the withdrawal of German competition, rising freight rates, increasing demands for war supplies, and so on, were largely favourable to the commercial fortunes of the Company. Even some small measure of shipbuilding could proceed, for Holts were able to place orders in Hong Kong both for themselves and for the Straits Steamship Company.

Another feature of Holts' wartime experience is that although there were many instances of damage and tragedy in the fleet's wartime service record, the number of total losses was remarkably light. The British merchant marine as a whole lost one third of its tonnage, while the Blue Funnel fleet was depleted by less than one-fifth. None of the larger twin-screw steamers was lost and until Germany launched in February 1917 the campaign of unrestricted submarine U-boat attacks, Blue Funnel losses were minor indeed. Prior to May 1917 only four ships had been lost. As a result the Blue Funnel fleet was able successfully to complete a large number of voyages throughout the war years, and due to wartime acquisitions from the takeover of the Indra and Knight lines, the company ended the war with more ships and a greater tonnage than at the commencement of hostilities (Table 7.3).

Table 7.3   Blue Funnel fleet, 1913–18

Year	Voyages	Number of ships	Gross tonnage
1913	152	71	467 940
1914	136	73	487 202
1915	152	75	492 394
1916	113	80	523 640
1917	128	79	534 086
1918	133	72	492 934

Source: 'The Blue Funnel Line: Alfred Holt & Co.', *The Manchester Guardian Commercial*, 24 January 1924.

A man like Richard Holt could not be immune from pangs of conscience at the large profits brought by war. On 15 September 1916, he wrote in his diary, 'Affairs at India Building are flourishing marvellously. No wonder those who are suffering severe loss through the war are casting envious eyes on the shipowners and the other small classes who are making great profits.' A few months later, recording 'remarkable' financial results, he added 'No wonder those who suffer in money as well as personal sacrifice feel grudgingly towards those to whom the war has brought a fortune.' Not that Richard Holt distanced himself from public duty or the affairs of the less fortunate. He was a member of parliament throughout the war, and served on an Advisory Committee to the Admiralty on merchant shipping affairs. Both personally and through the business he was active in various Liverpool war charities. Only three months after the declaration of war he converted his house at 54 Ullet Road into a hospital with forty beds for wounded soldiers and did not take up residence there again until peace had returned. Already at the beginning of 1914, before the war, the Managers had set up a special fund 'for charitable or semi-charitable purposes'. This money was to come from any surplus over £50 000 which accrued to the Managers under their agreement with Ocean (which incorporated a percentage of profits), and was established with £7000. In 1915 a further £10 000 was added to the Trust which, the shareholders were told, 'has been very valuable and will be needed after the War'. Further amounts totalling £25 000 were added over the next three years, and the fund was used to support a large number of Liverpool war charities during and after hostilities. There were two further trusts established in these years which have continued to the present day. The first, known as the P. H. Holt Trust, was set up as a fund founded under Philip Holt's will 'for necessitous servants'. This fund, to which the Company added from time to time, was used for general charitable as well as staff benevolent purposes. The second, the Holt Education Trust, was established by the Company in 1915 upon the death of Philip Holt as 'a memorial to the late Alfred and Philip Holt'. The Philip Holt Trust (as it was originally called) was set up with £20 000 for 'education in the Liverpool district in memory of the Brothers Holt'.

While the huge wartime profits gained by shipowners may have been viewed by many with distaste, it is interesting that Holts at any rate did not push their profits to the limit. As Richard Holt explained to shareholders early in 1916:

> The managers have not thought it wise to try to extort the last farthing from the regular shippers, and on the outward berth have been contented to charge very moderate freights while mounting the most regular service which the requisitions of the Admiralty and the difficulties due to shortage of labour in the ports have permitted.

At this stage outward freights had been raised by only about 50 per cent on their pre-war levels, despite higher general price and cost increases, while inward freights had risen about twice this figure. On the Australia to England run, for example, freights had been lifted in the first weeks of war from £2 7s 6d to £5 5s a ton, and were raised on several occasions thereafter. Richard Holt argued that the company should 'contribute towards the stability of the national finances' by assisting exports, a rather curious example of financial myopia since in the last analysis wars can be won by imports but not by exports.

High wartime profits were a natural consequence of rising freight

Table 7.4   Holt dividend payments, 1913–18 (£)

1913	106334
1914	106334
1915	106334
1916	148867
1917	212668
1918	212668

Source: Ocean Archives: Shareholders'
Annual Meetings.

rates and remunerative requisitions at a time when demand for shipping services was greatly in excess of supply, and supply could not be readily increased. Shipowners could always point out, and they did, that war-time taxes were high (Holts paid around £3 million to the exchequer in excess profits tax), and were fully aware that fleets were deteriorating and replacements likely to be far most costly than normal rates of depreciation on pre-war values would cover. Many shipowners were profligate in distributing profits, but Holts, in keeping with the company's traditional social attitudes and financial caution, showed a measure of self-denial in wartime dividend payments (see Table 7.4).

Dividends in 1917 and 1918 were paid at the rate of 50 per cent but while this showed a doubling over 1913 levels this was no more than increases in the general price level. In real terms, therefore, dividends did not rise at all and it could be argued that shareholders did not participate in Ocean's immense wartime profits.

Tempered by taxation and moderate freight rates as they were, wartime profits nonetheless rose dramatically. Table 7.5 shows how steamers' earnings reached nearly £3 million in 1917, whereas before 1914 a figure of £500 000 had been considered remarkable. Even the figures in Table 7.5 do not give the full picture, since delays in government accounting for requisitioned and part-requisitioned voyages meant that the totals returned were incomplete. Doubtless the rounded totals for 1915 and 1916 reflect this uncertainty, while the 1917 figure, despite its precision, can only be an approximation. For the year 1918, when

Table 7.5   Net steamer earnings 1914–17 (£)

1914	482072
1915	1250000
1916	2500000
1917	2876647

Source: Ocean Archives: Managers' Minute
Books.

nearly all the voyages were requisitioned at certainly very adequate rates, no detailed figures are available.

These figures do not include the earnings from the small joint West Australian service, but these too more than trebled in wartime, profits rising from just under £15 000 in 1914 to over £46 000 in 1917. A breakdown of these aggregate figures among the various trades is not very useful since the wartime voyage records were incomplete and requisitioned journeys formed an increasing share of the total. Moreover during the war new trades between the Far East and New York were acquired, so that comparisons with pre-war years cannot readily be made. All trades, though, flourished, with profits in the China and Straits trades rising from £490 000 in 1915 to £1 055 000 in 1917, and the Java trade even more strikingly from £186 000 to £513 000. And in 1917 the seven voyages on the Panama Canal route earned profits of nearly £65 000 a trip. This new route had been developed at the beginning of 1915, running via the Panama Canal (opened in 1914) to and from New York and the Pacific Coast of North America. It was combined with a monthly Trans-Pacific service between Manila, Hong Kong, Japan, British Columbia, and Puget Sound. As a result of this 'round the world' service, the old China Mutual route via Suez to Japan and then across the Pacific to British Columbia and Puget Sound was discontinued.

It would be tedious to detail the burgeoning Holt accounts for the war years, especially since the available figures are incomplete and subject to many qualifications arising from factors such as taxation, war insurance, government compensation, and requisition payments. But a few additional statistics will show vividly enough Holts' flourishing fortunes. The reserves in December 1912 had stood at £2 500 000. By December 1919 they had grown to £5 500 000. At this latter date the building suspense account, arising from war damage, stood at £1 250 000. Investments, meanwhile, were £8 million, largely converted to War Loan, and were bringing in over £200 000 compared with under £20 000 a year before the war. The book value of the fleet had stood at £2.4 million in 1914, and had been written down to £1.8 million by 1918. And in each year around £1 million had been carried forward to the year following after all expenses had been met.

Faced with large profits the Managers sensibly followed a policy which might be described as cautious expansion where this was possible. As far as the fleet was concerned, significant new building was out of the question in view of the huge pressures on all dockyards. The company was fortunate to have embarked upon a substantial building programme just before the war. Five vessels were delivered to Ocean in 1914; none in 1915; one, the *Tyndareus*, in 1916; four more in 1917, including the 7500 ton *Autolycus* from Swire's Taikoo yard in Hong Kong; and none in 1918. In 1915 two small ships were purchased by the

NSMO for transhipment of cargoes between the United Kingdom and continental ports. These were the *Veghtstroom* (1353 tons) and *Elve* (899 tons). In 1915, also, three ships were ordered in Hong Kong for the newly acquired Straits Steamship Company and were delivered the following year.

Fleet and route expansion also took place as a result of the purchase of two small Liverpool lines. In August 1915 Holts acquired the Indra Line and its seven steamers of 38 072 gross tonnage. The Indra Line had been engaged in the trade between China and New York and the purchase consequently brought Holts a place on the China–New-York Conference. The deal was arranged by Richard Holt, after he had been approached by Sir Thomas Royden, son of the founder of the line which had started in 1854. In his diary, Richard Holt recorded the purchase 'for a prodigious sum of £750 000. This deal I negotiated with Tom Royden at the Admiralty on his overture and trust it may be successful for us.' The shareholders, incidentally, were not told the full terms of the deal but were informed that the price was 'reasonable'. Following the purchase the ships were duly dignified with 'Homeric' names, and one of them the *Euryades*, built in 1913, remained in Blue Funnel service until 1948. For a time the American trades were, indeed, very successful, but with the return of peace constant conference problems, low freight rates, and the impact of the great slump on America after 1929 all combined to make the American venture far less profitable than Richard Holt had hoped. Immediately after the war the New York service was expanded with a joint arrangement with the two Dutch Mail lines for a service between New York, the Straits and Java via the Cape.

The second purchase was of four ships, each of around 7000 tons, from the Knight Line. This purchase cost Holts £600 000. The Knight Line had been started in 1878 by Greenshields, Cowie and Co., and like the Indra its ships had been largely engaged in Far Eastern services across the Pacific. At the beginning of the war the four Knight ships, which were little more than superior tramps, had been chartered by Ocean. In the summer of 1917, with an eye to post-war demands, Richard Holt decided to purchase the vessels outright. The names *Knight of the Garter, Knight of the Thistle, Knight Templar* and *Knight Companion* remained unchanged (perhaps the Managers did not feel the ships worthy of Blue Funnel names), though the latter ship, despite being torpedoed in June 1917, stayed in service until 1933.

As well as adding to the fleet by these acquisitions, the Company was able to continue the pre-war development of its overseas properties during the war, and some major extensions to existing sites were undertaken in Hong Kong, Shanghai, Hankow, and Sumatra. In Hong Kong a second wharf and new godowns were built, while in Shanghai Holt's wharf at Pootung was doubled in size during 1917 with the purchase of adjacent land from the China Navigation Company for £100 000. When work was completed in 1923 the wharf at Pootung could berth four large ocean liners. In 1918 land was bought at Belawan for a wharf and godowns. By 1919 the Kowloon and Shanghai developments were valued at £4 750 000. Also in 1919 a major new office development was started in Singapore on the site of the existing office, a site described to shareholders as 'one of the finest in the East'. For this project Ocean engaged the well-known Liverpool architects, Briggs and Thornley.

The Company also took steps to expand its own accommodation, as the firm's thriving business outpaced efforts to improve and alter the existing India Buildings offices. In 1916, therefore, the Managers decided on a long-term plan to rebuild India Buildings when war was over, and £149 757 was paid for 'all the property between Water St, Chorley St, Brunswick St and Drury Lane' for the future construction of a new India Buildings specially designed for the Company's needs. Shortly after the war, in 1921, Holts further added to their Liverpool amenities by purchasing playing fields for the use of employees.

A new and, with hindsight, misguided policy was started in 1916. This was the purchase of shares and control in a number of companies connected with shipbuilding and operation. The aim was explicitly to gain some measure of control over their management, and consequently of business with Holts. In April 1916 Holts bought an interest in the Caledon shipbuilding yard, a small Dundee concern established in 1874 which had previously undertaken a limited amount of work for Blue Funnel. In 1917 there followed investments in another shipbuilding company, Scotts of Greenock, and in the Rea Towing Company of Liverpool; in 1918 came holdings in Lester and Perkins, a London engineering and ship repairing company; in 1919 and 1920 Holts acquired shares in Jenkins Brothers of Liverpool, ship repairers, in the North Eastern Marine Engineering Company of Wallsend, and in British Perlit Iron Co. in 1919–20. This latter year saw an end to the process when the sudden collapse of the post-war boom left Ocean with a number of over-capitalised and unprofitable concerns on its hands.

The rationale behind these acquisitions was that Holts would otherwise be unable to secure shipbuilding and engineering services when the expected post-war reconstruction boom took place. Part of the problem was that several large shipping concerns were integrating with shipbuilders while shipbuilders themselves were combining, so that as a meeting of shareholders was subsequently reminded, 'a possible monopoly threatened from a combination of shipbuilders and ship-repairers creating difficulties for those owners who were not integrated with builders and repairers'.

All the firms involved in this spate of acquisition had long associations with Blue funnel ships. Scotts of Greenock had, of course, a particularly close and long-lasting association. The yard had built the *Plantagenet* for Alfred Holt in 1859, and 83 ships later the last Scott vessel for Blue Funnel, the *Cyclops*, was delivered in 1948. The acquisition of Jenkins Brothers was precipitated by a bid for the company from Harland and Wolff, for the Managers were anxious to protect various Holt patterns for castings which the Birkenhead firm possessed. The initial investment in Caledon, 'a small and thoroughly well managed concern' was substantial, around £200 000 being subscribed for extensions and improvements. The investment in Scotts, originally £184 237, was for a one-third interest in the concern with an agreement for Ocean to buy complete control 'if necessary'. Henry Bell Wortley became a director of both shipbuilding companies.

The exact sums invested in these sundry concerns is impossible to estimate from surviving records, but the holdings in Scotts and Caledon were certainly the most substantial. By December 1922 total investments in the two concerns was put at £589 000, while investments in the other five enterprises was roughly about £125 000, one half of which was accounted for by Lester and Perkins. Almost without exception these investments proved unprofitable once the post-war boom collapsed, although in the case of Rea Towing, whose tugs on the Mersey were used exclusively by Blue Funnel, the holding proved a fruitful and enduring one. By 1924 Lester and Perkins had gone into liquidation, Jenkins Brothers had been reconstructed (and in 1928 Ocean was forced to write off an accumulated loss of £65 890), and the holding in both Scotts and Caledon had been substantially written down. To some extent Holts were able to protect their holdings in Scotts, Caledon, and North Eastern Marine by their orders for new ships, and these firms participated strongly in the post-war rebuilding programme. Caledon, in particular, recovered well from the world depression of the early 1930s. In July 1939 a note in the Manager's Minute Book recorded that in view of the 'very fine work done by Mr Henry Main in connection with the recovery of Caledon Shipbuilding and Engineering Co.', 6000 shares in Caledon were to be given to him free of charge.

The disappointments of these investments should be placed in perspective. Nearly all leading shipping concerns were engaged in similar acquisitions and the Holt management was by no means alone in believing that without such safeguards it might be impossible to obtain speedily adequate new ships with which to meet expected post-war demands. In this respect it is perhaps worth emphasising that Holts' investments look rather modest, and certainly did not saddle the company with the huge expensive acquisitions which was the fate of some shipping concerns. Another point is that many of Holts' wartime investments, viewed as a whole, were profitable and supportive of the company's business – investments such as the Far Eastern wharves and warehouses, Straits Steamship, the new Ocean building in Singapore, and others. By comparison with other shipping companies Holts dispersed their wartime profits with circumspection, and this was to stand the firm in good stead in the difficult times ahead.

We have yet to touch upon the wartime performances of the fleet itself, when British merchant ships were caught up in a major war for the first time since Napoleonic days. In the First World War, as in the Second World War, there were triumphs and tragedies, countless and uncounted acts of individual heroism and courage, stories of tasks fulfilled under the most arduous and dangerous conditions. A brief survey cannot begin to do justice to epic episodes, while the retelling of particular events inevitably distorts the full picture. Summarise and distort we must, but here perhaps we can add another point. Just as in wartime to focus on a particular incident, a sinking perhaps, or an escape, will do scant justice to others, so war itself brings its own distortions. For in peacetime, too, the business of merchant shipping can be perilous with a multitude of hazards, from storms to collisions, foreign wars to tropical diseases.

Being essentially a cargo line, few Blue Funnel ships were suitable for what might be called combatant service. Some, like the *Ajax*, became hospital ships, and others were used as store ships, but most were left to continue their normal business of ferrying cargoes. The Australian ships were different. Here the three passenger ships, *Ascanius*, *Anchises*, and *Aeneus* were ideal for conversion to troop carriers, and were requisitioned by the Admiralty almost at once. In 1915 the larger *Nestor* and *Ulysses* were similarly commandeered, and the Australian service was necessarily suspended, not to be revived until April 1920. The five ships contributed sterling wartime service, making many transport voyages, and were also used for returning troops after the war. Incidentally the five ships remained in Blue Funnel service throughout the inter-war years and were again requisitioned to serve once more at the outbreak of the Second World War.

Two aged Blue Funnel ships, the *Hector*, and *Menelaus* saw special war service. The *Hector*, after structural alterations, served as a balloon ship during the ill-fated Dardanelles campaign in 1915 and 'spotted' for the *Queen Elizabeth* at Gallipoli. The *Menelaus* was also used as a balloon ship, being sold to the Admiralty for extensive alterations in March 1916, and renamed *Davo*. As *Davo* she 'spotted' for monitors during operations on the Belgium coast. Holts received £50 000 from the Admiralty for *Menelaus*, incidentally, roughly what the ship had cost in 1895, though costs were rising rapidly during the war.

Most of the Blue Funnel ships not under Admiralty charter were requisitioned by the Controller of Shipping early in 1917. As Roland Thornton expressed it, the British merchant marine was effectively on charter to the nation, and by the end of 1917 as many as 78 ships of the Blue Funnel Line, which then totalled 83, were under requisition. Many of these requisitioned ships were now diverted from their customary Far Eastern routes to take troops and war goods from the United States, or to perform other war tasks. The requisitioned ships, including the Australian vessels, ferried huge numbers of soldiers across the seas. By the time peace returned in November 1918 some 117 000 troops of the American Expeditionary Forces had been taken to Europe, and a further 131 000 'Imperial and Colonial' troops taken to various theatres of war. There were also Portuguese troops brought to France and cross-channel operations ferrying troops and supplies. In addition many Chinese 'coolies', some 15 000, were taken across the Pacific and via Panama to work beyond the lines in France. And at the close of hostilities Blue Funnel ships were engaged too in the return of troops and Chinese labourers to their homelands.

Mention of the Shipping Controller and government requisitioning raises a curious episode which briefly brought Richard Holt and the Blue Funnel line before the public eye. The immediate background was the toppling in December 1916 of the Liberal Asquith government by the Liberal Lloyd George, who formed an alliance with Bonar Law and the Conservatives, with Labour support. An 'infamous conspiracy' was Richard Holt's diary entry, Holt remaining with the small Asquith rump in opposition. Almost at once Lloyd George addressed the question of shipping which he said 'had become the most vital and vulnerable point in the issue of victory or defeat'. He appointed a worthy but obscure Glasgow shipowner, Sir Joseph Maclay, as Shipping Controller and head of a new Ministry of Shipping. At the opening of 1917 Sir Joseph began the wholesale requisitioning of Britain's merchant fleet for government service. He thereupon wrote to shipowners that 'you are to continue to manage and run those vessels as if they were not requisitioned ... the only difference is that as you complete their voyages you will hand over to the Government the profits you make.'[11]

This was too much for Richard Holt. He considered such action by the Controller illegal, and he therefore refused the requisitioning order and took the Controller to court. The crux of Holt's case was not opposition to the requisitioning of ships as such, for the right of the government to act thus had never been questioned (and the Admiralty had already requisitioned many vessels). The issue was the requisitioning also of the shipowners' *services*, by compulsion under the mantle of the all-embracing Defence of the Realm Act. In Holt's view the government had no such compulsory powers, and, alone among British shipowners, refused to comply.

Holt's view was entirely consistent with the same principles which had led him to oppose conscription. The government, even in wartime, should not have complete command of the individual. Thus the services of shipowners should have been requested, not ordered, and proper blue-book rates offered. The courts, in the case brought as China Mutual Steam Navigation Company *v* Maclay upheld Holt's actions in a judgement delivered in November. This led the *Economist*, which was 'not sorry that Mr R. D. Holt should have successfully challenged in one instance the proceedings of the Executive', to point out: 'Nothing has happened except that the Shipping Controller is compelled to put his requisition of liners in a regular form, and to ask, rather than command the services of their owners in working the ships for him. These liner owners – not excepting the litigious Mr Holt – are as patriotic as the rest of us.'[12]

In reporting the incident to shareholders Richard Holt hinted that a certain rancour about the 'infamous conspiracy' may have influenced his actions. The new government, shareholders were told, had requisitioned nearly all the merchant marine 'in the pursuance of the undertakings given to the Labour Party, in exchange for their assistance in removing Mr Asquith and his friends from office'. Doubtless with tongue in cheek, Holt announced that since the Managers considered requisitioning would be more profitable than ordinary trading, and having established compulsory requisitioning of shipowners' services illegal, they had decided to fall in with the scheme.

Following his victory, Richard Holt then cooperated wholeheartedly with the Shipping Controller, and Blue Funnel vessels, often now sailing in protected convoys, played their full part in the dire conditions which prevailed in the period of unrestricted submarine warfare. With the return of peace, Holt was no doubt thankful to note in his diary in December 1918, 'probable we shall soon be free of the Shipping Controller', who, he magnanimously added, 'has really done his work very well'.

II

The cold statistics of the war years may be briefly related. In all, 16 Blue Funnel ships were destroyed, in addition to the two small NSMO ships purchased in 1915. Twenty-nine other ships were damaged, ten of them seriously crippled, by shellfire, mines, or torpedoes. Around 80 officers and men in these ships lost their lives. To these may be added 14 of the India Buildings Office staff who were among the 82 young men who joined the army or navy. The latter, shareholders were told in 1919, received 1 VC, 1 DCM, 7 MCs, and 2 MMs: 'this may be considered to reflect creditably on the young men – mostly boys – of whom the office staff is composed'. Rather surprisingly, and sadly, no record was given

Table 7.6 Ocean Steam Ship war losses, 1914–18

Ship	Built	Lost	Details
Troilus (1)	1914	19.10.14	Sunk by Emden, Indian Ocean
Diomed (1)	1895	22. 8.15	U-boat, off Scilly Isles
Achilles	1900	31. 3.16	U-boat in Atlantic, off Ushant
Perseus	1908	21. 2.17	Mined off Colombo
Troilus (2)	1917	2. 5.17	U-boat off Malin Head
Calchas	1899	11. 5.17	U-boat of Tearaght Island (SW Ireland)
Phemius	1913	4. 6.17	U-boat off Eagle Is. (NW Ireland)
Laertes	1904	1. 8.17	U-boat off Prawle Point (S Devon)
Veghtstroom	1902	23. 8.17	U-boat off Cornwall
Elve	1904	?.12.17	No trace after leaving Oporto
Kintuck	1895	2.12.17	Mined off Godrevy Lighthouse, St Ives
Eumaeus	1913	26. 2.18	U-boat in Channel, off Ile de Vierge
Machaon	1899	27. 2.18	U-boat off Cani Rocks, Tunisia
Autolycus	1917	12. 4.18	U-boat off Cape Palos, near Cartagena, Spain
Moyune	1895	12. 4.18	U-boat off Cape Palos, near Cartagena, Spain
Glaucus	1896	3. 6.18	U-boat in Mediterranean, off Sicily
Diomed (2)	1917	21. 8.18	U-boat off Nantucket lightship
Oopack	1894	4.10.18	U-boat in Mediterranean, off Malta

Source: Ocean Archives.

to shareholders of lives lost or decorations of those who served in the merchant fleet. The ships' losses are best expressed in a tabular form (see Table 7.6).

The eight losses in 1917 took a toll of 54 lives. Only four lives were lost in the seven sinkings in 1918, although a further 12 died from attacks on four ships which were able to make port. Many more were seriously wounded. The heaviest losses came with the sinkings of the two little Dutch steamers in 1917. Both were lost shortly after being taken over from NSMO by the Admiralty. The *Veghtstroom* was mined, while the *Elve* was never heard of after leaving Oporto and was presumably sunk by a submarine. In both disasters a total of 36 officers and crew lost their lives. Another tragic loss came when the *Laertes* was torpedoed, and sank with the loss of 14 lives. These sinkings stand out, for as emphasised earlier, Blue Funnel losses were relatively light. The majority of comparably large lines fared considerably worse than Blue Funnel, in casualties both of ships and men. Good fortune must have played its part, but so, too, did the high standards of training and discipline shown by Blue Funnel crews, by many examples of quite exceptional skill and courage, and also by the sturdiness and speed of the 'Holt class' vessels. There were many spirited escapes as ships drove off submarines by gunfire, most ships being armed with a gun mounted aft. The *Tyndareus*, in particular, owed her life to improvements in design made as a result of experience gained from the *Titanic* disaster.

18 Nurses on board *Charon*, First World War (*photograph courtesy of Liverpool Museum*)

Ocean's first lost was *Troilus*. This splendid new ship had completed her maiden voyage on 15 July 1914, and set out again for the Far East on 16 July. War was declared while she was in the east. On 19 October, on the home trip, *Troilus* was challenged by the notorious German cruiser *Emden* in the Indian Ocean. Typical of the rather gentlemanly early phases of these hostilities the *Emden* announced her presence and sent a boarding party to remove the crew and six passengers, who were allowed to bring with them their luggage and food. *Troilus* was then sunk. The captives were well treated and transferred to a ship bound for Cochin. Only a few weeks previously the Ben Line's *Benmohr* had been similarly caught by the *Emden* and the passengers deposited in Cochin also. At the end of October 1914 the *Emden* made a daring attack on Penang, disguised as a British cruiser with the addition of a dummy funnel. But a fortnight later the *Emden* was driven ashore in the Cocos Islands by HMAS *Sydney* and destroyed. A drinking vessel from the *Troilus* recovered from the wreck was long kept as a memento in India Buildings.

The second loss was more tragic when the unarmed and 20 year old *Diomed* was attacked in August 1915 by a submarine just south of Ireland. The *Diomed* was overhauled and sunk, but only after Captain

Myles and two of the crew had been killed by shellfire, and the chief mate badly wounded. Several others of the crew were killed by shellfire and drowned as the ship went down, including five Chinese engine-room ratings. The loss of the vessel, Richard Holt told shareholders, 'is a small matter compared with that of the brave men, British and Chinese, who died so unhesitatingly in the execution of their duty'.

A final 'war story' may be given as an instance of remarkable survival. The *Tyndareus* had been delivered at the end of 1916 and began her maiden voyage the following January under Captain George Flynn. On board were around 30 officers and 1000 men of the 25th Middlesex Regiment on their way to the Far East by way of the Cape, under command of Lt Colonel John Ward, a Labour Member of Parliament. On 6 February, off Cape Agulhas, the ship hit a mine on the starboard side. The huge explosion opened a gaping hole, and *Tyndareus* began to sink at the head. Perfect discipline was kept by Captain Flynn and Colonel Ward; the troops were ordered to boat stations, and despite rough seas and a strong wind, all were transferred safely to two ships which had answered her SOS, the Blue Funnel *Eumaeus*, and the hospital ship *Oxfordshire*. A naval vessel, HMS *Hyacinth*, and a South African government tug were sent out from Simonstown to give assistance, but in fact could give very little. The *Hyacinth*'s commanding officer considered *Tyndareus* a danger to navigation and ordered Captain Flynn to beach her. Flynn ignored the order, and resolutely took his sinking ship, stern first and in stormy conditions, to safety at Simonstown. There were two immediate consequences. One was a dressing down for Captain Flynn from the Admiral of the port for ignoring the naval officer, accompanied by the request of 'the honour of your presence at dinner tonight'. The second was a letter to Richard Holt from Viscount Buxton, the Governor General of South Africa:

> The hole is a terrible one, the bottom, plates and bulkhead twisted and torn. It is a wonderful thing that she was able to float at all and did not sink within a few minutes . . . You and your firm, the designer and the builders, are greatly to be congratulated on the result of the great improvements in regard to double bottom, bulkheads, etc., de-signed for the *Tyndareus* – These were, I presume, consequent on the loss of the *Titanic*. There is no doubt whatever . . . that without these well designed and costly improvements the ship would have gone to the bottom.

This same *Tyndareus* had a long and eventful history, which included further war service between 1939 and 1945. She ended her days as a specially adapted pilgrim-carrier after the war, and was finally sold to breakers in 1960 after a record 44 years as a Blue Funnel ship.

8
After the War: An Uncertain World

The Blue Funnel Line is so well known in China that it always comes as a surprise to the China hand on leave to find that everybody in England is not familiar with it.

(A passenger, *Manchester Guardian Commercial*, 1924)

When peace returned in 1919 Britain's shipping industry, although few realised it at the time, faced a new world. The period before 1914 had seen a long period of growth and prosperity, with quickly rising trade in primary products; the war years themselves had seen immense profits as earnings from merchant ships soared. But by 1921 the good years had gone, and for the remainder of the inter-war period there was to be a constant struggle against downward freight rates, rising costs, and intensifying competition. The inter-war years were bisected by the world depression of 1929–33, the worst slump in history, when problems for shipowners and shipbuilders reached new depths. In 1932 nearly one-fifth of Britain's ocean-going tonnage was laid up, while few of the voyages made were returning profits.

The end of the war had brought a quite phenomenal boom throughout Britain and most of the world, based on a natural desire to restock and rebuild after the long war years, and an unnatural, or at any rate unjustified, confidence that the prosperity of 1913 would return. The boom collapsed in the closing months of 1920, and in 1921 prices fell and unemployment reached levels never before recorded. This was the start of mass unemployment which stayed at more than a million (in 1932 more than three million) until the Second World War. Reality slowly forced itself upon the reluctant minds of Britain's businessmen. Exports would not quickly recover their pre-war buoyancy, and some markets, like many of Lancashire's pre-war textile outlets, were lost permanently. The labour market had changed too. Workers, more vocal and effectively organised than before 1914, could resist vehemently any attempts to reduce wages or increase working hours. Many gains had been made by labour during the years of acute manpower shortages. In the war years trade-union membership as a whole had doubled, and the shop steward's movement made considerable ground at factory level. Gains would not

171

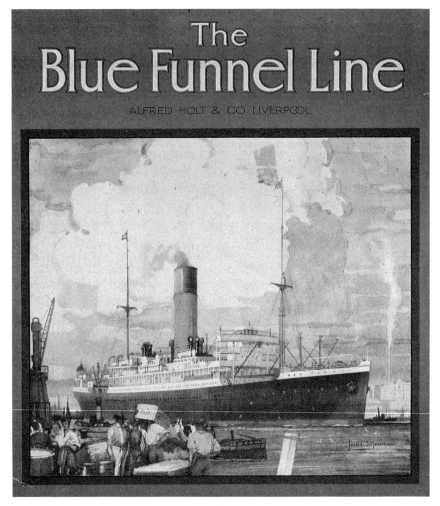

19 Illustration by James Mann used for publicity in the 1930s. (*photograph courtesy of Liverpool Museum*)

be lightly yielded. Moreover Britain's industries and of course its shipping lines faced great competition from powerful competitors. The productive power of the United States expanded dramatically between 1914 and 1919, while a number of newly industrialising countries were also becoming major competitors. The world was entering a phase of economic nationalism, with rising levels of protection, state subsidies to industries and shipping lines, and various other measures which assaulted the traditions of free trade which had nurtured Britain's pre-1914 supremacy.

A few facts and figures will illustrate this changed environment in which Britain's shipping industry found itself. Throughout the inter-war years world shipping tonnage was always well in excess of the volume of trade, by comparison with the immediate pre-war years. In 1914 world dry cargo tonnage was some 45 million, in 1920 54 million, and in 1929 60 million. Some nations had expanded their merchant fleets dramatically. Japan, for example, had doubled its tonnage between 1914 and 1920 to three million, and the tonnage was to double again by the late 1930s. The United States had expanded its fleet even faster, from 2 million to over 12 million tons between 1914 and 1920. And between the wars not only Japan but Germany (from a very low base in 1919 admittedly), the Scandinavian countries, the Netherlands, and others, added substantially to their fleets.

But trade nowhere matched these figures of tonnage. In 1920 the volume of world sea-borne trade was actually one-fifth *lower* than pre-war levels, and although by 1924 it had recovered to 1913 levels, the ratio of tonnage to volume always remained higher. In the great depression between 1929 and 1933, things became dramatically worse. In 1932, at the height of the depression, world trade was (in volume) about the same as in 1913; world tonnage – much of it laid up – was 50 per cent higher. It should be borne in mind, too, that much of the tonnage was considerably faster and more efficient than in pre-war days, so that the carrying capacity of world fleets was even more in excess of demand. At the same time some cargoes, like oil, required specialist vessels rather than conventional cargo liners. Technical progress in the First World War had been rapid, and the inter-war years saw a new generation of diesel-driven motor ships, many of them run by Britain's competitors, often with government subsidies.

If we turn from the world to the British shipping scene, the picture is no better. Many of the old export trades were depressed throughout the inter-war years, especially coal and textiles. It is a striking fact that although British tonnage was roughly at its 1913 level throughout the inter-war years (rather higher by 1929, lower during the subsequent depression), the total tonnage clearing with cargo from Great Britain was always at least one-third lower between the wars. Even if we exclude coal, we find that the volume of exports was almost the same in 1929 as in 1913, and in 1938 one-third lower. And this to be carried by Britain's similar-size fleet, the competitors' larger fleets, and all of them improving in capacity and efficiency. If we look simply at the port of Liverpool, total outward dry cargoes failed to increase in tonnage by comparison with the immediate pre-war years, a fact which underlines the significance to Holts of maintaining a monopoly of the west-coast berths.

No wonder, then, that the inter-war years generally saw intensified competition and downward pressure on freight rates. Freight rates, for instance, fell precipitously from their 1919–20 boom levels to an average of something like 10 per cent above their 1913 levels by 1922. By 1925 they had fallen to the 1913 levels, and by 1928 below them (though 1913 had been an exceptionally good year). Yet costs were rising. The cost of buying a ship, even after the 1921 price falls, remained greatly in excess of pre-war levels, while bunker coal remained costlier and of inferior quality compared with pre-war days. And many shipping lines had already reconstructed their fleets between 1919 and 1921 when ship-building costs were three times their 1913 levels and waiting times for delivery twice as long. Wage costs remained obstinately high and, despite sharp reductions in 1921 and further falls in 1925, were substantially above their 1913 levels both in money and real terms.

I

On the eve of the Great War Alfred Holt and Co. had been in an enviable position. Several years of very high profits and fleet expansion had left the Company with large reserves and one of the most powerful fleets in the world, more ships than even their old rivals, the P&O, and approaching the P&O tonnage. Nearly the tonnage, too, of the British India Steam Navigation Company, though this company ran twice the number of ships. Blue Funnel vessels were the largest single cargo line using the Suez Canal, the major trader between Europe and the Far East, the biggest individual shipping concern based in Liverpool.

In 1918 Blue Funnel was, in many respects, an even more powerful company. As we have seen, the war years brought quite unprecedented, almost indecent, profits; enlarged investments both at home and abroad in property and subsidiaries; and the acquisition of two small shipping lines and the New-York–Far-East route. At the close of the war the Holt fleet was of similar size to that existing at its outbreak. It had been, on the whole , a good war.

There were other respects, however, in which Blue Funnel's role within Britain's shipping was changing. Before 1914 many of Britain's great shipping lines were controlled by family dynasties, Booths, Brocklebanks, Cayzers, Ismays, Holts. By 1918 a different pattern had emerged. By now P&O (having amalgamated with British India and other concerns) possessed over 300 ships with a gross tonnage of over one and a half million. The Kylsant Royal Mail empire grew in an even more spectacular fashion, owning in 1929 140 companies and a fleet of over 700 ships with a tonnage of over 2½ million. Within this empire were to be found familiar names, such as Lamport and Holt, Glen, Shire, and Elder Dempster.

If the names of those controlling Britain's shipping industry were fewer, they were also grander. Peerages were not uncommon, and there was a sprinkling of viscounts and earls. The chairman of P&O was offered the throne of Albania six years before Richard Holt was offered the Chairmanship of the Mersey Docks and Harbour Board. Alfred Holt and Co, seemed, possibly, rather drab, with their cargo vessels having little of the glamour surrounding large passenger liners. Although Beatrice Webb had predicted an imminent knighthood for her nephew in 1913, Richard Holt remained unknighted until 1935, when he was getting on for 70. Lawrence Holt, it is true, had earlier declined a knighthood in 1925 (for reasons which remain unexplained; they may be connected with his father's refusal of a baronetcy in 1895, or they may not). But for the Holts there were no peerages. The long delay in Richard Holt's baronetcy, incidentally, remains another puzzle. His service within the wider shipping community was outstanding, and his contributions to Liverpool equally great. On the other hand he made no secret of his enmity with Lloyd George, whom he despised, and his brave battle against compulsory conscription in the First World War and defence of conscientious objectors brought him few friends and was a major factor in ruining his political ambitions.

Liverpool, though remaining a great and dominant port in the inter-war years, lost something of its national role. In 1920 Liverpool had handled about one-third of the nation's trade, a greater proportion even than London. By the end of the 1920s it was down to one quarter, and by the end of the 1930s well under one-fifth, by which time London was handling double Liverpool's trade (a trend caused partly by the growth of oil and other bulk cargoes outside Liverpool).

To portray Liverpool and Holts undergoing something of an eclipse in the inter-war years gives, of course, an exaggerated view. But such an exaggeration gives an interesting perspective, for we should consider this: Holts, Liverpool-based as they were, 'private' company as they remained, family-run as they continued, were successful. In the bleak inter-war years such success was remarkable. Never for one moment did Holts lose their reputation for financial soundness, the high quality of their fleet, and the excellence of their management. And as depression clouds moved across Britain's shipping skies, the few visible stars seemed all the brighter. Throughout the 1920s Holts were able to maintain to shareholders the dividend at its 1917 level. Yet Glen paid no dividend after 1923, Lamport and Holt none after 1925, while P&O were forced to reduce theirs in 1926. Nearly everywhere Britain's shipping lines were in considerable difficulties, some of the great ennobled empires above all. In 1930 the Kylsant group dramatically collapsed, Holts receiving an interest in Elder Dempster, and later salvaging Glen and Shire from the wreckage.

No wonder, then, that in the inter-war period the ethos of Holts became rather self-congratulatory, with a tendency to despise London, to cling to family traditions, and to survey the troubled shipping world with an intellectual condescension. Holts success in the inter-war years was not simply that they survived, but survived with style.

The company was fortunate in three particular respects. First, their 'portfolio' of trades gave them some the world's few regions where there was significant commercial progress, especially South-east Asia and China. This in itself is perhaps a little surprising in view of the great political turmoil in China at the time, and we will examine the causes of China's relative buoyancy later. Other trades, too, flourished from time to time. For example, both the Malaya and Java trades were profitable for much of the 1920s; between 1922 and 1928 Malaya's rubber exports to Europe (the bulk of which were carried in Blue Funnel ships) increased by one-third in volume, while those from the Dutch East Indies doubled.

Second, in contrast to many conferences, the Far Eastern Conference was quickly re-established after the war, and operated as an effective body throughout the period. This was of no little significance, for elsewhere, in the South American trades for example, there were long periods of bitter competition and collapsing freight rates. Not that friction within the Conference and threats from without were absent. Richard Holt was at times greatly troubled by the demands of German and Scandinavian lines in particular, and Rickmers reappeared after the war as a running sore until staunched in 1936. Conference problems took Richard Holt on a business trip to Holland in April 1928. His diary recorded:

> I think our negotiations were successful – certainly we arranged all our difficulties with the Dutch with whom we get on well, but the Germans are not pleasant; they want to get back their pre-war position which is impossible seeing that others, Scandinavians, Japanese, etc., have partly filled it and cannot be driven out. One of the disgusting features of our Conference is that both the Germans and Scandinavians freely break their agreements if necessary to get business for their lines.

Third, Holts' dependence on purely British trade lessened considerably. Most significant was the move towards continental loadings, started just after the war at a time when the German merchant fleet was non-existent. In the east, also, Holts were able to benefit from rising trade between third countries, for example with cargoes of raw materials to Japan, and of Japanese goods to South-east Asia.

In addition to a propitious pattern of trade and trade routes Holts were fortunate in other respects. If we compare Holts with rival concerns it becomes quite clear that enormous problems were created for some shipping lines by (a) investing in subsidiary companies in the war and post-war boom at ruinously high prices – this was especially the case where shipping companies absorbed shipbuilders and engineering firms, which then suffered from years of acute depression; (b) wholesale reconstruction of fleets at the height of the 1919–20 boom, again leaving a legacy of over-capitalisation; (c) excessive public issues of shares to finance such projects, or to take over other shipping lines. Borrowing was often made, too, to finance other schemes, such as improved port facilities or new office buildings. Share issues then put pressure on the company concerned to pay dividends to shareholders, and even more share issues or bank borrowing at high interest rates sometimes resulted.

Now Holts were not immune from these problems. Holts did invest unwisely in a variety of enterprises, and did purchase ships at excessive cost in the post-war boom, and did find themselves unable to finance the new India Buildings without bank borrowing. But what is striking is not only the relatively small nature of the investments, so that failure was not a disaster, but also that some of the investments proved in fact commercially rewarding. Many of the overseas properties, for example, which were built up during the war years, yielded a steady profit, while the Straits Steamship Company regularly returned a sum to Ocean which was about 8 per cent of net steamship earnings during the 1920s. Conservative accounting, moderate dividends (remembering the low capitalisation of the company) large accumulated reserves which enabled the writing down of assets to realistic levels, and a continued emphasis on self-finance all contributed to the healthy state of Ocean's balance sheet and profit-and-loss account in the difficult post-war years.

It is worth reflecting that Holts benefited in a rather special way from being a family-based company. There was no temptation to indulge in grandiose accounting displays to dazzle shareholders, nor to raid hidden reserves in order to improve the look of the balance sheet. In large measure the Managers were the owners (the Managers and their immediate families held over £50 000 worth of stock in 1930), and any such manipulation would be pointless. Moreover the traditional Holt accounting method for its shipping business, by calculating a net profit on each voyage after deducting for expenses and repairs, focused attention exactly on the areas of strengths and weakness in the business. Through Holts' network of agents, the Managers in Liverpool were able to observe in minute detail the varied fluctuations in trade conditions which could make all the difference between a profitable and an unprofitable voyage.

To say that Holts benefited from the structure of their trade, confer-ence relations, sound investments and cautious finance, is to a great extent the same as saying that Holts prospered because of its sensible and effective management. Not quite, for there were elements of good fortune too. But throughout the shipping world Holts had a reputation second to none for efficient fleet management and the quality of their employees both on ship and on shore. Led by Richard Holt, the small group of exceptionally able partners sustained a world-wide enterprise with knowledge and understanding of the most detailed aspects of the business. Richard Holt related in his diary how he met in Carlisle the Secretary of the local Labour Party who asked if he was related to the Holt of Blue Funnel. On being told that he was indeed that Holt the Secretary observed 'What strikes me about people like you is that you must know an awful lot of things and have to think about an awful lot of things.' He was quite right.

II

Before looking at the 1920s in any detail it is useful to glance at Holts' overall commercial record during the inter-war years. Table 8.1 presents such a survey, and, though it is far from giving a complete picture, it

Table 8.1   Blue Funnel: net steamer earnings,[1] 1919–39 (£)

Year	Net steamer earnings	Fleet tonnage	Earnings ton
1919	1438661	498740	2.88
1920	2161748	517973	4.17
1921	664643	543150	1.22
1922	560196	557902	1.00
1923	86655	681603	0.13
1924	560753	684952	0.82
1925	520768	687100	0.76
1926	564282	699489	0.81
1927	651376	686201	0.95
1928	844939	689847	1.22
1929	848829	705312	1.20
1930	423570	705156	0.60
1931	156579	684664	0.23
1932	483275	675655	0.71
1933	583683	654630	0.89
1934	446845	654383	0.68
1935	377821	622475	0.61
1936	567838	618001	0.92
1937	988620	621704	1.59
1938	1160953	651856	1.88
1939	959743	595199	1.61

[1]Figures exclude the West Australian trade.
Source: Ocean Archives: Managers' Minute Books and Shareholders' Annual Meetings.

20 Hon. Leonard Harrison Cripps, Manager,
1920–44

helps to isolate the main phases. Apart from the dismal performance in 1923 (which will be discussed in due course), we can see clearly the precipitous falls in 1921 after the post-war boom, the severity of the world slump, especially in 1931, and the very clear recovery which swept Blue Funnel's earnings per ton past pre-depression figures, reaching just before the Second World War the best returns since the post-war boom.

If the commercial environment was marked by upheaval and contrast, the managerial scene presented great continuity. By and large the business of the company remained at the end of these years very much as it had been at the opening, with around 70 Blue Funnel vessels serving routes and trades built up in an earlier period, guided throughout by the dominating figure of Richard Holt. Within the management and many of the departments there was a fortuitous continuity in the long service of senior figures. The inter-war years were the era of Richard Holt, but the era, too, of Lawrence Holt, Sydney Jones, John Hobhouse, Leonard Cripps and Roland Thornton, all of whom served the company continuously throughout the 1920s and 1930s. It is hard to resist the conclusion that the long periods served by this tight-knit group of senior managers during a very distinctive epoch did much to cement Holts' reputation, both inside and outside the firm, as a highly individual and unique organisation.

21 Sir John Richard Hobhouse, Manager,
1920–57

III

Holts, like other shipping concerns, faced both an uncertain furture and
the immediate tasks of post-war reconstruction at the close of 1918. In
purely financial terms the company had never been in a stronger
position. At the end of the war the reserve fund stood at £5 million and
investments (including subsidiaries) at some £8 million. In addition the
building expense account, provided by the Government to cover war
losses, stood at nearly £1.5 million. The fleet, at 74 vessels of around
500 000 tons, was of similar strength to that existing before 1914,
although ironically, in view of the world surplus tonnage soon to
become apparent, Richard Holt was worried that the new service
between New York and the Far East would strain the Company's
capacity. The end of war brought back to India Buildings many familiar
faces who had served in the war. Company policy had been to continue
full pay to those serving their country, and all were offered their jobs
back when they returned. In addition all salaries were increased to the
points which those who had left might otherwise have attained.
Symbolic, perhaps, of the many breaks with the past which the First
World War brought was the death in 1918 of the last of the original

22 Roland Hobhouse Thornton, Manager,
1929–53

shareholders, the redoubtable Captain Turner Russell who died at the age of 84. Turner Russell had joined Alfred Holt in1860, commanded the first *Achilles* until 1877, and was subsequently Marine Superintendent until 1889 and then Nautical Adviser before retiring in 1916.

Holts' most pressing need in 1918 was the need for reconstruction and renovation of the fleet. During the war years virtually all new construction had ceased, and the average age of the fleet was four years older than it had been in 1914. The position was even worse than this indicates, for the war years had seen exceptionally rapid technical advance in ship design and engines, while levels of maintenance had been at a minimum. The fleet was therefore in need of overhaul and modernisation, and new work had at once to be put in hand. Unfortunately this was at a time when British shipyards were working at full capacity, prices were soaring, and delays inevitable.

Accordingly, towards the end of 1918, nine new steamers were ordered from Caledon, Scotts, Hawthorn Leslie and Workman Clark. These were for five of the *Diomed* class, and four of the improved *Bellerophon* class. At this time Lord Inchcape of P&O was desperately trying to sell his 156 standard ships which he had taken over from the government in the expectation of a ready sale in boom conditions. He

23 Launch of *Diomed* by Mrs. R. D. Holt (*left*), June 1917 R. D. Holt is second
from right. (*Photograph courtesy of John Swire & Sons*)

approached Holts, but the Managers turned down the offer in favour of
their traditional policy of building ships to their own design. In any
event they cannot have been anxious to contribute funds to a rival. As
the Managers informed the shareholders in January 1919 'it appears
wiser to concentrate the Company's resources on acquiring vessels of
the highest character and most efficient type, rather than to meet what
will probably be keen competition with an equpiment which is neither
efficient nor cheap'.

The original contracted cost of these new vessels was more than
£4 million, some three times the cost of equivalent vessels by comparison with 1913. Eventually the final costs were reduced by negotiation
with the shipbuilders concerned once prices and costs had started to
tumble at the end of 1920. Nevertheless, the cost of these ships was
well in excess of their earning capacity.

Designs for the new ships, which in terms of efficiency, reliability,
and quality, were in accordance with Holts' traditional standards of
excellence, owed much to the work of H. B. Wortley. Wortley's death in
1919 left an unfillable void in the senior management. Although John
Hobhouse and Leonard Cripps became Managers in October 1920, they
had neither the experience nor professional expertise to match Wortley's

contributions at management level. The managerial changes produced a not insignificant shift towards 'family' control (both Cripps and Hobhouse were cousins of Richard Holt), and towards the 'amateur manager'. Thus Lawrence Holt now became responsible for fleet policy, but, as we have noted before, Lawrence Holt was not a technical expert, and he became beholden to the Naval Architects and Marine Engineers.

During 1919 the Company, with cooperation from the Shipping Controller in releasing the Blue Funnel vessels, was able to re-establish something like its normal service on most routes, while the five Dutch ships (seized by the British government in 1917 to stop them falling into enemy hands) were all returned to the Dutch flag and a full service restarted. Government requisitioning of steamers ended formally on 15 February 1919, and as each requisitioned vessel completed its voyage the ship was then handed back to Ocean's control. Many of these ships needed a complete overhaul, and the Managers set up a programme of systematically overhauling every ship in the fleet, a process which took two years to complete.

The new orders for vessels placed in 1918 all suffered considerable delays, due in the main to continued shortages of labour and materials, and also to a series of damaging strikes. The *Achilles*, ordered earlier in 1917, was more than a year late when finally delivered from Scotts in 1920 at a cost of £545 000 (a 12 knot, 11 400 ton ship; in 1912 Scotts had built the 13 knot, 10 200 ton, *Talthybius* for Ocean for £117 000). Originally Holts had hoped that six of their new vessels would be ready for service in 1920, but in the event only two were delivered, and the programme was not completed until 1923, though in fairness it should be added that the Managers themselves had called a halt to construction during the sharp recession of 1921.

Competition quickly made itself felt on most Blue Funnel routes, and the Mangers felt it was imperative both to restart their full services as soon as possible, and to revive the conferences which had lapsed during the war. There was particular concern about Japanese competition, since any increased Japanese presence would almost inevitably affect the trades in which Blue Funnel was involved. We know how effectively Japanese lines had established themselves in the war years. In 1916, for example, NYK started no fewer than three new services, including a monthly service between Japan and New York via the Panama Canal, one between Formosa (Taiwan) and Java via Singapore and Hong Kong, and a serivce between Japan and Australia. Mitsui Busan Kaisha also started a monthly service between Japan and Australia in October 1916. In 1919 the Japanese established themselves on Blue Funnel's own home ground, the NYK running a regular monthly service between Liverpool and the Far East, and a little later entered the Java trade. The enlarged United States fleet, too, was running a competitive service between New

York and the Far East, and was threatening to invade the trade between Europe and the Far East. With active support from Warren Swire and Richard Holt the Conference was quickly reestablished in 1919. Interestingly the Secretary of the Conference was paid a salary of £1150 a year according to a letter written in 1919 by Warren Swire, to which sum the conference contributed £700, Holts £200, and Butterfield and Swire £250.

The year 1920 was a prosperous one for Blue Funnel with the post-war boom in full swing for most of the year. The trans-Pacific trade with America booming and competition not yet intense was particularly favourable. In that year commercial voyages produced a net profit of £1 523 532, and it is notable that the main China-Straits trade, though accounting for most of the voyages, produced only one-third of the net earnings (Table 8.2).

Table 8.2   Net voyage earnings, 1920 (£)

Number of voyages	Trade	Net earnings	Earnings per voyage
14	New York	444 427	31 745
49	China	553 908	11 304
14	Java	38 923	2 780
9	Australia	214 572	23 841
3	Trans-Pacific	271 702	90 567

Source: Ocean Archives: Voyage Accounts.

These immediate post-war results, however, were exceptional. Indeed, the voyage returns for the rest of the decade showed an almost exactly opposite picture, with the American, Australian and trans-Pacific trades doing badly, and the China and South-east Asian trades proving successful.

The enormous profits made on trading in 1920 were put to good use, and displayed the admirable caution which served the Company well. The surplus was used to write down the value of the new steamers, the Managers being increasingly concerned at the sharp fall in prices which occurred in November and December that year. Richard Holt told the shareholders in January, 'it is difficult to believe that any ships can be profitable at their present cost', and he argued that these high costs had the same root causes as the company's 'apparent prosperity'. 'It is only proper', he said, 'to use this prosperity now to avert definitely the evil consequences at a later date of the present inflated prices.'

During 1920 the Company was able to restore nearly all services on its routes to pre-war levels, and, as we saw in Chapter 3, was soon able to extend operations in a number of directions. Taking advantage of the eclipse of Germany's merchant navy as a result of the war, Blue Funnel services were developed between European ports and the Far East. Monthly sailings from Hamburg, Bremen, and Rotterdam were added to the main line Far East voyages, and also prior loadings at Hamburg and Bremen on the Amsterdam–Java line. In these trades the Hamburg–Amerika line and the North German Lloyd Line became the Company's agents in Germany for trade between Germany and the Far East, and DADG (German–Australia Line) were appointed as agents for the Germany–Java trade. These were important moves, for they added considerable flexibility to the outward cargoes available for the Far East, and helped compensate for the difficulties experienced by Britain's export trades. Another development was an arrangement with the Canadian state-owned line for a joint service across the Pacific, with Butterfield and Swire as agents for the Canadian line. The arrangements included through rates on the Canadian National Railway, matching the services already enjoyed with American Railways at Seattle. In addition a joint service was run with the Dutch Mail lines between Java and New York via the Cape.

The appointment of the German agencies reflected a conciliatory attitude towards an erstwhile enemy which was typical of Richard Holt's belief in the peaceful purposes served by trade. After the Second World War, too, the company was notably ready to readmit German and Japanese lines to their former trades. An emergency meeting of the Conference held in September 1921 at India Buildings agreed the terms for readmission of the North German Lloyd and Hamburg–Amerika Lines and, as a gesture of goodwill, Richard Holt arranged for both lines to be represented by Mansfields in the Straits until the lines could make their own arrangements. In November, 1921 Richard Holt informed Mansfields that 'both companies have been most friendly in their negotiations and both offered us the agency quite spontaneously.' He also warned 'we must anticipate that the personnel on the German ships – masters, officers, crew – will have a very cold reception in all British ports – they have a good deal to live down, but you will of course see that these people do get justice and fair play, and you will protect them from insult or persecution.' Richard Holt was, of course, a businessman as well as a diplomat. He added 'We at Holts have decided that the best chance of retaining for the Blue Funnel a share of the trade between Germany and the Far East lies in cordial cooperation with these German Lines while we still have something substantial to offer.'

As part of Ocean's plans for post-war development Richard Holt made an extensive tour of the Company's routes and major agencies between November 1919 and June 1920. He visited Singapore and Malaya, Sumatra, Java, Manila, China, Japan, British Columbia, and Puget Sound. Part of the purpose of the visit was to gain first-hand experience of the problems and opportunities in the various areas served by the company, and in particular to pinpoint any weaknesses in the organisation of its agency system. He later felt able to report that 'looking at things broadly it is safe to say that the company is extremely fortunate in its representatives abroad', but privately he was concerned about the relationship between the company, Mansfields, and the Straits Steamship Company. The following years saw some major changes in the structure and organisation of Ocean's representation in the Straits.

Between 1921 and 1925 the company experienced some difficult years, and was forced to shelve earlier plans for expansion. In 1921 Richard Holt noted in his diary that trade was 'shocking', and forecast no improvement without a 'drastic reduction in wages'. This fight between costs and prices was to become a constant theme of the inter-war years. The spring and summer of 1921 brought a bitter coal strike, which in turn greatly increased the costs of coal-fired steamers. This strike probably determined the Managers once and for all to dispense with coal for new ships; from henceforth all new ships ordered by the company were either diesel-powered motor ships, or oil-burning steam-ships. In addition the company began to convert a few coal-burning ships to oil-burning. At the end of the coal strike Richard Holt noted in his diary, 'we just bought coals at Glasgow at 39s a ton, thrice the pre-war price, less than half that of a year ago.' Indeed, at the height of the 1919–20 boom Holts had been obliged to pay 140s a ton for coal in Liverpool.

Despite bad trading conditions in the 1921 slump, the company was not only able to maintain its 50 per cent dividend, but add £70 000 to the credit balance of the profit and loss account. Nonetheless, the slump in prices and declining freight rates persuaded the Managers to slow down the rebuilding programme. For most of 1921 building ceased on all the Holt vessels under construction, and was recommenced only at the end of the year.

In this new economic environment, the Managers nevertheless decided to push ahead with plans to build ships to carry passengers to the Far East, a scheme mooted prior to the war. Accordingly in 1921 two twin-screw cargo passenger steamers were ordered from Scotts and Cammel Laird, the vessels both ready in 1923, being capable of carrying 155 first class passengers at 15½ knots, and having a crew of 80. These were the *Sarpedon* and the *Patroclus*. The Managers thought the passenger traffic 'not likely to pay', but was being introduced 'to secure the valuable goodwill of the planter and merchant communities'.

The close of 1921 found Richard Holt in an optimistic mood, for he thought the sharp slump of that year would have a salutory effect on the labour market, and squeeze out much of the competition which had mushroomed in the hot-house atmosphere of wartime and post-war inflation. Blue Funnel's competitors, wrote Holt, 'are all burdened with very costly fleets and most of them have still to learn their business'. Holts, by contrast, were better placed 'for a return to pre-war conditions', and the company 'does not require to borrow money at 7 per cent'. The following year, in fact, despite reasonable steamer earnings and a 'satisfactory level of cargo', the company found it necessary to borrow heavily from the bank for the first time. The borrowing, primarily from Lloyds Bank, was to cover shipbuilding costs, and by the close of 1923 the level of indebtedness to Lloyds stood at £1.6 million, though this was only a small proportion of assets.

Trade conditions in 1923 were appalling, especially in Java and Australia, where every Blue Funnel voyage recorded a net loss. In the Australian passenger trade, Holts found it necessary to join with White Star on the outward service, giving alternate fortnightly sailings. The first advertised 'joint' sailing was that of the *Ascanius* in March 1924 and later that year the Australian Government Commonwealth Line service from Glasgow to Australia joined the arrangement. In 1926 the Aberdeen Line joined as well, and a joint service was maintained until 1939 (with Shaw Saville taking over White Star's Australian interests in the early 1930s) when Holts then withdrew. Steamer earnings in 1923 were catastrophically low, so much so that Richard Holt felt unable to divulge them to the shareholders. The Java trade was hard hit by troubles in Europe (1923 was the year of the French occupation of the Ruhr, and saw an astronomical inflation in Germany), so that European demand for Java products was low. Amsterdam was an important centre for the onward despatch of goods to European markets, and troubles in Germany and central Europe during the inter-war years had an immediate reaction on the Java trade. Australia's problems were different, but no less damaging. For much of the 1920s Australia's economy was bedevilled by a series of bitter dock strikes and sharply rising costs at the major ports, and when as in 1923 such endemic troubles were joined with exceptionally severe weather the result was disastrous.

The best results in 1923 came from the China trade, where 32 voyages returned profits. In December 1922 Richard Holt had recorded in his diary:

business is not easy at present. The steamers are making some money, at any rate more than covering out of pocket expenses, but there is a restlessness everywhere. The various steamship interests, both British and foreign, all seem to want more than they have got and I fear cast covetous eyes on the China trade which has been sensibly managed and has a good volume of business. It has been a constant difficulty to avoid any break up.

24 Yokohama after the earthquake, 1923 (*photograph courtesy of Liverpool Museum*)

The reference to 'sensible management' was, of course, the operation of the Conference which was largely under the control of Holts and Swires, and which was able to maintain freight rates at satisfactory levels.

At the close of 1923, though, the China trade was boosted by a disaster, the great earthquake which struck Japan in September of that year. Following the devastation came an immense rebuilding programme, which was felt by Blue Funnel through greatly increased cargoes to Japan from Europe and the United States throughout 1924. Blue Funnel vessels were also involved in the disaster. The earthquake struck on the morning of 1 September 1923, destroying much of Tokyo and virtually the whole of the port of Yokohama. In all nearly 25 000 inhabitants of Yokohama perished, out of a total population of around 440 000, and most of the buildings were destroyed either by the earthquake or subsequent fires. Two Blue Funnel vessels were in port that morning, the *Philoctetes* and the *Lycaon*, both managing to survive the disaster without serious mishap. The former, under the command of Captain Propert, narrowly escaped collison with an OSK liner, and the Master decided in an action subsequently criticised to leave port and head for Kobe. The *Lycaon* remained in Yokohama and, along with other vessels,

25 *Sarpedon* at Liverpool landing stage, 1923 (*photograph courtesy of Liverpool Museum*)

was able to give assistance to some of the many thousands of injured and destitute left in the ruined city.

The year 1923 was also notable for the delivery of the two Far Eastern passenger steamers, the *Sarpedon* and the *Patroclus*, both coal-fired. The following year the last coal-fired steamer *Asphalion* was delivered and the Managers recorded that 'nothing but the uncertainty which exists as to the supplies of oil and the future comparative price of oil and coal causes the Managers to doubt that the internal combustion engine will supplant the steam engine using coal as fuel for ordinary propulsion at sea.'

The technical development of the Blue Funnel fleet after the First World War calls for some comment. Undoubtedly the main development in marine engineering in the inter-war years was the possibility of substituting internal combustion for steam power in the form of efficient diesel engines of various kinds. Now it is frequently stated that British shipowners were dilatory in modernising their fleets by comparison with those of their rivals, especially the Scandinavians and Dutch. In 1925, for example, only 3.9 per cent of the British merchant fleet was equipped with diesel propulsion, by comparison with 21.4 per cent in

Sweden. By 1933 the proportion so driven in British ships had grown to 14 per cent, but in Sweden it was then 33.3. per cent, in Holland 26 per cent, and in Norway 43 per cent. It is often forgotten, though, that the fleets of both Germany and Japan, which proved such powerful competitors in many trades, had scarcely a greater share of motor ships in their fleets than Britain.

The motor ship had made its appearance in cargo ships, pioneered by the Danish East Asiatic Company, just before the First World War. By 1920 marine diesel engines had been developed sufficiently to make them a practical possibility for liner owners. Diesel ships had a number of significant advantages over traditional coal-fired vessels, and one major drawback. The big advantage was that for a given power, far less fuel was used, so that the space required for fuel bunkers was much lower and more cargo could be carried. In addition, diesel ships needed to carry fewer engine-room ratings and the conditions under which the engine crew worked were far less arduous. The main drawback was that diesel fuel was expensive. Moreover the capital costs of diesel ships were rather higher than for conventionally-powered ships, while the engines themselves were still in an early stage of development in the 1920s. They were generally more efficient in small, relatively low powered vessels, than in large fast ships. By the 1920s, too, a long series of experiments with Parsons' steam turbine principle had produced an efficient steam engine (which might use either coal or oil of a much cheaper sort than high quality diesel), which provided an alternative to the motor ship. From around 1920 there developed a long and fluctuating battle between the steam and diesel engines, and it was not until a series of major improvements in the latter in the 1950s that it became appropriate to use diesel engines in large high-powered vessels.

It will be obvious even from the above brief discussion that many factors might influence a shipowner's decision as to whether to order a steamship or a motorship, even were he not hidebound by tradition. The relative price of fuel would be an important factor, fuel in the inter-war years being the largest single item in the running costs of a long-distance liner. If coal was cheap and of good quality, as from the Welsh coalfields, this might be decisive. The Scandinavian countries, on the other hand, who needed to import all their fuel, would face a different situation. Then again the shipowner would need to consider the costs of fuel en route (oil, for example, was cheap in Aden and Sumatra). Another factor would be the relative savings of crew costs in a motorship. The employment of Chinese engine-room ratings, whose wages were less than half those of European ratings, would reduce the significance of any savings brought by the use of diesel. The size and speed of the ship, too, would need to be considered.

How did Holts react to the new technical possibilities? We may conveniently examine the shape of the Holt fleet in 1934, since by this time the last of the Blue Funnel ships built between the wars had been delivered with the single exception of the small *Charon*, a diesel-powered ship of 3700 tons which arrived in 1936.

In May 1934, the Blue Funnel fleet stood at 78 vessels with a total gross tonnage of 621 360 (average 7966 tons). Of these 78 ships, 30 had been built before or during the First World War and 48 subsequently. The Fleet Book classified the fleet into three groups by size. At that date there were 24 'large twin screw ships' each with a gross tonnage over 9000. Then came a group of '*Lycaon* and succeeding classes' of 26 ships, within the range 7500 to 7900 tons. Finally came 28 'smaller ships', from 3066 tons *Centaur* to a cluster of vessels around 6700 tons of various classes. Table 8.3 classifies the 78 ships of 1934 by size and by steamer (coal or oil-fired), or diesel motor ship:

Table 8.3   Steamers and motorships in the Blue Funnel fleet, May 1934

| Class (tons) | Steamers | | Motorships |
	Coal	Oil	Diesel oil
Up to 6700	10	3	15
7500–7900	16	2	8
9000 and above	22	2	0
	48	7	23

Source: Ocean Archives: Fleet Books.

Interestingly, coal-fired ships still predominated in 1934, but this pre-dominance was concentrated very much in the larger classes of ships (which included the pre-war Australian passenger vessels). The pattern is even clearer if we look just at the 48 post-war ships (see Table 8.4).

Thus nearly half the ships built between the wars were diesel-powered, although the proportions of total tonnage was far less than this, at around 25 per cent. Nevertheless, this ratio was considerably larger than that for British shipping as a whole, and indicates that Holts were actively aware of the possibilities held out by the new form of propulsion.

Table 8.4   Post-war steamers and motorships, 1919–34

| Class (tons) | Steamers | | Motorships |
	Coal	Oil	Diesel oil
Up to 6700	1	3	15
7500–7900	9	2	8
9000 and above	8	2	0
	18	7	23

Source: Ocean Archives: Fleet Books.

Looking a little more closely at the technical aspect of fleet development, it is evident that the Company designs showed no clear pattern
between the varied claims of motorships and steamship before 1925.
This was reflected by divisions within the Company itself, the two
principal engineering superintendents each advocates of different
methods, while Lawrence Holt favoured steamships and Leonard
Cripps motorships. None the less, Holts had been among the earlier of
British lines to build diesel ships, and Lawrence Holt had played an
important part in arranging in 1921 for diesel machinery to be fitted by
the subsequently famous Copenhagen firm of Burmeister and Wain.
The first diesel ship delivered was the *Tantalus* in 1923, followed by the
Medon the same year, with similar but improved machinery. In 1926
there followed the *Orestes* and *Idomeneus*, also with Burmeister and Wain
engines. When delivered in July that year the *Orestes* was the highest-
powered diesel-engined cargo liner afloat, and her design pioneered
what later became standard practice, the fitting of cast steel crankwebs.
The next major programme of motorships consisted of five *Agamemnon*
class ordered at the end of the 1920s. In a successful effort to reduce the
large amount of space taken up by these early diesel engines Holts
decided to increase pressure by supercharging. A Holt engineer, A. G.
Arnold, later recalled how Burmeister and Wain 'were not anxious to
adopt supercharging but the owners were determined to try it, and
experience since that time has shown how right they were.'[1] Less
successful was the earlier experiment with the 1926 *Stentor*, whose 4-
stroke double-acting engine from North Eastern Marine Engineering
developed insufficient power and was replaced with a Burmeister and
Wain engine in 1930. Experiments were also tried with Scott–Still diesel
engines. The small *Dolius*, delivered in 1923 attracted attention throughout the world because of its fuel efficiency. The *Dolius* had a speed of 11
knots, and Holts were sufficiently encouraged to proceed with a 14 knot
Scott–Still diesel ship, the *Eurybates*, in 1928, though the engines of this
ship, built to a wholly novel design, needed several modifications before
they performed entirely satisfactorily. Noteworthy also were the four 15
knot ships fitted with supercharged North Eastern Marine–Werkspoor
engines built for the Java trade in 1930. Again these engines, commonly
called 'Water Street engines', had a number of novel features and, after
initial modifications, performed well until all four ships met premature
deaths in the Second World War.

These, then, were the diesel ships added to the Blue Funnel fleet by
the time of the great depression. The new steamers were mostly powered
by efficient double or single reduction geared turbines, the last ones
before the Second World War being the *Antenor* and *Polydorus*, delivered
in 1925. The year before had seen the last coal-fired ship to be added to
the Blue Funnel fleet, the *Asphalion*, a steam turbine ship of 6300 tons. By

then the age of the coal-fired reciprocating triple expansion engine had passed too, the 1916 *Tyndareus*, the 1917 *Elpenor*, and the 1920 *Machaon* being the last representatives of a form of propulsion which had taken Blue Funnel to the forefront of British shipping in the decade before the First World War. The year *Machaon* was delivered saw the arrival of the *Achilles*, the first with a double reduction geared turbine. The Managers noted after a few months that 'although due to labour troubles the gearing of the *Achilles* is defective in workmanship, the results on the *Achilles* are very good, much better than the sister ship *Tyndareus* with a different eingine'. Thus had the old era passed.

We have seen that Holts were far readier in the 1920s to introduce diesel engines into the smaller vessels than into the large ones. This focuses on one of the fundamental problems with the marine diesel engine, not solved until around 1959, the relatively low output per cylinder in early engines. This meant that if large amounts of power were required for driving larger or faster ships the engines would become large and complicated, with many cylinders. The improvements which were achieved after the Second World War are shown in Table 8.5 which illustrates the sevenfold rise in b.h.p. (brake horse power) per cylinder achieved between *Tantalus* in 1923 and *Priam* in 1966.

Increased power was not the only improvement. There were considerable space savings in the later engines, while the weight in relation to b.h.p. developed fell 75 per cent between 1923 and 1962. There was too, a 25 per cent saving in fuel consumption, added to which it became possible after 1960 to burn low-grade boiler oil in the diesel engines, due to work initiated by Holts in the 1930s and pursued vigorously after the

Table 8.5   Development of diesel engines

Year	Ship	Number of Cyl.	BHP/Cyl.	Total BHP
1923	*Tantalus*	16	300	4800
1925	*Eurymedon*	16	240	3800
1926	*Orestes*	16	390	6000
1929	*Agamemnon*	16	540	8600
1930	*Stentor*	16	900	5400
1936	*Glenearn*	12	1000	12000
1947	*Calchas*	8	860	6800
1949	*Ascanius*	7	1000	7000
1959	*Menelaus*	6	1420	8550
1962	*Glenlyon*	9	2000	18000
1966	*Priam*	9	2100	18900

Sources: A. G. Arnold, *Development of the Marine Diesel Engine in Messrs Alfred Holt & Company's Fleet, 1921–56* (A. Holt & Co., 1957); Sir Stewart MacTier, *Shipowning and the Marine Engineer* (Liverpool, 1962); and Ocean Archives: Fleet Books.

war by W. H. Dickie, a saving of 40 per cent on fuel costs in the early 1960s.

Arising from this discussion we can perhaps make a few points in summary. First, there is no simple connection to be made between 'efficient' shipowners adopting motorships, and conservative owners remaining with steam. In fact both diesel and steam technology was in flux, and the decision which to adopt depended on a great many factors. One factor was the size of vessel to be propelled and the power required. Holts showed that where their smaller ships were concerned they were very willing to adopt diesel propulsion, and also demonstrated that far from being conservative they were ready to experiment, to make mistakes, and to learn from these mistakes. In pioneering the use of the Burmeister and Wain engine in Britain, in experimenting with the Scott–Still and North Eastern Marine engines, and in encouraging the use of supercharging in diesel engines, Holts showed a spirit of innovation often thought to be lacking in Britain's shipping industry. If there was conservatism in Holt ships, it was to be found in design rather than propulsion, in the preserve of the naval architect rather than of the marine engineer.

IV

By 1924 there were signs of a general improvement in Holts' business which was to be sustained until the onset of the great depression. The Australian trade remained bad, with several voyages losing money. A long seamen's strike there took place throughout the summer and autumn, Richard Holt describing it as 'mutiny' rather than a strike. The Australia service suffered additionally from a seamen's strike in South Africa, and the average profits from each voyage was little over £1000. The main results are shown in Table 8.6.

The China–Straits trade was, in consequence, roughly twice as profitable as any other branch. Less satisfactory was the liquidation in 1924 of Lester and Perkins Ltd, a shipbuilding and engineering company in

Table 8.6   Net voyage earnings, 1924 (£)

Number of voyages	Trade	Net earnings	Earnings per voyage
55	China	329763	5996
29	Java	84069	2899
12	Australia	14425	1202
26	New York	83194	3199

Source: Ocean Archives: Voyage Accounts.

which Holts had acquired an interest in the heady days of 1918. The firm had been in difficulties as soon as the post-war boom collapsed, and Holts had written down its valuation so that in 1924 they actually received £23 000 more than the stock valuation. This additional sum they put towards writing down their holding in Caledon.

During 1924 the Managers had established a company, India Buildings Ltd, in which the Ocean Company took all the ordinary shares in return for the site for a new office building, previously acquired by a series of purchases. The purpose of this device was to raise capital for the building, which was to have ample accommodation for letting to tenants, by the issue of debentures. Half the debentures were immediately issued, but in the depressed state of Liverpool property it proved impossible to sell them except at ruinously high rates of interest. Accordingly, the company had to raise money from the bank, and during 1925 the overdraft with Lloyds Bank rose to over £2.5 million, raised both for shipbuilding and India Buildings.

By 1925 political troubles overseas were having a marked effect on the affairs of Ocean. A war in the Holy Places stopped the pilgrim trade for the entire year, leading to a loss for the company of £100 000, while there were renewed strikes of seamen in South Africa and Australia. Most threatening was trouble in China, where anti-foreign riots took place against a background of disintegrating central authority. This is not the place to detail these problems, but it may be briefly recalled that in 1911 a revolution led by Sun Yat-sen had successfully overthrown the Manchu dynasty and established a republic. The 'unequal treaties' remained in force, causing growing resentment, especially in the post-war years when Bolshevik propaganda aroused 'anti-imperialist' sentiment. In 1925 Sun Yat-sen, the one slender thread of national unity, died. The country plunged into a long period of war-lord factionalism, strikes, boycotts, murders, and rival communist and nationalist groupings leading to virtual civil war. By 1928 China had, at least nominally, become unified under Chiang Kai-Shek. But in the 1930s the situation disintegrated rapidly as Japan invaded Manchuria, and eventually launched a full-scale war against China in 1937.

Anti-imperialist sentiment in the 1920s focused against Britain, and a 15 month strike against British goods and against trade with Hong Kong lasted from June 1925 to the autumn of 1926. The trouble spread to Hong Kong, and Butterfield and Swire had a torrid time there arranging for Blue Funnel vessels to be loaded and unloaded and despatched on time. As usual, Holts' agents managed efficiently, and Blue Funnel vessels were little inconvenienced by the troubles. In July 1926 a general strike occurred in Hong Kong, but Swires could write to Liverpool assuring the Managers that the *Patroclus* 'was got away and other Blue Funnel Ships are being handled as they come along ... with Chinese crews brought up from Singapore'. At times such letters arrived in India Buildings almost every day, but usually with the same message: by dealing with each situation on a day-to-day basis Blue Funnel business was carried on with a minimum of inconvenience. Swires were also able to reassure Liverpool that the Holts wharf staff 'performed well' during the strikes in what must have been very uneasy circumstances. Later, at the beginning of 1927, came the 'China Affair', when Chinese rioters threatened the international settlement in Shanghai, and mobs took over British concessions in Hankow. An international force of 40 000 troops were sent to defend Shanghai, and several Blue Funnel vessels were employed in the transport of these troops.

By 1926 Holts had spent over £100 000 on the Hankow wharves and other facilities, and although the property was not damaged in the riots, the Company was unable to use it for many months. The Managers felt that they had over-extended themselves in China, and decided to dispose of the Hankow and other small properties which were at the mercy of civil disturbance. Incidentally it is interesting to see the valuation put on some of the larger overseas assets held by the Company in 1926:

	£
Pootung	776 771
Kowloon	473 445
Singapore	278 503
Dutch East Indies	454 284

Against this background one would expect the China trade to collapse. But remarkably it did not. Time after time Richard Holt expressed his astonishment. In May, 1927, he wrote 'in spite of all the disturbances in China trade is still alive on a larger scale than anyone could have expected', and in January 1929, 'The trade with China still goes on remarkably considering the chaos in the government of that country.' Reviewing the year 1929, Richard Holt told shareholders that the China trade was remarkable; it was 'almost wonderful that it should go on as it does'. Probably the key factor explaining this resilience is that China should not be considered as a single country. The very political disintegration of the country meant that conditions in one region would not necessarily be repeated elsewhere. Circumstances in one, or two or ten 'treaty ports' might be hazardous, but in 1928 there were over 90 treaty ports (some dozen served regularly by deep-sea liners). There were, too, some underlying favourable factors. The enormous population represented a great potential market, while the various main China exports, tea, silk, the 'China produce' like beans and bean oil, bristles, egg products, hides, skins, and other miscellaneous goods, could be

Table 8.7 China's foreign trade (exports
+ imports), 1920–8 (million Haikwan
Taels)[1]

Year	Value
1920	1304
1921	1507
1922	1600
1923	1676
1924	1790
1925	1725
1926	1988
1927	1931
1928	2187

[1] The tael fluctuated between about 2s 6d
and 3s 6d in these years.

*Source: Hsiao Liang-lin, China's Foreign
Trade Statistics, 1864–1949* (Harvard University, 1974) p. 24.

obtained from a variety of outlets and their production could not easily
be diminished by civil unrest (not like a steel mill or an oil refinery, for
example, where a single strike could close production completely).
Moreover the continued decline in the value of China's silver-based
currency boosted exports. Even the effective boycott against Hong Kong
in 1925 and 1926 brought a compensating increase in the direct trade
between China and the Dutch East Indies, the Straits, and Singapore.
Table 8.7 shows the total value of Chinese foreign trade (exports plus
imports).

The point made earlier that the China trade was not a single trade but
a variety of trades which could compensate for one another was
confirmed in a report written to the Department of Overseas Trade by
the Commercial Counsellor to the British Legation in Peking. He wrote
in 1928:

China is such a vast country that a falling off of revenue of trade in,
say, southern ports, is more than compensated for by increased
returns from the northern ports. In 1927, for instance, figures for the
foreign trade of Shanghai and the Yangtze ports show an all round
decrease as compared with 1926, which is largely compensated for by
the increased trade done at Tientsin, Dairen, Tsingtao, Antung and
other northern ports.[2]

A fascinating insight into conditions at Holt's Wharf during the inter-
war years has been recorded by Captain W. J. Moore, who became
Assistant Manager at the Wharf in 1937 and was later variously Manager

there, a Blue Funnel Master, and Godown Manager in Shanghai with Butterfield and Swire.[3] Captain Moore noted the crucial role played by Chinese 'compradores' who acted as essential intermedaries for large European companies in their dealings with the Chinese community. In the 1930s at Shanghai:

> our compradore was Yang Wei Ping, and he was an important man, maintaining his own office and staff of book-keepers and shroffs. He had put up a substantial bond to get the position and his terms of remuneration as well as his liabilities were set down in a solemn legal contract. He had political contacts and he was consulted by the wharf manager on all purely Chinese matters. All cash was handled by his men, who were called 'shroffs', and he was held responsible for any thefts from godowns. The godown men were appointed and guaranteed by him and all sheds and godowns had his locks on them in addition to ours.[4]

(The word 'shroff' derived from a term which originally referred to those who could 'scent' out by touch debased coins from those with their full silver content).

Captain Moore also gave a colourful picture of cargo-handling at Holt's Wharf:

> Although the ordinary porter's hand-truck had been introduced back in the 1920s (it was said that the coolies threw the first lot into the Whangpu), prior to 1945 practically all movement of cargo on the wharves of China was effected by bamboo coolies. These men were specialists, and ours anyway were known as 'Hupeh men', presumably because they came originally from that province. Each pair of men had their own special trusted bamboo. The tractor, the fork lift, and the mobile crane had not yet come on the scene, which rang and echoed with the characteristic 'Yo-ho, oh-ho' chant used by the coolies as a help in keeping step and synchronising their efforts. It was quite a sight, and one not to be seen today, when a hundred or more men clustered around a heavy boiler or other piece of machinery discharged from a ship and weighing perhaps forty tons, criss-crossed their bamboos and bits of rope, then gradually took the weight on their shoulders and, step by step, like ants carrying an egg, walked it off to wherever it was required. These men lived in the villages of Pootung, close to the wharf, and hired their services out at a daily rate (which was pretty low). A contractor or 'coolie boss', called Wong, provided ours as necessary. He was another middleman, essential to us as well as to the labour, with whom we could not otherwise communicate.'[5]

If we can stand back from the year-to-year events, it is clear that for all vicissitudes and vexations there was a definite sense of advance and recovery after 1925. Certainly political troubles remained a constant source of anxiety and threat, and in these years Richard Holt and the other senior managers were in constant touch with Warren Swire and the agents in Shanghai and Hong Kong. Politics and strikes continued to influence conditions throughout the second half of the 1920s. At the height of the China boycott, the Australian strike, and the cessation of the pilgrim trade Richard Holt had hurried back from holiday in 1925 to take control of affairs in India Buildings. He recorded a little testily:

> I don't like leaving my partners alone to face all these disturbances, and they are serious for us involving a considerable loss of money. It is remarkable how one's private affairs are affected by circumstances over which one has no control. Strikes in Australia, ferment of crude nationalism in China, the religious effervescence of the Wahabis stopping the pilgrimage to Mecca – all these things react unfavourably on my holidays and reduce my income.

The following year the pilgrim trade was resumed and some 30 000 pilgrims were carried in Blue Funnel vessels. Richard Holt wrote in his diary,

> I find it impossible to avoid a quiet chuckle at the thought that a middle class English household in Liverpool has its material welfare seriously affected by the religious differences of Mahometans. The world is a funny place. The interdependence is much greater than most people suppose and it matters more to me and many most respectable fellow citizens how the Sultan of the Wahabis behaves than what happens in football, cricket, or golf matches although no one would suspect this from reading the newspapers.

The commercial results for 1926 were satisfactory, especially consider-ing that the year saw a prolonged strike and a 9-day General Strike in June. The strikes, and continuing difficulties in raising money on the capital market for India Buildings, led to a sharp increase in bank borrowing, the debt at one stage in the year rising to £2 785 726. The General Strike itself made little impact on Liverpool's shipping business, and all Holts' vessels were loaded and unloaded without difficulty. The year saw higher net steamer earnings than at any time since the post-war boom had collapsed in 1921. The improvement continued. True, the Australian trade, plagued by labour disputes and rising costs, remained depressed, but the Java trade, aided by a steady increase in the Mecca pilgrims and growing exports of rubber and copra, prospered. The

American trades and the trans-Pacific also remained depressed, but the China trade, as we have seen, made relatively good progress. The result was that net steamer earnings rose, and the company was able to pay off £1.4 million of its overdraft between 1927 and 1928, so that by the end of the latter year it stood at £1.3 million, half its peak level two years before.

Two major events in the late 1920s transformed the company's day-to-day operations in Liverpool. One was the move to new India Buildings, the other the transfer of discharging to the newly-opened Gladstone Dock. The year 1927 saw both the start of demolition at old India Buildings and the opening of the magnificent new Dock. The new India Buildings was the culmination of work planned during the First World War, and work had begun on the new headquarters in 1924. Saturday, 7 January 1928 was the last day at the former offices which had housed the Ocean Steam Ship Company since its inception, and the following Monday the staff moved to its new quarters. The building was not finally completed until 1932, by which time it had cost nearly £1.5 million. The architect was H. J. Rowse, responsible also for designing the Mersey Tunnel and the Liverpool Philharmonic Hall, and his new steel-framed building, with its Portland stone exterior and spacious marble-clad interior was in its day one of Liverpool's most impressive landmarks. Inside, such matters as lighting, heating, and the lift services were among the most advanced available, but not everything looked forward. On the seventh floor was replicated the open-plan 'quarter-deck', so preserving the Holt tradition of a management accessible to one another, and in touch with other staff.

From the early years of the twentieth century Blue Funnel ships had discharged in Liverpool at Queens Branch Docks, but by the First World War the facilities available there were becoming insufficient for the Company's needs. Ocean planned to move their discharging berths to the vast new Gladstone Dock complex, on which work had started as early as 1905. Due to wartime and other delays the construction of the new Dock was not completed until 1927, and in July of that year Blue Funnel ships were able to move to their new and permanent berths. Fittingly, Richard Holt had succeeded to the Chairmanship of the Mersey Docks and Harbour Board at the beginning of 1927, for he had been a prime mover in pushing forward the construction after the war. And as chairman he welcomed King George V and Queen Mary at the official opening of the new Dock on 19 July. The new facilities, where Holts' employed their own stevedores and other dock staff, provided opportunities for more efficient and extended operations. A new venture was the establishment of Holt's Warehouse, the Company adapting the upper storeys of a large shed alongside the berths as a warehouse. Ocean employed there a staff of specialists to deal with various aspects of storage, despatch, and sampling of cargoes, and Holt's Warehouse not only improved the efficiency of the company's own handling operations but provided a source of revenue from other users of the facilities. Soon additional space was acquired as stored cargoes, especially raw rubber, increased. In 1933 Dunlops installed special storage tanks alongside the Warehouse for the reception of latex pumped directly from the ships' tanks. Managerial responsibility for both the Warehouse and latex ventures devolved largely on John Hobhouse, and no small part of the success of these endeavours was due to his initiatives.

By the close of the 1920s there was a mood of confidence in the new India Buildings which was reflected in Richard Holt's diary note in January 1929, 'The Blue Funnel enterprise is flourishing.' With growing earnings from the major trades (the Java trade accounted for over one-quarter of total net voyage earnings in 1928) the company decided to embark on another round of fleet building. In 1928 the Managers ordered five new vessels to take advantage of the better conditions they expected. Two vessels, of 7800 tons, were for the China trade; three, of 6500 tons for the Java trade. The success of such an investment required some years of moderately good trade. At worst Richard Holt had hoped in January 1928 'that we may be allowed 12 months at least without disturbance, or misfortune'. But well within these eleven months the Japanese had invaded Manchuria, the long American boom had come to an end, and the Wall Street stock market crash in October signalled the start of the worst international depression ever to hit the modern world.

9

The 1930s: Collapse and Revival

All your Managers can do is to keep up a brave front and be ready to meet whatever may befall with forethought and resolution.
(R. D. Holt to shareholders, February, 1930)

The volume of international trade in the year 1933 measured in terms of gold was hardly more than a third of its value in 1929. Those are the bare statistics of the great depression, but behind those statistics stand the want and misery and malaise which surrendered to the desperate remedy of war.[1]

The words are Sir Compton Mackenzie's, and he does well to remind us that the Great Depression was not simply an economic catastrophe, but a political disaster as well. No depression before or since has struck with such intensity, spread its tentacles so wide, or had such profound and long-lasting effects. One side of the coin was falling prices, mass unemployment, and an avalanche of bankruptcies. The other side was the political response; a rise everywhere of beggar-my-neighbour policies of protective tariffs, trade quotas, currency depreciation, and debt defaults. If the curtain of depression began to lift in 1933 in some countries, and was lifting nearly everywhere by 1936, it was to reveal a very different world. Far from the pre-1914 ideals of free trade, the gold standard, and economic interdependence, which Britain had vainly tried to re-create in the 1920s, the 1930s was a period of economic nationalism, of low international investment, and low levels of world trade.

The depression and consequent changes mark an obvious turning point in Ocean's history, and this makes it the more important to recall again the many elements of continuity in company affairs, in both management and fleet operations. To some extent depression reinforced continuity since uncertain trading conditions were hardly propitious for major changes. Thus, with minor exceptions, no new vessels were ordered for the Blue Funnel Line after the pre-depression programme was completed in 1931. Moreover there were few changes at the management level. Only one new Manager, Roland Thornton, was appointed. This was in 1929, and in Thornton's opinion long overdue.

Only one Manager left, W. C. Stapledon, who retired in 1930 after a period just short of 30 years with the company. Managerial succession was safeguarded, however, by the appointment of two recruits to the Liverpool Office in 1932, Sir John Nicholson and George Palmer Holt. At the departmental level too, long service and continuity remained characteristic in many sections. In particular, H. Flett, (Chief Naval Architect), S. Townley (Inward Freight), and F. Jackson (Steamship) long dominated their respective spheres.

As a great trading nation Britain was naturally hard hit by the depression, and shipbuilders and shipowners were, of course, in the eye of the storm. Overall, Britain's shipping tonnage in the 1930s declined both absolutely and as a proportion of the world total. World trade fell sharply, and at the low point of the depression the volume of such trade was less than two-thirds of its former level. World shipping tonnage fell, but not so fast as trade. The 1930s was, in consequence, a decade of fierce competition and falling freight rates. Unfortunately too, in the environment of strident nationalism, a number of countries built up subsidised fleets which added to the problems. Freight rates fell by around 25 to 30 per cent, and at the worst of the depression, in July 1932, around 3.5 million tons of British shipping, some 17 per cent of the fleet, was laid up.

For Alfred Holt and Co. the optimism of 1929, a 'goodish year' Richard Holt had termed it, quickly gave way to pessimism. Suddenly and dramatically the company found itself faced with falling earnings, declining profits, laid-up ships, and mounting bank debt. The year 1930 saw 'extreme world depression in industry and trade', as Richard Holt told the Annual Meeting at the beginning of 1931, and he warned that, at current levels of earnings, dividends could not be maintained. Phrases from the Annual Report for 1930 tell their own story. The China trade: 'now badly affected by the collapse in the price of silver'; Japan: 'even more severe'; Malaya and the Dutch East Indies: low prices of tin and rubber 'have greatly reduced the volume of trade, the natives are poorer and less able to pay for pilgrim passages'; Australia: 'difficulties grow and may have to deteriorate before they can improve'; and the United States, where 'trade from all ports of the Far East to America has been greatly depressed'.

That was 1930, a year when Holt's voyage profits fell to only half of their 1929 levels. And then followed 1931, far, far worse in every respect. Steamer earnings now plummeted to only about one-third of even their depressed 1930 levels. Ships loading at Birkenhead were fortunate to load fifteen hundred tons of cargo; before the slump they had regularly loaded six or seven thousand tons. Parallel with lower steamer earnings went falling revenues from almost all other sources, such as dividends from investments, rents from India Buildings, and returns from overseas properties. During the 1920s investments had regularly yielded around £151 000 a year, by 1933 the figure recorded only a bare £31 000. In March 1931 the Company could pay only a reduced dividend, and the Managers sent a circular to shareholders warning them that further reductions might be necessary. A few months later shareholders were told that the Managers 'feel it their duty to conserve the resources of the Company to the utmost until the present extreme depression of trade has passed away', and the usual October dividend instalment was passed altogether. This was a sad occasion for the company, the first since the earliest days that no dividend could be paid. Explaining the situation to shareholders Richard Holt rather disingenuously laid the blame on 'low earnings, the political crisis, and the high bank rate'. In fact the Managers had already decided to pay no dividend long before the political upheavals of September 1931, which brought the formation of Ramsay MacDonald's National Government, the collapse of the gold standard, and a crisis level of Bank Rate. What Richard Holt did not say also was that with the Company in the throes of a vigorous economy drive, where every employee had taken a pay cut, and when some staff were being declared redundant, a dividend to shareholders was unjustifiable in the circumstances.

The financial crisis brought few compensations, but one came at the low-point of mid-1931 when a Blue Funnel ship took a most unusual cargo of £500 000 worth of gold bullion to America. This was at the time when panic was sending gold streaming to the United States in such quantities that all the normal Atlantic liners had their maximum permitted loads already.

Blue Funnel's financial situation at this juncture was dire, and remained so throughout 1932. Not only had the slump struck unexpectedly but the company was in many ways ill-equipped to counter it. Before 1900 a not inconsiderable element in Holts' success had been the ordering of new tonnage in times of depression, when prices and borrowing costs were low. But the inter-war years saw an unfortunate reversal. As we saw earlier, expensive new ships were acquired in the early 1920s at prices reflecting the immediate postwar boom. The ships then had cost around £500 000 each, compared with perhaps £130 000 for a similar vessel on the eve of the First World War. Yet the Holt fleet still maintained many elderly and increasingly obsolete vessels, some over 30 years old. The Managers, faced with the faster and newer ships of competitors, decided to inaugurate an expensive building programme just as the depression clouds were gathering. In 1928 and 1929 no fewer than 9 new diesel ships were ordered at an average cost of nearly £300 000 each, and the new ships entered service, in the midst of depression, in 1929 and 1931. These nine ships consisted of 5 *Agamemnon* class vessels of 7500 tons, capable of a speed of 16 knots, and four smaller ships of 6300 tons. Ironically, in placing their orders for the new ships, the Managers had feared long delays, as they had experienced already with the *Eurybates*, delivered several months late in 1928. So with unfortunate prudence they decided to contract for their new building programme further ahead than they would normally have contemplated.

At times of boom and rising prices the costs of fleet modernisation may to some extent be offset by a buoyant secondhand market for discarded vessels. Indeed, Alfred Holt's old policy of ordering in depressions made sense largely because it was additional tonnage being ordered, without the need to sell existing ships at the depressed rates. In the early 1930s Holts had the worst of both worlds. They had bought ships whose earnings could not cover their high costs, and they were faced with disposing of obsolete ships on a glutted world market. Holts did, indeed, sell 13 old ships in 1930 and 1931 at very depressed prices, but many ships had none the less to be laid up (at one time as many as one in five) while others had to make part of their voyages in ballast.

The timing of the rebuilt India Buildings was equally unfortunate, constructed at high prices in the 1920s and ready for letting just as depression struck. Demand for office space in a depressed Liverpool was almost non-existent, and the company was forced to leave some parts of the building without tenants, while a lease for what the Managers thought was an unfairly low rental was acquired by Lloyds Bank. The Lloyds Bank episode rankled with Richard Holt, who felt that the bank was exercising undue power as the company's major creditor (although Lloyd's in turn could argue that they were paying a competitive rent in a very depressed market).

The seriousness of the company's situation was reflected by mounting bank debt and the inability of the company to sell its shares. At the onset of the slump at the end of 1929 Holts was already in debt to Lloyds Bank for around £2 million. During 1930 the overdraft rose by more than £1 million and at one point in 1931 exceeded £4 million (although the end of year balance sheet showed around £3 million). These loans were secured largely against the value of ships, but since the real worth of the fleet was declining in line with earnings, Lloyds became increasingly reluctant to lend further. Indeed, in September 1931 Holts were obliged to deposit securities with the bank against a further loan. Lloyd's policy in asking for these guarantees was part of a general bank policy towards ship-building loans, for the banking world had been shocked by the failure of the Kylsant group and by the falling values of ships against which they were being asked to lend money. Even so, Lloyds' reluctance to extend Holts' overdraft was a severe blow to the management. The problem was not so much that Holts had a particularly large outstanding debt relative to assets (the company was not highly geared in present-day terminology), for the value of investments, property, and the fleet of more than 50 ships was worth many times more than borrowings. Rather, the problem was that in 1931 ships were a particularly unattractive asset, while low stock-market values and the difficulty of disposing of Ocean's own stock added to the uncertain environment. The sterling crisis of August and September 1931 brought high levels of Bank Rate, with consequent falls in the values of the gilt-edged securities in which company reserves were held. In fact, from 1932 interest rates fell rapidly, and this then brought rising values to the bonds (especially government bonds) in which, under Holt's conservative financial management, the reserves were largely held.

For a time though, the financial clouds were dark indeed, and much disquiet was generated by the reluctance of investors to buy Ocean stock which came on the market increasingly as shareholders themselves sought to turn their holdings into cash. On a large number of occasions in 1931 and 1932 the minute books record the tenders of stock for which no bids were received, while at other times the Managers' Minute Books noted that sales could only be made at very low prices.

Richard Holt fought the depression with weapons based on the principles of Victorian financial orthodoxy. One step, taken in 1932, was to reduce the reserve fund by more than £1 million and to use this sum to write down the book value of the fleet. This was a necessary recognition that the depressed earnings of the ships were no longer adequate to pay off depreciation at the pre-slump historic costs, though it was a bold use of reserves at a time of great financial uncertainty. With further sums later written off as well as normal (8 per cent) depreciation, and with little new building, the book value of the Blue Funnel fleet declined steadily, as Table 9.1 shows.

The Managers applied the principles of sound finance not only to the balance sheet; they also instituted a regime of strict cost-cutting and rigid economies throughout the organisation, at home and overseas, on shore and on land. In the emergency it was Leonard Cripps who directed most of the cost-cutting operations. Cripps showed both zeal

Table 9.1   Book value of fleet, 1932–9 (£ million)

Year (1 Jan.)	Number of ships	Fleet book value
1932	54	5.67
1933	54	4.18
1934	52	3.83
1935	52	3.52
1939	51	2.62

Source: Ocean Archives: Managers' Minute Books.

and resourcefulness in undertaking a thorough overhaul of Holts' financial and accounting practices, and in a host of ways made his influence felt in varied spheres of company affairs. Since Cripps can rightly be regarded as one of the architects of Holts' survival in depression and recovery thereafter, it is appropriate at this stage to consider his contributions in a little more detail.

Leonard Harrison Cripps was born in 1887, son of Alfred Cripps and Theresa Potter. Alfred Cripps, later the first Lord Parmoor, was a parliamentary barrister, his mother one of the remarkable Potter sisters and sister, therefore, of Richard Holt's mother Lawrencina. The talented Cripps children included Stafford, two years younger than Leonard, later, as Sir Stafford Cripps, to become Britain's post-war Labour Chancellor of the Exchequer. Leonard was considered less academically gifted than other members of his family, and took up a military career with the cavalry before the First World War. Badly and permanently wounded in the retreat from Mons he subsequently managed a munitions factory and joined Holts at the end of the war, and was made a Manager, together with John Hobhouse, in November 1920.

Cripps' contributions as Manager were many and varied, although undoubtedly his main achievement was to implement economy measures and overhaul accounting procedures during the years of the great depression. Like all the Holt Managers, he was given, or assumed, particular areas of responsibility, and most notably for Cripps these areas lay in Java and in the Far Eastern Conference. He was for nearly 20 years Chairman of the British Chamber of Commerce for the Netherlands East Indies (a body which Holts had been instrumental in establishing in 1919), and with his useful contacts and knowledge of the Java trade he was able to work fruitfully with the Dutch Mail lines, with whom Blue Funnel ran joint services after 1920. He helped improve and rationalise the pilgrim trade, which became very depressed in the early 1930s, promoting Ocean's acquisition in 1931 (jointly with its Dutch partners, SMN and Rotterdam Lloyd) of International Agencies, Jeddah. The chief purpose of this agency was to streamline handling of the pilgrim traffic, and Ocean's interest in the agency lasted until 1959 when ownership had perforce to pass to Saudi Arabian nationals. Cripps also represented Ocean at the Far Eastern Conference, alongside Richard Holt, and the success with which the Conference countered its immense problems owed a great deal to Cripps's skills as a negotiator.

We cannot, of course, attribute all the cost-cutting achievements which Holts instituted during the slump to Leonard Cripps, for many individuals, including department heads as well as Managers were involved. Yet there is no doubt that Cripps's influence on policy here was large and perhaps decisive. This we may glean from three factors. First, a number of specific areas where Cripps made largely individual and effective contributions. An example was Cripps's control of fleet expenditure which he inherited when Lawrence Holt replaced Stapledon as ships' husband in 1931. Cripps immediately investigated all aspects of voyage costs and instituted new forms of voyage accounting. One innovation was to have Masters send by the new airmail services repair indents forwarded from each ship's last homeward port. These were then analysed by superintendents at the weekly meetings over which he presided, and any obvious cases of overcharging could be quickly dealt with. He also set up a Voyage Accounts Department in 1931 to examine overseas expenditures which were settled on Holts' behalf by agents (usually on the authority of the Master). This brought to light a number of areas of loose financial control and even fraud; it was discovered, for example, that suppliers of dunnage in Shanghai had for years added one or two noughts to their totals. Attention to shipping costs was overdue, for Holt vessels were traditionally expensively crewed and operated. By the 1930s not only was the company faced with having to save money, but Holt ships were facing growing competition from countries with lower wage costs and subsidised fleets. Cripps personally introduced a number of changes. These ranged from covering ships' hulls with red lead to save paint, to cutting down on the breakfast eggs given to the crew.

Cripps also made some significant improvements to commercial operations. Most notably he instituted a programme of vigorous canvassing for outward cargoes. He also initiated the new trade from Poland, through the port of Gdynia, a significant cost-cutting venture in view of the cheap bunker coal available there. The opening of the splendid new port of Gdynia at the beginning of the 1930s is a sombre reminder of political realities in the inter-war years. Poland needed access to the sea for her survival, yet German hostility made it impossible for Poland to depend either upon German sea-ports or upon the Free City of Danzig (Gdansk). Accordingly, the Polish government embarked upon the construction of a fine modern port in 1925, and the Blue Funnel line was among the first foreign lines to use the port regularly.

The second manifestation, as it were, of Cripps's influence was the resentment which his many intrusions into all aspects of company business aroused. His attempts to improve boiler efficiency annoyed engineers; his investigations into repair costs annoyed shipwrights; his overhaul of accounting arrangements annoyed financial departments; his efforts to improve freights annoyed the freight departments and the agents; and so on. He was, though, fully supported by Richard Holt and also by those departmental managers who realised the importance of the changes introduced.

The third element in Cripps's impact stemmed from his own personality. He certainly showed a tremendous energy, a grasp of detail, and an imperviousness to the unpopularity which his measures and his overbearing manner sometimes brought. More significantly, though, he was a passionate believer in the importance of making economies and cutting out any form of extravagance or waste. In short, it is hard not to believe that Cripps, in the slump, was in his element, and it was Ocean's good fortune that Cripps should have found an opportunity for his talents at the appropriate time. He displayed in India Buildings a large design of a thermometer, reminiscent of church fund-raising appeals, with the vivid red 'mercury level' displaying Holts' overdraft at the bank. His cost-consciousness had an almost philosophic basis, as his published address on 'The Elimination of Waste', delivered in May 1933, showed. This address, and Cripps's austerity regime at India Buildings, cannot but recall Churchill's famous 'strength through misery' jibe against brother Stafford's post-war budget policy.

Before leaving the contribution of Leonard Cripps we should mention also his role with Elder Dempster. When in 1932 Richard Holt became Chairman of the line, Cripps also joined the Board, and from the mid-1930s West African trades absorbed a growing share of his time. As Richard Holt's health began to decline Cripps largely withdrew from Blue Funnel affairs, and in 1941 succeeded Holt as Chairman there.

The particular economies instigated by Leonard Cripps were, of course, but part of a wider spectrum of measures taken throughout the length and breadth of Blue Funnel operations. Older ships were sold, or laid up. At one time around a dozen ships of the fleet were lying idle. Some services were pruned, such as the Java and American schedules. In 1929 the Company had operated 33 Java voyages; in 1933 only 22 were made. Similarly in 1929 30 New York voyages, outwards and homewards, were completed; in 1935 only 13 sailings were made. The Australian passenger service, too, was reduced, the fortnightly joint service via the Cape being reduced to a monthly one in 1931, while for a few months during the worst of the depression the service was completely suspended. Thus between May and October 1931 mail and passengers for South Africa from Australia were obliged to journey to Colombo to await a connecting ship for the Cape.

Successive reductions in the victualling allowance for ships resulted in considerably lower expenditures. During 1931 the rate for cargo and 'semi-passenger' ships was reduced from 1s 7.33d a day to 1s 4.75d. The following two years saw reductions to 1s 3.91d and 1s 3.74d. As a result total victualling costs for these ships fell in the depression from £133 855 in 1931 to only £98 853 in 1935.

No member of the company was immune from the effects of the depression. A series of emergency measures taken in the summer of 1931 included a 10 per cent pay cut throughout the organisation (for sea as well as shore staff), staff reductions, and redundancies. The famous 'redundancy letter' dated 18 August 1931, and sent to every Master and mate, deserves to be quoted in full:

Dear Sir,

The grave and continued contraction in the Company's Trades has reached a point where it becomes necessary for the Managers to take steps to reduce the number of Masters and Officers (as well as Engineers, Stewards and others) in the Company's service. The Company cannot continue to stand the financial burden of a surplus of seagoing employees, even with the aid of enforced holidays, or half-pay, hitherto so cheerfully accepted; it is moreover imperative to meet the fierce competition of present day by the most rigorous economies in all directions.

The selection of those who will be called upon to fall out of the Company's service is a most painful task. It will probably be necessary to terminate the employment of at least 12 Masters, 12 Chief Mates, 12 2nd Mates and about 30 3rd and 4th Mates. In doing so due consideration will be given to age, health and record of service.

Should any Master or Officer be in a position to retire voluntarily, with or without aid from the Company, he is requested to communicate with us verbally through the usual channels. There is no need to acknowledge this letter.

Redundancies were, in fact, greater than heralded by this letter. More than sixty pursers were made redundant and clerical work formerly undertaken by them was henceforth the duty of the wireless officer.

I

Partly due to the cost-cutting schemes pursued by the Ocean management, there were some significant reductions in voyage costs in the 1930s. External circumstances, too, were lowering a number of costs which was a natural consequence of extreme depression. For example, port handling charges tended to decline, especially overseas where there were some sharp falls in labour costs, while reduced trade levels meant less port congestion and quicker turnaround times. Lower fuel and victualling costs also helped, while falling silver prices in the east lowered wages and other charges in the Far East when silver-based currencies were converted into sterling (although sterling itself was

26 Cargo for the *Centaur* arriving by a team of 13 camels at Carnarvon, Western Australia, c. 1920, after a journey of over 200 miles (*photograph courtesy of Liverpool Museum*)

devalued in September 1931 when Britain left the gold standard). It is impracticable to apportion the contribution of these varied elements to reductions in total voyage costs, nor is it easy to estimate the overall reductions which were obtained. But if we look at the individual voyage accounts which ran on similar routes and with similar levels of cargo both just before and during the great depression, it appears that total voyage costs fell by at least 20 per cent, and often by more. A few representative examples are given in Table 9.2.

The actual figures in the table are illustrative only, since the costs of each voyage were an amalgam of a multitude of factors, such as the cargoes carried, ports of call, and so on. Comparisons of voyage costs between ship and ship, or of different voyages by the same ship, is a hazardous business. But the story told by these examples is a fair reflection of the general pattern of voyage costs after the depression, except when special repairs had to be undertaken. Since, while costs fell, earnings rose after 1931, overall profitability recovered significantly.

Holts' recovery from the depths of depression in 1931 was remarkable, for British shipping as a whole remained sadly depressed for much of the decade. We saw in the previous chapter that net steamer earnings (and earnings per ton) picked up sharply after 1931 and, after a short setback in 1935, there followed several years of strong growth in which both net earnings and earnings per ton rose substantially above pre-depression levels.

Another indicator of better times was the speedy resumption of

Table 9.2   Selected voyage costs, inter-war years

Ship	Voyage number	Year	Voyage costs (£)	% decline
Asphalion	12	1929	36549	
,,	22	1934	24454	34
Rhesus	37	1929	42074	
,,	51	1937	26647	37
Stentor	5	1928	33003	
,,	25	1937	26692	20
Idomeneus	6	1929	44685	
,,	20	1935	30555	32

Source: Ocean Archives: Voyage Accounts.

dividend payments. As early as March 1932 the Managers felt justified in declaring a half-yearly dividend at the rate of 20 per cent of capital stock per annum. This rate of dividend was maintained until 1936 when an additional December bonus was paid (Table 9.3).

The figures for steamer earnings and dividend payments may together be taken to indicate a gradual recovery from the depression as early as 1932 and a full recovery after 1935 (especially when it is recalled that Ocean had in that year acquired the Glen Line at a bargain price). Of course the actual level of dividends, which look high, mean very little since they were paid on a capitalisation unchanged since 1902. Nevertheless, when we compare Holts' performance with other companies the significant point is that dividends were paid at all. To pay dividends in the early 1930s and to increase them in 1936 was not the typical performance of major British shipping lines. P&O, for example, paid no dividend at all between 1931 and 1935, and Cunard and many other

Table 9.3   Alfred Holt and Co., dividends paid, 1924–39 (percentages)

	%		%		%
1924	50	1929	50	1934	20
1925	50	1930	50	1935	20
1926	50	1931	20*	1936	20 + 5
1927	50	1932	20	1937	30 + 10
1928	50	1933	20	1938	30 + 10
				1939	30 + 10

*Half yearly dividend reduced in March and passed in October.
Source: Ocean Archives: Reports to Shareholders.

Table 9.4   Net earnings, 'China' trade 1927–39 (£)

	Net earning	% of total	Voyages	Earnings per voyage
1927	248716	38	45	5527
1928	346509	39	38	9119
1929	372619	44	41	9088
1930	283189	67	49	5779
1931	86811	56	46	1887
1932	231069	48	47	4916
1933	327628	56	46	7122
1934	292213	65	50	5844
1935	266167	70	56	4753
1936	314583	55	43	7316
1937	500026	51	55	9091
1938	571796	49	47	12166
1939	431854	45	43	10043

Source: Ocean Archives: Voyage Accounts.

lines were similarly unable to pay dividends for some or all of these years.

Holts' striking rate of recovery from the world slump deserves some comment, since it runs counter to the general experience of British shipping. The lower voyage costs, other internal economies, and the writing down of fleet values were part of the story, but only part. Table 9.4 suggests that the main Far Eastern trade was relatively prosperous for much of the decade. This many-sided trade remained the backbone of Holt operations, and as the table shows, the number of China voyages was maintained at a similar level throughout the period.

The relative buoyancy of 'China' earnings is shown by the much higher proportion of total earnings contributed by the Far Eastern trades by comparison with the late 1920s. In some years the share rose to more than two-thirds, and remained throughout the 1930s higher than in the pre-depression years. Thus a good part of the explanation for Holts' revived earnings must be in the performance of the Far Eastern trades, whether to China itself or the Straits and other regions served.

Table 9.5 looks in more detail at net earnings from the various trades through the depression years.

The uneven and patchy performance of steamer earnings in the various trades should warn us against laying too much emphasis on management policies by themselves as a way out of depression. Had Holts' fortunes rested solely on the American trades, no amount of managerial competence and dedication could have prevented bankruptcy. The pattern of trade was obviously also of significance and here, as in

27 This advertisement for the Java trade appeared in the *Java Gazette*, 1933, at the nadir of the Great Depression (*photograph courtesy of the Liverpool Museum*)

Table 9.5   Net earnings per voyage (£), 1929–36

	China	Java	New York	Australia	Trans-Pacific
1929	9088	9081	4315	4193	2640
1930	5779	3745	350	1991	−1345
1931	1887	2206	−925	4995	−1270
1932	4916	4488	245	11427	2243
1933	7122	3037	957	14097	1504
1934	5844	2925	−2200	10579	−826
1935	4753	1447	−1619	8822	−1281
1936	7316	1979	4444	14237	−193

Source: Ocean Archives: Voyage Accounts.

the 1920s, the company was most fortunate to conduct trades with some of the few relatively buoyant areas in the 1930s.

As Table 9.5 demonstrates, the most enduring features of the 1930s were the depressed Java and American trades and the significant revival of the Straits, Far Eastern and Australian trades. The situation in the Dutch East Indies, where the 1920s had seen considerable prosperity based on rising rubber, tin, copra, and other exports, and a flourishing pilgrim traffic, was especially bleak for much of this period. Table 9.6 shows how precipitous was the drop both in Blue Funnel freight receipts and the profit remaining after costs had been deducted from those receipts.

As always, depressed conditions were reflected in a low volume of pilgrimages, and Richard Holt reported to shareholders that in 1931 passengers in this trade were the lowest ever recorded except in wartime, and earnings were 'negligible'. While succeeding years were to see some improvement in various parts of the Far Eastern trades, especially after 1933, the Java trade remained depressed. The main reason was that Holts' trade with the Dutch East Indies was based essentially on carryings of bulky primary commodities to Europe. Almost alone in the region, the Dutch continued to cling to the gold standard and a high exchange rate throughout the depression. In

Table 9.6   Java voyages, 1929–31 (£000)

Year	Total freights	Net earnings
1929	646	267
1930	436	116
1931	284	59

Source: Ocean Archives: Voyage Accounts.

consequence, exports from the Dutch East Indies were increasingly expensive for those countries, like Britain, which had devalued. Low export earnings, of course, reduced the incomes of planters and wage-earners alike, and this factor, coupled with intense competition from Japanese manufacturers, reduced outward cargoes to Java also. The result was that in each successive year until 1936 total freight returns from the Java trade declined. In 1933 some £160 000 was realised (less than one-quarter of average pre-depression earnings); in 1934 the figure was £103 000; in 1935 only £85 000; and in 1936 a miserable £31 000. Only from 1937, following long overdue devaluation, was there some small improvement, and by 1939 there were signs that the trade was once more about to flourish. By this latter date the carriage of pilgrims had revived to its 1920s levels of around 10 000 a year between Jeddah and the East Indies, whereas in 1936 the number carried had slumped to less than 2000.

American trades also remained depressed throughout much of the decade. This was in part due to the depth and length of the slump in the United States, which greatly reduced that country's capacity to import. It was due also to considerable competition among shipping lines and to the lack of effective conference organisation in both the New York and Pacific trades. There was competition from Japanese and American vessels, both countries running subsidised fleets and hence able to furnish ships considerably faster than commercial considerations alone would allow. In July 1930 Holts wrote to fellow members of the Pacific West Bound Conference threatening withdrawal, and stating the company was, 'much dissatisfied with the conditions under which the trade from North America to the Far East is being conducted'. Holts complained of competitors being allowed to undercut conference rates with impunity, and were incensed that Blue Funnel ships routed to Singapore could take only limited amounts of cargo destined for Siam. 'The arrangements about Siam are unsatisfactory and unfair to us', the letter added. The enduring problem in the American trades, though, was over-tonnaging by subsidised fleets. In 1933 Richard Holt commented that 'national prestige appears to be more important than procuring a good trading profit', while the following year was 'deplorably bad', with still 'no advance towards harmonious working'. In 1935 Ocean withdrew one of their four ships from the trans-Pacific trade and considered abandoning the service altogether, but thereafter there was some perceptible improvement in the organisation and consequent earnings on both New York and Pacific routes.

In contrast to the Dutch East Indies and American trades, the main Australian services with the United Kingdom proved a perhaps surprisingly bright feature of Ocean's business. In 1931, the worst year for Holts, Australian voyages produced easily the highest average levels of net earnings per voyage, £6560 compared with £2650 in the China trades and £1260 in the Java trades. This result was mainly due to the ability of the well-structured conference to maintain freight rates, and it was also due to a stimulus given to Australian exports by a sharp devaluation of the Australian pound in 1930. In 1932 the net earnings of Holts' Australian steamers were described as 'decidedly good', and in 1933 'excellent', with returns of £226000 from 21 voyages, helped by a 'remarkable advance in the price of wool', as shareholders were told.

Up to this point Holts had run their Australian service jointly with White Star, continuing an arrangement which had been set up in 1923. In 1934, however, as one of the many by-products of the Kylsant collapse, White Star merged with Cunard and sold its Australian interest to Shaw Savill and Albion Co, Cunard having its own Australian interest with Port Line. This move necessitated some changes. A joint service with Shaw Savill was maintained, and the old Liverpool firm of Gracie, Beazley & Co. was appointed as sole loading broker for both lines on the outward trade from Liverpool to Australia. For inward discharging Ocean handled Shaw Savill ships and managed all berthing and dock handling for both lines, both loading and discharging.

In 1933 Australian affairs were disrupted by the entry into the Australia–United Kingdom trade of the powerful Vestey shipping group with their Blue Star Line. The problem here was that Britain's new policy of imperial preference, following the Ottawa Agreements of 1932, led to losses for Vestey in their traditional business between the Argentine and the United Kingdom. As a result Vestey ships were diverted to the more lucrative Australian trades. The threat for conference ships was that the fast, purpose-designed Blue Star ships could carry chilled beef from Australia to Europe, whereas hitherto chilled beef had only been a minor cargo for the British ships since, as we saw earlier, the length of the voyage was too long for conventionally chilled beef. The Australian Conference initially refused Blue Star entry, but Vesteys were able to contract with local meat exporters and so ensure sufficient cargoes. Within days Blue Star ships were admitted to the conference to avoid a costly rate war. The move spurred Holts to experiment with new ways of chilling beef, and it was directly as a result of the Blue Star threat that the *Idomeneus* made her pioneer voyage with gas-chilled beef in 1933. Blue Funnel's Australian earnings were little damaged from the new competition. Net earnings in the trade increased from £26000 in 1930 to £50000 in 1931 and to £137130 in 1932. By 1937 profits had reached £168000, which was four times the pre-depression figure.

In the Australian trade Holts benefited from a combination of relatively firm freight rates and a strong conference. As we have seen, the regulation of the Australian trade had been reorganised in 1929 with the setting up, under Australian government auspices, of the Australian

Table 9.7   Index of Australian Conference
freight   rates,   greasy   wool,   1922–37
(July 1922 = 100)

October 1925	90
April 1926	81
April 1932	86
July 1933	81
July 1937	72

Source: K. Burley, *British Shipping and
Australia 1920–39* (Cambridge University
Press, 1968) p. 354.

Oversea Transport Association. This step signalled Australian approval
of the closed conference system, and in the ensuing years the stability
of freight rates and rationalisation of trade was a significant feature
of Australian shipping. For Holts, wool freights were of paramount
importance since wool was far and away the most important Australian
cargo carried in Blue Funnel ships, making up around 30 to 40 per cent
of Australian cargoes in the inter-war years. Table 9.7 shows how freight
rates for the carriage of wool remained virtually unchanged through the
depression (1929–33) and this was a significant factor in the profitability
of the Australian trades.

    While the main Australian trade was developing steadily in the 1930s,
the small specialised West Australia service with Singapore was also
returning useful profits. Except for the period 1929–32, when Holts ran
only one ship on this service, Blue Funnel had traditionally furnished
two, with their partners, the West Australian Steam Navigation Company, providing a third. In 1935 the West Australian Company's single
ship, *Minderoo* became a total loss. This, coupled with the entry of the
West Australian State Shipping Service to the Singapore–Fremantle
trade, decided the WASN Co. to give up the service, which they had run
jointly with Holts since 1891. Blue Funnel had only recently taken
delivery of the new *Gorgon* for the trade in 1933 to join *Centaur*. They
now considered a third ship 'imperative', and ordered a new vessel,
Charon, which was delivered from Caledon in 1936. The West Australian
trade thus had the distinction of being the only Blue Funnel trade to
benefit from new ships after the last delivery of the pre-depression
orders in 1931. Although only a minor part of Holt operations, the West
Australian trade provided a profitable service in small specialised ships,
designed for the highly individual carriage of people, animals, and
refrigerated products which the trade served. Table 9.8 records the
profitability of the service during the inter-war period, and, apart from
the single very depressed year of 1931, the ships always made useful
returns.

Table 9.8   West Australian trade, 1922–39, net earnings (£)

1922	13149	1927	10314	1932	23142	1937	26009
1923	10000	1928	19621	1933	17646	1938	42550
1924	9900	1929	31157	1934	14127	1939	30710
1925	13377	1930	24464	1935	31382		
1926	27983	1931	533	1936	24830		

Note: From 1922–8 2 ships were run; 1929–32 1ship; 1933–5 2 ships; 1936 onwards 3 ships.
Source: Ocean Archives: Voyage Accounts.

In China itself Blue Funnel ships were able to sustain a surprisingly high level of trade despite the political turmoil, civil unrest and hostilities with Japan. To some extent the conflicts with Japan actually benefited Ocean, for the Japanese invasion of Manchuria in 1931 (and open war in 1937) encouraged Chinese merchants to boycott Japanese ships and goods not only in China but also in Hong Kong and throughout South-east Asia. Moreover as Japan's merchant vessels were diverted to carry war supplies, especially after 1937, the proportion of normal trade carried by British ships increased. At the same time Japan's rapid industrial growth in the 1930s, and fast-growing trade with her Asian neighbours, brought additional cargoes for Holts.

Table 9.9, drawn from Chinese customs returns, shows the rather surprising trends which we have noted. Although only a very rough guide, the table does bring out the continued expansion of British shipping by comparison with the years before 1914, and shows how even in the world slump (due partly to the boycott of Japan's ships) British shipping tonnage increased. We cannot deduce prosperity, of course, for the tonnage may simply indicate fierce competition. However, we know from other sources that Holts were able to maintain a satisfactory trade with China both in volume and in net earnings after the singularly depressed levels of 1931, and that the Conference was able to sustain reasonable freight rates.

The Straits homewards trades also recovered significantly. Staples of

Table 9.9   China: British tonnage entered and cleared, 1909–34

Year	Tonnage
1909–13	35500
1920–4	47500
1925–9	50000
1930–4	58000

Source: Hsaio Liang-lin, *China's Foreign Trade Statistics* (Harvard University Press, 1974) pp. 232–3.

the trade continued to be rubber and tin, supplemented by a vast range of other primary commodities such as copra, palm oil, natural dyes, hardwoods, sugar, coffee, and many others. Rubber and tin typically accounted for around two-thirds of Malaya's total exports by value, and nearly half of Blue Funnel's freights between Singapore and Europe. During the late 1920s trade in both commodities had boomed, spurred by America's apparently insatiable appetite for tinned food and motor cars. But then came the slump, and prices collapsed. At the low point in 1932 prices of both tin and rubber were barely one third of their former levels. Thereafter there was some recovery, helped by international restriction schemes, and by the late 1930s most regions of South-east Asia were returning to more prosperous conditions. However, when we consider the low prices obtained for primary products in the 1930s we should remember that freight rates were levied upon volumes of cargoes rather than values. Thus, up to a point, the very low prices of primary commodities in the great depression were more of a problem for the shipper than the shipowner. Indeed when, as happened on many occasions, producers were forced to sell even at ruinously low prices, this could mean good cargoes for the ships. Also, in the main liner trades in which Blue Funnel were involved, freight rates did not fall as quickly as prices. Throughout the depression the Far Eastern Conference held together despite general world over-tonnaging and the constant threat of competition. In explaining the relative strength and cohesion of the Conference in the 1930s a significant factor was British dominance, led by such powerful groups as Holts (with four votes after 1935) and P&O, which made agreement relatively easy. Holts were certainly skilled and experienced conference negotiators, and Richard Holt and Leonard Cripps were able to safeguard the company's position during this difficult period.

Another factor in Ocean's strong showing in the 1930s was the relative weakness of Japanese and other foreign competition in the direct trades with the UK, itself evidence of the success of the closed conference system in these trades. Table 9.10 shows the proportions of trade carried between the Far East and the United Kingdom in the year 1937 (by

Table 9.10   British share of UK–Far East trade, 1937

	UK exports to	UK imports from
Malaya	96	89
China	97	85
Japan	60	72

Source: Imperial Shipping Committee, *British Shipping in the Orient* (London, 1939) p. 20.

28 *Maron* embarking troops, Kowloon Bay, 1937 (*photograph courtesy of Liverpool Museum*)

value), and it can be seen that even in the direct trade with Japan British shipping took a preponderant share in both directions.

The strength of Britain's trade with the Far East is also shown by figures of the Suez Canal traffic. Interestingly, and perhaps surprisingly, the tonnage of British shipping passing through the Canal was actually one-third higher in 1936 than for the average of the years 1909–13; moreover Britain's share of the total had scarcely fallen at all, standing in 1936 at 58 per cent compared with 62 per cent before the war.

Thus the China and Straits trades, the backbone of Holts' operations, made a substantial contribution to Ocean's recovery in the 1930s. Already in 1933 the results showed that for the same number of voyages as the previous year, 71, profitability had increased by 26 per cent from £371 000 to £469 000, an improvement due largely to better homeward cargoes from the Straits. During the next few years the trend continued, with significant recoveries in shipments of rubber and other plantation products. By the closing months of 1936, Richard Holt was able to tell shareholders that 'there has been nearly as much cargo as can be lifted comfortably', and for the following year trade was 'good in every branch'. That year saw the outbreak of war between China and Japan which reduced the outward trade of both countries, but war generated trade in some commodities, and the diversion of Japanese shipping from normal activities provided considerable opportunities for Holt vessels trading to Hong Kong and South-east Asia. Thus in 1938 earnings per voyage for the Far Eastern services were around one third higher than in the immediate pre-depression years, and, though reduced somewhat in 1939, remained significantly higher than a decade earlier. However, trade with mainland China was increasingly affected by the worsening conflict. In August 1937 the Japanese attacked Shanghai and the port was closed to international trade for three months. In perilous circumstances the families of the European staff at Holt's Wharf were evacuated on the *Stentor*, which was lying alongside the Wharf when the attack began. But despite the fighting, no Holt employee was injured and the Wharf itself remained undamaged during this 'incident', which was, of course, but a prelude to the larger conflict about to engulf the region and the world.

Passenger trades did badly in the slump, and throughout the inter-war years generally never fulfilled their pre-war promise. Fares on Blue Funnel passengers services in the interwar years showed no tendency to rise, and in fact the trend was downwards as Table 9.11 shows.

The same pattern was evident in the Far Eastern passenger services, the first class single from Liverpool to Singapore falling from around £88 in 1929 to £74 in 1938. The main cause of the downward trend was overtonnaging and consequent intensified competition, especially during the great slump at the beginning of the 1930s. Blue Funnel, like other lines, attempted to alleviate the problem by introducing a wide range of special fares, excursions, and cruises, and these became a familiar feature of the 1930s. For example in 1935 the company advertised a trip on the *Patroclus* to arrive in time for the flowering of Japan's cherry blossom, at a first class return fare of £135 (the normal fare was £161). Similar concessionary return fares were offered to South Africa, China, Australia, Ceylon, and elsewhere, in fact to virtually every principal region of call. Passengers could leave Liverpool for Cairo in 1939 on the *Antenor*, for example, returning to London on the *Aeneas* 29 days later for a fare of £45 which included landing and embarkation expanses at Port

Table 9.11   Average first-class single fares (£) 1913–38

	UK–Sydney	UK–Cape Town
1913	81	20
1921	110	62
1928	96	60
1938	78	56

Source: Ocean Archives: publicity brochures.

29 Typical advertising poster in the inter-war years

Said, quarantine taxes, first class return rail fares between Port Said and Cairo, accommodation with full board at the Continental Savoy Hotel, Cairo, and 'gratuities to hotel servants'.

In an attempt to make full use of otherwise empty space, Blue Funnel, like other shipping lines, began to develop well advertised cruises. Especially successful were the *Ulysses* cruises, started in 1932, which for a cost of £135 took passengers on a few months venture to Australia which included visits to the coral islands of the Great Barrier Reef in motor launches. In addition, in the mid-1930s, returning Far Eastern ships were taking cruising passengers from London to Glasgow via

Scotland's western isles, with a two-day stop-over in Rotterdam (six days, 8 guineas). Other special cruises were organised – for example, to Madeira and the Canary Islands – while rather inauspiciously, with Hitler's war plans reaching a crescendo, in mid-1939 Holts were advertising a cruise to Poland of all places. The 'Special Cruise to Norway and the Baltic Sea' on the *Ulysses* cost 20 guineas, and landed passengers in Gdynia from where they were transferred to a hotel in Zoppot, 'the popular seaside resort situated within the Free City of Danzig'.

Depression and over-capacity, coupled with the extensive building programme inaugurated in 1928–9, dictated that there would be few additions to the Blue Funnel fleet, as we have seen. A consequence was a significant ageing of the fleet in these years, and, despite the sales of some of the older vessels, the average age of the Blue Funnel fleet rose from 14.7 years in 1934 to 18.4 years in 1938. At the same time there were disquieting signs that even the best ships were becoming less competitive in comparison with those of some of the company's rivals, especially overseas competitors. In 1929, for example, the Hamburg–Amerika Line built a class of six 15 knot, 7900 ton vessels for the Far Eastern trade. These ships, powered by single reduction geared turbines, carried crews of only 64. By contrast the *Agamemnon* class 14 knot diesel vessels launched at the same time carried crews of around 80 (50 of whom were Chinese). As the 1930s progressed the Managers became increasingly concerned by competition from faster rivals. Such competition was most evident in the American and Pacific trades, for the main UK–Far-East trade was still very much a British-dominated conference preserve. Thus in 1937 British ships still took between 85 per cent and 95 per cent of UK trade to and from Singapore and China, while even in the UK–Japan trade trade 60 per cent of exports and 72 per cent of imports were handled by British vessels. Elsewhere, though, competition was intense, with new generations of fast cargo ships operated by German, Japanese, Italian, and Scandinavian lines. For their New York service, for example, NYK introduced six 18 knot freighters in the 1930s. British shipowners complained that such competition was often unfair, being sustained by government subsidies and low-paid crews. Blue Funnel vessels also faced severe competition in some trades from British rivals, and in the Australian trades at the end of the 1930s two new Blue Star ships capable of 17½ knots were carrying chilled beef between Australia and the United Kingdom.

Holts, though, could not be accused of complacency in the 1930s. As well as their vigorous economy drives, energetic outward canvassing, and moves to develop new trades, the company planned to maintain its position by first strengthening the Glen fleet, and then building a new generation of ships, similar to the *Glenlyon* class, for Blue Funnel. That such a vision was not realised was a consequence not of economics but

Table 9.12   Straits Steamship profits,
1932–9 (Straits $)[1]

1932	332765
1934	330000
1936	364750
1938	500050
1939	680953

[1]Straits $ = 2s 4d
Source: K. G. Tregonning, *Home Port Singapore* (Singapore: Oxford University Press, 1967) p. 162.

of politics, as war preparations and war itself dictated the course Blue Funnel was to take.

An interesting sidelight on commercial activity in South-east Asia is provided by the performance of the Straits Steamship Company. The feeder and coastal services provided by the company are a good reflection of the overall economic conditions of the region, and these services certainly belie a picture of unrelieved gloom and depression in the 1930s. In fact after 1934 profits rose strongly, and Tregonning, in his history of the company, wrote 'we cry in vain if we weep for the Straits Steamship Company, for it went from strength to strength. Illogical though it may appear, it finished the thirties in many ways in better shape than when they began'.[2] The same might be said of Alfred Holt and Co. and it was largely under the guidance of the former Holt man, C. E. Wurtzburg, who had taken over as Chairman in 1932, that the profits of Straits showed sturdy growth (see Table 9.12). Wurtzburg, incidentally had joined Mansfields from India Buildings in 1921 as a direct result of Richard Holt's on-the-spot examination of affairs in Singapore and decision to improve and streamline operations there. Wurtzburg was destined to have a brilliant career both in the commercial and wider public world in Singapore, and he became Chairman of both Mansfields and Straits Steamship.

By the late 1930s, despite evident short-term prosperity in most branches of Holt trades, a darkening political situation in Europe and the east was casting a long shadow over the company's future. True to his Gladstonian Liberal traditions, Richard Holt watched with horror the advance of political and economic nationalism in the wake of the great depression. Holt's creed was essentially that free trade and the flow of international investment promoted prosperity, world peace, and harmony. The whole post-war morass of trade and currency restrictions, reparation demands, and economic nationalism had made impossible the re-creation of the sort of harmonious economic environment which had existed before 1914. Richard Holt's philosophy was well expressed to shareholders at the beginning of 1932 when he pointed out that the prosperity of the company depended upon the prevalence of peace and goodwill in world affairs, and it was essential that:

> a final settlement of reparations and war debts should be made on such terms that the debtors can pay what is due from them without exhausting their vitality, and that the barriers to trade, established by monstrous tariffs, prohibitions and other interferences should be done away with and swept away. The prosperity of the world as a whole is essential if the business of any great steamship company is to be truly profitable: the profit of the shipowner can only be obtained when there are many people able and willing to travel, and many people able and willing to buy from and sell to those dwelling in all parts of the habitable globe.

Tragically, such hopes were dashed by the rise of aggressive nationalism in Europe and Asia. The month of September 1938 saw Britain brought to the brink of war as Hitler made a series of humiliating demands upon Czechoslovakia. Blue Funnel ships ceased calling at German ports for fear of seizure should war break out, and on 26 September a telegram was sent from India Buildings to the Far Eastern agents to withdraw ships from dangerous zones and send them home if possible. If this were not possible, and seizure imminent, the agents were given authority to sink the ships rather than let them fall into enemy hands. The following day, 27 September, the British fleet was mobilised and war seemed inevitable. At the last moment war was temporarily averted by the Munich Agreement, but few among the leaders of the western democracies now doubted that conflict would eventually come, and from that time rearmament and war-preparedness became a constant theme of British life.

At the beginning of 1938, an already ailing Richard Holt (he was nearly 70) warned shareholders that 'the uncertainty of conditions in the Far East, the financial problems of Mr Roosevelt in the United States, and the general unrest caused by the political situation practically all over the globe prevent any view being taken of the future of trade with any certainty of its fulfilment'; at the beginning of 1939 he described the political outlook as 'precarious'; at the beginning of 1940, Britain, the Company, and most of Europe were at war.

II

The notable acquisitions of the Glen Line and a substantial holding in the Elder Dempster Group came in the aftermath of the dramatic collapse of Lord Kylsant's shipping empire in 1930. Although the detailed history of the two shipping enterprises lies beyond our scope here a few words should be said about these historic concerns, both of which later came to form an integral part of the Ocean group.

Just prior to the First World War Sir Owen Philipps's (created Baron Kylsant in 1923) fast-expanding Royal Mail Steam Packet Company had absorbed the Shire and Glen Lines, both trading to the Far East, and Elder Dempster, trading to West Africa. First of the trio was Shire, in which Philipps took a half share in 1907 and bought the remaining half in 1911. The previous year, 1910, had seen the somewhat spectacular purchase of a group of companies registered as Elder Dempster and Co., and it was through Elder Dempster that in 1911 Philipps acquired the old-established Glen Line.

We have encountered the Glen Line at several stages in this history. The line dated from 1869, the year of the opening of the Suez Canal, founded by James McGregor, a partner in the shipowning and shipbrok-ing firm of Alan C. Gow and Co. The Scottish origins were permanently preserved in the 'Glen' prefixes to all the line's ships. The firm was later registered in 1880 as McGregor, Gow and Co. James McGregor was one of those perceptive individuals quick to see the prospects opened up by the Suez route to the east, and his first steamer, the *Glengyle*, made its maiden voyage to China from London in 1871. By 1882 the Glen Line was operating 15 steamers (Blue Funnel then had around 23), and was giving fierce and enterprising competition to Ocean and other eastern carriers. The Glen ships were noted for their speed, and in 1874 the *Glenartney* had returned to London from China in a then record time of 44 days. In the 1890s Glen proved readier than Ocean to build large steel ships with triple expansion engines, and a great deal of John Swire's exasperation with the Holt brothers was due explicitly to the unfavour-able comparisons he drew between Glen and Ocean vessels. Another area of initiative was in the adoption of diesel engines after the First World War, Glen again outpacing Ocean. Glen had an enterprising conference history, being like Ocean and Shire a founder member of the first China Conference, was the defendant in the famous 'Mogul Case' in 1889 which determined the legality of the conference system, and initiated the first China–New-York Conference. Already by the 1880s, though, the momentum which had seen Glen emerge as a formidable competitor to Blue Funnel was slackening. A combination of changing trading patterns (for Glen, with its speedy but expensive ships was hit particularly by the declining tea trade) and an unresponsive manage-ment led to steady decline. By 1911 poor trading conditions in China had put the firm in difficulties, and this provided an opportunity for the aggressive Philipps to secure a controlling interest in Glen (incorporated as a limited company the year before).[3]

Shire like Blue Funnel and Glen, had also originated in the expansive decade of the 1860s. The line was founded in 1861 by a Captain David Jenkins, trading as D. Jenkins & Co. Initially engaged in the Caribbean trade with sailing ships, Jenkins started trading with the Far East in the mid-1860s. His first steamer, the *Flintshire*, (all ships being named after Welsh counties) was delivered in 1872 and made her maiden voyage not only to the Straits and Hong Kong, but on to Nagasaki, Kobe, and Yokohama. Jenkins was one of the leading pioneers to trade with Japan, and he advertised his line as the 'Japan Line of Steamships' as early as 1872. By 1890 Shire steamers were making regular services from Hamburg, Antwerp and London, to Singapore, Hong Kong, and Yokohama, with intermediate stops at Penang, Manila and elsewhere as demand dictated.

Like Glen, Shire built some fast new steamers in the 1870s and 1880s, and presented often formidable competition for Blue Funnel services. Early in the twentieth century, however, the line suffered a series of setbacks, including substantial losses during the Russo-Japanese war of 1904. In 1906 an arrangement was made with the Liverpool shipping company, T. and J. Brocklebank, Brocklebanks taking a half share in the Shire Line and transferring five of their own ships to Shire routes. Further changes came in April 1907, when Philipps's Royal Mail also purchased an interest, so that Jenkins, Royal Mail, and Brocklebank were equal third partners in the newly reconstructed Shire Line of Steamers. Only a few months later, in July, Royal Mail bought out Jenkins's interest and finally, in 1911, bought the Brocklebank holding too, to become sole owner.

Already at the beginning of 1911 the Royal Mail Group had, through Elder Dempster, secured control of Glen. Glen's Far Eastern services overlapped those of Shire, so that it was a logical development for Philipps to rationalise and combine the services of the two lines. In 1920, consequently, the sailings of the two lines were completely integrated under the title 'Glen and Shire Lines', and the funnels of the vessels painted in Glen's red and black.

A necessary step towards integration was the fusion, which took place in 1912, of the two broking firms, McGregor Gow and Co. (who in effect were managers of the Glen steamers), and Shire's Norris and Joyner. McGregor Gow had been acquired by Elder Dempster at the same time as Glen, while Norris and Joyner were subsequently bought by Philipps as a prelude to his intended rationalisation of the Far Eastern services. The rather clumsily named new firm of McGregor, Gow, Norris and Joyner was simplified in 1912 to McGregor, Gow and Holland, upon the appointment of Charles Holland as Managing Director. McGregor, Gow and Holland became managers of the integrated Glen and Shire Lines in 1920.

Elder Dempster and Co. had been formed by Phillips and Lord Pirrie,

Chairman of the giant Belfast shipbuilders Harland and Wolff, in 1910. Since 1901, Pirrie had been Chairman of the African Steam Ship Co., one of the major components of a West African shipping empire built up by Sir Alfred Jones. Jones's group included also the British and African Steam Navigation Co. and Elder Dempster and Co., both companies involved in the West African trade. Upon Sir Alfred Jones's death in 1909, Philipps and Pirrie were able in a remarkable deal to buy all the Jones interest, which included 109 ships, totalling over 300 000 tons for only £500 000. Philipps became Chairman of the new concern, and henceforth Elder Dempster was used as a springboard for even further expansion, including the purchase of Glen.

The collapse of Lord Kylsant's group in 1930, and the subsequent trial and jailing of Lord Kylsant himself in one of the most publicised court cases of the twentieth century, is now an indelible part of modern commercial history. But however ill-advised Kylsant's ambitious plans, and however dubious the group's accounting procedures, two points deserve some emphasis. First, it was the world slump which precipitated the collapse, and Kylsant's companies were by no means the only ones to suffer bankruptcy or near-bankruptcy in these years. Second, many of the companies in the Royal Mail group were well run, and several were notable for their adoption of new technology and improved management methods in an industry often blamed for conservatism and complacency. Both Glen and Elder Dempster were among the more viable and attractive companies within the group, Elder Dempster in particular carving out some lucrative and well-managed trades in West Africa.

During 1929 Kylsant's difficulties became known to a small group, including members of the government and the Governor of the Bank of England. No one at that stage was aware how grave these difficulties were, nor how extensive were the ramifications, but as the months unfolded it gradually became clear that a leading British enterprise was facing collapse and liquidation. This, in the circumstances of deepening depression, was unthinkable to the government, and in the middle of 1930 a full-scale rescue operation was launched. Kylsant was forced to relinquish active control of his companies and Walter Runciman was appointed deputy chairman of Royal Mail and director of the main subsidiary companies. Runciman's appointment was made to reassure the public, for he was a respected figure in shipping circles, a prominent Liberal MP, and was President of the Board of Trade during the First World War – a post he was to hold again between 1931 and 1937. He was also a close friend of Richard Holt.

The rescue plan set up a three-member group of Trustees, of whom Runciman was one until he joined the government in November 1931, to oversee the management of all the Kylsant companies and to arrange the best possible salvage terms on behalf of investors and creditors.

Richard Holt, it so happens, had been on an extended tour of the Far East as the Kylsant drama unfolded. He was, though, kept informed of events by Runciman and other friends, Runciman writing to Holt in June 1930 shortly before his return, 'I wish you had been here during the past two months, for I should have wished to consult you repeatedly.' Richard Holt was quickly aware of the prospects opened up for Ocean, and was attracted at once by the possibility of acquiring Glen, with its east-coast berth, and also gaining a foothold in the West African trade. Already in July 1930 Holt met Runciman to stake Ocean's claim, and to moot the purchase of Glen and Elder Dempster. In August he met Sir Robert Waley Cohen, chairman of the United Africa Company (the leading West African shipper with around half of the total trade) to discuss the possibilities of a joint Ocean–UAC shipping company based on Elder Dempster. Nothing came of these early moves, for the Trustees were still in the process of unravelling the complicated knots of Kylsant's interconnected operations. But it is indicative of Richard Holt's shrewdness and opportunism that he was among the first of a great many shipowners to make contact with the Trustees and he lost no time in pushing Ocean's interest in the affair.

The detailed history of the long and involved salvage operation is irrelevant here, but it is necessary to outline the course of events leading to Ocean's involvement with Glen and Elder Dempster. As part of the rescue plan the Trustees in 1932 formed two companies to operate respectively a South American and a West African group of lines. The Trustees were obviously anxious to find as Chairmen shipowners who would bring confidence to the business world and to the public at large. They chose Sir Frederick Lewis (created Baron Essendon in June 1932), Chairman of Furness Withy, to head the South American group, and Richard Holt the West African one. Both men had considerable reputations as eminent and reliable shipowners and both, accordingly, had the full support of the Treasury and the Bank of England. After informal soundings the Governor of the Bank of England, on behalf of the government, formally invited them to accept the chairmanships in March 1932, and both then formally agreed. *The Times* noted the confidence such appointments would bring, commenting 'It would have been impossible to find two shipowners better qualified, on the grounds of reputation or experience, to take over the great fleets forming the Royal Mail Group; indeed, the active cooperation of such men in itself affords a guarantee to all the many interests concerned that thair affairs will be in the best possible hands.'[4] *The Economist* in similar vein thought that 'the willingness of these two gentlemen to accept a task which is bound to make extensive calls on their time and energy is a testimony to their recognition of the national importance of the Royal Mail Scheme'.[5]

By the summer of 1932 various meetings of shareholders and creditors had approved the schemes, and Richard Holt accordingly became Chairman of Elder Dempster Lines Ltd, as the new group was called, in July 1932. Leonard Cripps joined the Elders board at the same time.

Under Richard Holt's leadership Elder Dempster, operating initially 55 ships, proved a flourishing concern despite generally depressed world trading conditions. The 1932 arrangement was essentially a temporary one, with a moratorium placed on certain debts and obligations until the company could be launched on terms which would be fair to creditors. Holt and other Ocean Managers were at this stage concerned with the growing likelihood of competition in the Far Eastern trades, especially from Japan. They were also aware of the growing strength of moves for tariff preference within Empire markets. The Managers were therefore attracted by the prospects presented by Elders for broadening their own base in Empire markets, and Sir Richard as he then was (he became a baronet in 1935) let the Trustees know of his interest in a shareholding in Elder Dempster which he coupled with a fixed term management agreement with Ocean. To expedite the scheme, Holt met Alan Tod, Barings' Liverpool Agent, and requested assistance. Barings promptly investigated the management and financial position of Holts, and concluded that it was a well-run, conservatively financed enterprise which had established a high reputation for efficient shipping management. Barings noted as an instance of its effective management that 'the partners sit on a dais in the general office of their shipping company and are in personal contact with all their employees'.[6] Barings now worked closely with Richard Holt to construct a scheme acceptable to the debenture holders of the Elder companies, and from these dealings emerged a lasting link between Ocean and Barings. After some months of intense negotiations between the various parties a new holding company, Elder Dempster Lines Holdings was formed on 4 June, 1936 to acquire all the interests in Elder Dempster Lines and some associated companies. The various component companies of the former group were now promptly liquidated, and the new company issued 1 560 000 ordinary shares of which Ocean took 675 000 £1 shares at £1 0s 11¼d, an investment of £750 937, giving Ocean a powerful stake in the holding company. At the same time a fixed-term management agreement was drawn up whereby Ocean would manage Elder Dempster's shipping operations for seven years for a sum of £200 000 a year. As well as Richard Holt and Leonard Cripps, Lawrence Holt also joined Elder's Board. Justifying the investment to shareholders Richard Holt explained that not only was the West African business good and that the investment should prove a profitable one, but that Ocean should take a greater interest in inter-Empire trade, and, typically, that Elders was 'essentially a Liverpool business'.

By this time Richard Holt had also safely pocketed Glen. From the outset of Kylsant's difficulties Holt was deeply concerned for the future of Glen, attracted both by the possibilities of the line's east-coast berth and conference position, yet fearful of its absorption by another powerful company. Towards the close of 1930, despite rapidly gathering depression and falling steamer earnings, Richard Holt contacted Runciman with a firm offer of £1 250 000 for the line. The Trustees were not at that stage ready to sell, and thought the price too low. As a result they rebuffed both Ocean and a number of other suitors who included Furness Withy and P&O. But by 1934 the Trustees were sufficiently far advanced in their reconstruction plans to negotiate the sales of some of the shipping lines. Already in 1933 Glen had purchased the two remaining active ships of the Shire fleet, so that the whole of Glen and Shire operations were now legally under the full control of Glen. In April 1934 the Trustees wrote to Richard Holt suggesting that he pursue once more his overtures for Glen. Holt offered to buy 50 per cent of the company, but this was rejected by the Trustees who were only interested in a complete sale.

In November 1934 matters moved further. Holts arranged a loan from Lloyds Bank for £500 000 for the purpose and in December put in a bid of £450 000 for Glen's fleet and a further £100 000 for McGregor, Gow and Holland. At this juncture Holts' most active rival for Glen was Lord Inverforth, Chairman of the United Baltic Corporation, who had previously had a bid of £250 000 rejected by the Trustees. While negotiations with Holts were underway Inverforth suddenly threatened that should Holts acquire the Glen Line, United Baltic would attack Ocean's existing trades. Richard Holt at once asked the government and Bank of England, through the Trustees, to put pressure on Inverforth to withdraw his threats. On 18 January he wrote to Sir William McLintock, Chairman of the Trustees:

> Within the last few days Lord Inverforth had threatened us with his displeasure if we buy the Glen Line and has informed one of my partners that he will attack us in our existing trades if we stand in his way. This is very surprising – seeing how largely he is financed by the Government through advances under the Trade Facilities Act. We should expect some undertaking from the Government – through the Bank of England – that they will use all their influence to protect us from an attack on our trade by a disappointed competitor for the Glen Line.

Later, on 6 February, Richard Holt wrote to McLintock of Lord Inverforth:

> we regard this person's threats as serious and must ask you to provide us with some guarantee that if we buy the Glen Line we shall not as a

consequence be the subject of an attack. If he wants the Glen Line he can buy it by offering better terms.

Reassurances were quickly forthcoming and the matter was fully settled by the end of February. Meanwhile Richard Holt had submitted a revised offer on 28 January 1935 and after further negotiations over Glen's large office building in Shanghai, which the Trustees pressed on a reluctant Holt, final terms were offered on 2 March and were accepted. This final bid of around £675 000 included the original offer of £450 000 for the Glen vessels and goodwill, and the remainder covered McGregor, Gow and Holland, the Shanghai building, and various other assets. Commenting on the deal to shareholders in 1936, Richard Holt explained that 'the Managers believe that after a short period of reorganisation the purchase will be a valuable asset and will strengthen our position in the China Trade'. The purchase brought Glen's ten vessels, on average 15 years old and aggregating 88 035 gross tons, under Ocean's wing. The price paid was only half what Holt had offered in 1930, a striking example of the way the depression had eroded shipping values, and it represented a brilliant bargain which opened a new window on the east-coast trades.

The formal date of acquisition was 31 March 1935, and on 29 April the first of a series of informal meetings was held to arrange details of the takeover between Richard Holt and John Hobhouse for Ocean and Cameron McGregor and Ernest Hills, joint General Managers of McGregor, Gow and Holland, for Glen. Although wholly owned by Alfred Holt and Co., the Glen Line was maintained as an independent operating company, with its own London-based board of directors and with the fleet, under its traditional flag and colours, managed separately by McGregor, Gow and Holland, where Cameron McGregor and Ernest Hills remained as Managing Directors. However, while Glen's London head office remained responsible for handling cargoes Liverpool took over responsibility for the design and maintenance of ships, for routes, and for crewing. Crew arrangements were integrated with Blue Funnel, the changeover meaning for Glen sea-staff better rates of pay, more paid leave, full pay when standing by ships laid up in periods of depression, and better treatment generally.

Other changes involved the rationalisation of agencies, with Butterfield and Swire, Meyers, and Stapledons taking over as agents for much of Glen's homeward trade, although Bousteads, Jardine Matheson, and other former agents continued in some areas. The purchase of Glen might well have led Holts to change their London outward brokers, Killick Martin. Glen, as an east-coast loading company was in direct competition with the Ben Line, whom Killick Martin also represented. However, Holts remained loyal to Killick Martin until the fall of Singapore in 1942 forced a break in the Far Eastern services, and when peace returned new arrangements were made.

The impact of Holts upon the new acquisition was quickly felt. Richard Holt became Chairman of the Glen Line and four Blue Funnel vessels were promptly transferred, two each to the Glen and Shire fleets. In 1936 the financial structure of the company was completely reorganised, with the capital of the company reduced from £2.3 million to £500 000, the amount being covered by the issue to Ocean of 471 250 shares of £1 bringing Ocean's total holdings to 499 894 shares. At the same time Glen's operations were able to show a small profit in 1936 and in the succeeding years were able to produce a small dividend for the owners.

In 1936 a programme of modernisation for the Glen fleet was launched with orders for 8 new ships known as the *Glenearn* class. These motor ships were of around 9000 gross tons, with a service speed of 18 knots. They were among the largest and fastest in the Far Eastern trade, the 'fast Glens' Winston Churchill later called them, and were to provide a fortnightly service between London and continental ports and the Far East. The ships were ordered from yards in Britain, Holland, Denmark, and Hong Kong, for rearmament was already putting pressure on British yards. The first, the *Glenearn*, was delivered from Caledon in 1938, but only two more, the *Glengyle* and the *Denbighshire*, had been delivered when war broke out in September 1939.

Parallel with the reconstruction of the Glen fleet were some significant managerial changes. In 1937 C. E. Wurtzburg returned from Singapore upon appointment as Managing Director of Glen and Chairman of McGregor, Gow and Holland. He remained to guide Glen through both the war and its aftermath before his premature death in 1952. He was then succeeded as Managing Director of Glen by Sir Herbert McDavid, another former Holt employee who had joined India Buildings in 1915 and was transferred to Glen in 1936. The various managerial and fleet changes promised much for Glen in the late 1930s. Sir John Nicholson considered that the company was poised to dominate the east-coast trade at a time when P&O showed few signs of an effective response. But of course it was not to be. The new *Glenearns* arrived not to bring commercial rewards, but to play a famous role in the war which coincided with their birth.

10

The Company at War

It was the London staff that most delighted me. Without Blue
Funnel background they were as keen and self sacrificing as the
Liverpool men and their courage was something to marvel at.
                    (C. Cameron Taylor, Straits Steamship Co., 1945)

The Company entered the war in a sound financial position, with an
ageing though still largely robust fleet, and an ageing though still largely
robust management. As far as the Blue Funnel ships were concerned,
fleet development had been virtually stagnant since the last pre-
depression orders, *Memnon* and *Ajax*, were delivered in 1931. The only
exceptions were the little *Gorgon* (1933) and *Charon* (1936), built for the
Singapore–West-Australian trade. The major building programme had
belonged to Glen. Of the eight *Glenearn* class vessels ordered under the
programme initiated in 1936, only three had been delivered by the
outbreak of the war, but others soon followed under the spur of national
emergency, four more being launched before the end of 1939 and the
last one, the *Glenartney*, was completed in 1940. These were fast,
powerful ships, and not surprisingly, four were requisitioned by the
Royal Navy within the first few months of war to be converted into Fleet
supply ships under the White Ensign. The *Breconshire* became one of the
best known of the heroic ships which sustained Malta during the seige
of 1942.

New tonnage for the Blue Funnel fleets had been under consideration
for some time, and the company had carefully nursed its resources in
preparation for what shareholders in 1938 were told would be a 'considerable building programme'. Already in 1936 Richard Holt had
written to John Nicholson, then on his Far Eastern tour, 'if you were
building new boats for the London berth homewards would you make
them bigger or faster than the *Agamemnon* and if so to what extent?'
This, incidentally is an excellent example of the sort of large and
responsible question with which Richard Holt would suddenly confront
his younger assistants. No building programme was forthcoming,
however, because of the overcrowded conditions of Britain's shipyards
as war approached. Only in 1939 were new orders placed with Caledon
for two *Glenearn*-type vessels for Ocean, the *Telemachus*, which was
taken over on the stocks by the Admiralty and completed as an escort

carrier, HMS *Activity*, in February 1940; and *Priam*, which was completed as planned in 1941. Subsequently, to replace their former ship, Holts were allowed a replacement which was delivered in 1943, and which was also named *Telemachus*.

So, in the autumn of 1939, Holts were in a somewhat fortunate situation in that their carefully harboured reserves had not been dissipated upon costly soon-to-be-destroyed tonnage. The fleet was fully covered by war insurance and government compensation agreements, and investments had been built up through the traditional conservative policy of dividend restraint and prudent husbandry of resources. The strength of investments is indicated by the growth of their yield, from only around £40 000 a year in the period 1932–4, to more than £234 000 in 1939. By the time war broke out the reserve fund stood at £5 million, and the special building fund at £324 776.

In the early hours of 3 September 1939, the British ultimatum to Germany expired, and the two countries were at war once more. For Alfred Holt and Co and other British shipping companies the circumstances were vastly diffrent to those at the beginning of the earlier war. Then, in what retrospectively looks like an age of innocence, there was a great deal of 'business as usual', even the sinking of merchant ships conducted in a way which observed certain standards of decency, and no Ministry of Shipping to regulate every movement made by the nation's merchant fleet. By 1939 much had been learned by both sides, and much of the old-world innocence lost. The vastly stronger German fleet of surface raiders and submarines was now fortified by the powerful *Luftwaffe*. New weapons like the magnetic and acoustic mines had been developed. And Hitler's orders were to strike and sink without warning. Long before the outbreak of hostilities the Admiralty, too, had laid its plans for the defence of Britain. These included once again the wholesale 'chartering' of merchant ships in the service of the nation, and on 27 August 1939, even before the formal declaration of war, the Admiralty assumed control of merchant shipping. Also, and in contrast to the arguments and vacillations of the earlier war, a convoy system was at once operated. As Captain Roskill has written:

> As the lights went out that evening in the darkened India Building in Liverpool every man or woman in the Holt organisation, from the Senior Manager down to the youngest office boy and the most junior typist knew that control of their great fleet had been taken out of their hands for an indefinite period: and that their ships, though remaining in the company's theoretical ownership, thenceforth formed a part of a greater organisation – comprising the whole British Merchant Navy mobilised for war.[1]

30 British troops being evacuated by Bellephoron from Dunkirk, 1940

The epic part played by the Holt fleet in the struggle is on record. It is told with memorable skill and great authority by the distinguished naval historian Captain S. W. Roskill in *A Merchant Fleet at War, 1939–45*. In this chapter we can do no more than trace the skeleton of a story which is fortunately preserved for all time in Captain Roskill's vivid prose. Even this skeleton is dramatic enough. When war broke out the Holt Blue Funnel fleet consisted of 77 ships. When peace returned only 36 of this great fleet remained, though two ships under construction in 1939 were delivered during the war. A further three Glens were lost. In the intervening five years the ships' size, speed and excellent construction, together with the fine seamanship and splendid discipline of their Masters, officers and crew, made them unenviable favourites for the most hazardous operations and functions – Malta convoys, armed merchant cruisers, landing ships. Even in the Atlantic convoys a Holt ship was likely to find herself in the post of honour, which was also the post of danger, as leader of a column or as Commodore's flagship.[2]

In addition to the Blue Funnel and Glen ships, some 40 vessels were placed under Holts' management, ships which ranged from captured Vichy-French ships to new 'Liberties'. Eight of these managed ships, too, were lost. The Blue Funnel losses, totalling 41 ships, included 26 from the Ocean fleet, 12 of the China Mutual, and 3 NMSO ships. This represented no less than 370 363 tons of the Blue Funnel fleet, and a further 28 179 tons of Glen ships were destroyed. The death toll was horrific, 324 lives in the Blue Funnel ships, and many more from the Glens, managed vessels, and losses of Holt seamen sailing under the White Ensign. Not a few, including some Masters, died under the intolerable strain of service in wartime conditions. And to the list must be added the countless maimed and wounded, or those who in one way or another had their lives shattered as a direct result of the war.

Those in Holts' service, high and low, gained a full measure of honours and awards as a recognition of their wartime contributions. These included 4 CBEs, 29 OBEs, 29 MBEs, 18 DSCs, one MC, 3 Albert Medals, 8 DSMs and 33 BEMs. In addition, J. R. Hobhouse was created a Knight Bachelor and Sir John Nicholson was made a Companion of the Most Eminent Order of the Indian Empire. To this list could be added decorations awarded by foreign governments and a very long list of Mentions in Despatches and Commendations.

In sharp contrast to the previous war, there was no lengthy period of relative calm in the troubled waters of 1940. The first vessel to fall was the *Protesilaus*, mined in the Bristol Channel on 21 January 1940. Before the end of that year a further seven Blue Funnel ships had been destroyed, and in 1941 nine were sunk. In 1942, the worst year of all, no fewer than 18 ships were lost, but by the end of that year the tide of war had started to turn. In 1943 only four Blue Funnel ships were lost and in 1944 the final two. By coincidence the last casualty, sunk by a U-boat in August 1944, was the *Troilus*, a strange echo of the first Blue Funnel victim in the First World War.

Were we to plot Blue Funnel casualties on a map of the world we would see how dispersed were the theatres served by these vessels. The English Channel, the Atlantic, off the coasts of the United States, North Africa, West Africa, the Philippines, Australia, India ... the list of casualties rings the world and reminds one how all-encompassing was the conflict and how the ships, often in their convoys, would be hunted alike by submarines, cruisers, or aircraft. The complete list of Blue Funnel casualties is given in Table 10.1.

This, in essence, was the experience of the fleet in wartime. It was an experience which unfolded against a background of events which flared in Europe and steadily engulfed the world. In April 1940 the Germans suddenly attached Denmark and Norway, in May Belgium and the Netherlands, and in June German troops entered Paris. In August that

Table 10.1   Holt war losses, 1939–45

Date		Ship	GRT	
21 Jan.	1940	Protesilaus	9577	Mined in Bristol Channel
25 Jan.	1940	Meriones	7556	Bombed off Yarmouth
29 Jan.	1940	Eurylochus	5722	Sunk by surface raider off the coast of West Africa.
17 Feb.	1940	Pyrrhus	7417	Torpedoed off Cape Finisterre
11 Mar.	1940	Memnon	7506	Torpedoed off West Coast of Africa
7 May	1940	Ixion	10263	Torpedoed in North Atlantic
17 June	1940	Teiresias	7404	Bombed off St Nazaire
2 July	1940	Aeneas	10058	Bombed off Plymouth
3 Sept.	1940	Titan	9034	Torpedoed in North Atlantic
25 Sept.	1940	Eurymedon	6223	Torpedoed in North Atlantic
11 Nov.	1940	Automedon	7528	Sunk by surface raider in Indian Ocean
8 Jan.	1941	Clytoneus	6273	Bombed off Ireland
14 Jan.	1941	Eumaeus	7472	Sunk by gunfire from submarine off Freetown
28 Feb.	1941	Anchises	10000	Bombed off Ireland
21 April	1941	Calchas	10304	Torpedoed off West Coast of Africa
26 Dec.	1941	Tantalus	7724	Bombed at Manila
11 Jan.	1942	Cyclops	10253	Torpedoed off American Coast
6 Feb.	1942	Talthybius	10253	Bombed and sunk at Singapore, later seized by Japanese.
3 Mar.	1942	Helenus	7366	Torpedoed off West Coast of Africa
27 Mar.	1942	Breconshire	8982	Bombed off Malta
2 April	1942	Glenshiel	9415	Torpedoed in Indian Ocean
5 April	1942	Hector	11198	Bombed and burnt out at Colombo (Armed Merchant Cruiser)
6 April	1942	Autolycus	7621	Sunk by Japanese cruiser in Indian Ocean
10 April	1942	Ulysses	14646	Torpedoed off Palm Beach
3 May	1942	Laertes	5825	Torpedoed off North America
17 May	1942	Peisander	6224	Torpedoed off Nantucket
26 May	1942	Polyphemus	6269	Torpedoed in Atlantic
28 May	1942	Mentor	7382	Torpedoed off Key West
9 Aug.	1942	Medon	5444	Torpedoed in Atlantic
12 Aug.	1942	Deucalion	7514	Bombed and torpedoed (aerial) in Mediterranean (Malta convoy)
13 Aug.	1942	Glenorchy	8982	Torpedoed off North America
5 Sept.	1942	Myrmidon	6278	Torpedoed off Freetown
11 Oct.	1942	Agapenor	7391	Torpedoed off West Coast of Africa
27 Oct.	1942	Stentor	6148	Torpedoed off North West Africa
13 Nov.	1942	Maron	6487	Torpedoed in Mediterranean
27 Nov.	1942	Polydorus	5922	Torpedoed off West Africa
3 Feb.	1943	Rhexenor	7957	Torpedoed in Atlantic
5 May	1943	Dolius	5506	Torpedoed in Atlantic
14 May	1943	Centaur	3222	Torpedoed off Brisbane
19 Dec.	1943	Phemius	7405	Torpedoed off West Coast of Africa
16 Jan.	1944	Perseus	10286	Torpedoed near Madras
31 Aug.	1944	Troilus	7421	Torpedoed in Indian Ocean

Source: Ocean Archives: War Records.

year the U-boat campaign against Britain's merchant fleet intensified. The submarines, hunting often in 'packs', sank great numbers of allied ships undertaking the perilous Atlantic convoys, especially in the year 1942. Meanwhile the war had spread to the Balkans, Russia, and beyond Europe to Africa and Asia. Japan, already at war with China and voicing support for the German cause with increasing stridency, attacked the American fleet at Pearl Harbour on 7 December 1941. Simultaneously Japan launched attacks throughout the east. Britain's Asian empire crumbled. Hong Kong fell on Christmas Day, 1941, Singapore on 15 February, 1942. In country after country the victorious Japanese flag replaced that of the former colonial powers. Manila fell on 2 January 1942, the Dutch East Indies capitulated on 8 March 1942, Burma was overrun and occupied in April. Meanwhile in China the Japanese had quickly controlled Canton, Hankow, Swatow, Amoy, and all the major ports on the China Sea following their full-scale attack on Shanghai in August 1937.

In a few short months virtually all the strongholds of Holts' Asian empire had fallen into Japanese hands. Mansfields in Singapore, Butterfield and Swire in China, Hong Kong, and Japan itself, and the various other agencies and staff in a multitude of ports and cities throughout the region had now to face the consequences of the invasions. All too often this meant internment, torture, and sometimes death. The upheavals in Asia precipitated a certain amount of trouble among the large numbers of Chinese seamen who formed a large and vital part of the Blue Funnel crews and those of other lines. One cannot but feel sympathy for their plight, risking their lives day after day in a cause for which they had no direct involvement, and suddenly deprived of any way of returning to their homelands and families. It speaks much for the good relations established between the company and its Chinese crews that many of them served so loyally and so uncomplainingly. Although some lines, for example Ben, had trouble with their Chinese crews in the opening months of war, there was little unrest in Holt ships until the end of 1941. But the fall of Hong Kong and Singapore dealt an irreparable blow to Britain's prestige throughout the east, and in 1942 cooperation between British and Chinese governments (with considerable and effective work by John Hobhouse) and higher rates of pay were used in an effort to improve morale. The role of the Chinese crews in Britain's war effort should certainly not be forgotten. Their fortitude in one incident was recorded in a letter from Captain D. L. C. Evans to Lawrence Holt after his ship *Glenartney* picked up survivors from a fire-stricken escort vessel in the Atlantic in April 1941:

The Chinese crew were truly excellent, working to the point of exhaustion ... As an illustration of the spirit prevailing, the Chinese boys made it clear that any attempt on the part of the survivors to offer any reward or gratuity would be most offensive to their feelings, and would be met with disdainful refusal. I can only say with all the sincerity that I possess that I am proud to have been in command of such a ship, manned by such excellent officers, midshipmen, and crew.

At home the depleted company organisation also struggled in arduous conditions. The war naturally took away the services of nearly all the younger staff, while the more senior, too, were often called upon to play a significant role in various spheres of the war effort. The company no longer had the services and experience of Richard Holt, whose death in March 1941 occurred only weeks before India Buildings was destroyed. Under the circumstances of war and ship requisitioning, ship management was necessarily very much a day-to-day affair, with little opportunity for any longer-terms considerations. Broadcaster Norman Lee visited Holts in 1942 and was given access to confidential papers about the war exploits of Blue Funnel ships. He later wrote to the company,

> I *thought* I knew something about owners ... Since my visit to Liverpool I have learned a lot; but it is different to what I had imagined. I didn't know they were very human people who give consideration to Masters and cabin boys alike. Where else in British industry will a millionaire owner spend an hour with a Chinese coolie labourer to find out the things that worry him?

Lee went on to pay a striking tribute to the work of Lawrence Holt whom he termed 'unofficial Commodore of the Merchant Navy', and who 'personally fashions great seamen, builds officers, creates Masters'.[3]

Let us chart briefly the main course of Company affairs. With the departure of Richard Holt, Lawrence Holt became senior partner, but in practice a great deal of the detailed running of company business devolved upon the shoulders of Roland Thornton. This was largely because more senior managers had their energies diverted elsewhere. Lawrence Holt, as Sir John Nicholson recalled, 'was wholly absorbed in maintaining the strength and morale of our seafarers and did a great deal to ensure official understanding of their problems through contact with his cousin Stafford Cripps, a leading Cabinet Minister, and the agency of Malcolm Glasier, then our London Marine Superintendent'. Others of the senior management were also more or less wholly diverted with war or other duties. John Hobhouse was frequently absent as a Regional Commissioner and as North Western Port Director, in this latter capacity doing much to retain the loyalty of the Chinese crew

members. Leonard Cripps had devoted an increasing amount of his time
to Elder Dempster after the mid-1930s, and as Chairman after 1941 was
wholly occupied there. Sydney Jones, who was 68 when war broke out,
was Lord Mayor of Liverpool until 1942, and had many other civic
duties. And Wurtzburg, McDavid and MacTier were seconded to the
Ministry of War Transport. Of the younger assistants, John Nicholson,
who had been appointed Assistant Manager in 1938, was away on war
service throughout the war, as was G. P. Holt.

For Thornton the war brought a measure of responsibility which had
too often been denied earlier. It was sad that this should have been the
case, for Thornton was a man of undoubted intellectual brilliance. We
may at this point appropriately turn to consider his contributions to the
company in a little more detail.

Roland Hobhouse Thornton joined Holts in 1919, after distinguished
war service. Educated at Marlborough and Balliol, Oxford, he had a
brilliant mind coupled with a capacity to master detail yet never lose the
essentials of an issue. He had a deep understanding of Britain's
shipping industry and its problems, and in 1939 published a small
excellent book on the subject.[4] Yet despite his brilliance, or perhaps
partly because of it, his role and situation in India Buildings was long an
uneasy one. Thornton had little in common with the partners except
distant relationship to most to them. To some extent he was, and
remained, an outsider, 'cousin of a cousin' rather than a *bona fide*
member of the Holt clan. Richard Holt, for example, always wrote 'Dear
Thornton'; elsewhere it was 'My Dear Jack' or 'My Dear Leonard'.
Unfortunately too, the division of responsibilities among the senior
Managers left little outlet for Thornton's talents in work which was
sufficiently challenging. Richard Holt was unwilling to let him play a
part in conference affairs, while the territories of the various Managers,
jealously guarded as they were, meant that Thornton's efforts to go
beyond routine frequently brought him into conflict with Cripps or
Hobhouse, or especially with Warren Swire.

Thornton felt, not without justice, that his progress to partnership was unduly retarded. While Hobhouse and Cripps were made
Managers in 1920, Thornton was not made an Assistant Manager until
1927, and raised to full partner in December 1929. Curiously Thornton's
Who's Who entry, and the company's official obituary notice in 1967,
claim 1927 as the year Thornton became a Manager, but the record
clearly states 'Special General Meeting, 19th December 1929, R. H.
Thornton appointed Manager'. Thornton's bitterness was certainly
exacerbated by his lack of private means and the company's traditional
parsimony. Thus he was getting £750 a year as Assistant Manager while
the five Managers were sharing around £25 000. All this meant that
Thornton was a difficult colleague, and at times considerable friction

existed between himself, Hobhouse, and Lawrence Holt. He was often at loggerheads with Warren Swire, who felt Thornton was treading on preserves of Butterfields and Swire in his efforts to improve aspects of the Far Eastern trade.

But if Thornton's abilities had been underused between the wars, thereafter he came to play a significant and indeed dominant role in the company's affairs. In wartime Holts was largely dependent upon Thornton for the maintenance of its day-to-day business, while in the post-war period it was Thornton, rather than the more senior partners, who fashioned the revival of the fleet and the rebuilding of Blue Funnel business.

I

When war erupted, the Holt fleet was well prepared. Already in February 1938 the Managers' Minute Book recorded that 'defensive armour' was being fitted to certain ships, and in the succeeding months most of the fleet, at Admiralty expense, was so protected. With the outbreak of war two long-planned actions were immediately put into effect. One was 'in view of the grave additional marine risks' the inevitable decision to insure the fleets under war risk insurance terms. The second was to separate the management of the six NSMO vessels whose organisation was constituted with three Dutch directors to conduct its business from Amsterdam. When the Germans invaded Holland the Amsterdam organisation ceased its brief existence, but the one ship then in Holland managed to escape German capture. Operation of the six vessels was then transferred to the United Kingdom, with the head office being relocated in Curacao. Table 10.2 gives in bare form the company's commercial performance.

Not that the company had any significant measure of control over its commercial operations since from the discharge of cargoes in February

Table 10.2   Ocean accounts, 1939–45

	Earnings from steamers (£)	Reserves (£)	Building reserve (£)
1939	957 742	5 000 000	324 776
1940	1 151 970	5 750 000	1 095 908
1941	1 911 346	5 750 000	1 903 190
1942	653 831	6 000 000	3 610 817
1943	406 023	6 200 000	4 264 484
1944	432 506	6 200 000	4 338 766
1945	460 118	6 200 000	4 588 766

Source: Ocean Archives: Managers' Minute Books.

1940 all voyages were requisitioned, with shipowners paid at official blue book rates. These rates were not generous. They were lower than pre-war levels and, as wartime costs rose, tended to decline in real terms. Yet, as shareholders were told in 1941, the company had 'no reasonable ground for complaining of the rates paid by the Government'. For the first two years of war dividends to shareholders were maintained at the pre-war rates (30 per cent plus a 10 per cent bonus), but from 1942 the 10 per cent bonus was not paid. Thus the financial returns to shareholders, bearing in mind wartime inflation, represented significantly less buying power in 1945 than they had done in 1939.

Under the circumstances of war the seven-year management arrangement with Elder Dempster which expired in 1943 was not renewed, and the Managers decided not to seek the management again when peace returned in 1945. With the retirement of Leonard Cripps from Elder Dempster in March 1945 the links with Blue Funnel were lessened for a generation, though Ocean remained a substantial shareholder.

A few words should be said here about the actual terms upon which the government took control of Britain's merchant shipping. For a short period the government attempted to distinguish between its own business and other trade, though with controlled freight rates. From February 1940, however, all ships were directly requisitioned and chartered at freight rates which were supposed to allow owners a return of 5 per cent on capital. The management of vessels, except those in naval service, remained in the hands of established shipowners, who were also assigned various additional ships such as captured vessels, or tramps which had no appropriate shore organisation.

The figures shown in Table 10.2 reflect the inevitable decline in fleet earnings as enemy submarines, mines, and cruisers took their toll, and it shows the concomitant increase in reserves as the company received war compensation for its fully insured vessels. Indeed, to the building reserve had been added by 1945 over £1 million from the Government tonnage replacement scheme. The company, too, maintained substantial liquid reserves. On the darker side, many investments had been written off completely, including assets in China, Singapore, Hong Kong and elsewhere, while the depleted fleet was getting steadily older and battle-scarred. While there can be no question that the First World War was a time of record prosperity and significant expansion, there can similarly be no doubt that in the Second World War no such gains occurred.

II

As with the previous conflict we can do no more than glimpse the exploits of Blue Funnel vessels through the inadequate lens of a few episodes. Although a representative sample from unique events is hardly possible we will look in turn at Blue Funnel participation in two key theatres of war, the Far East and the seige of Malta, and also record the tragic fate of the little hospital ship *Centaur*.

In December 1941 the sudden Japanese onslaught against the American fleets was paralleled by an attack on the British colony of Hong Kong and a simultaneous sweep through South-east Asia. When the attacks were launched, a good many Holt ships were in the Pacific and Indian Oceans, while the *Tantalus* and *Ulysses* were refitting in the Taikoo dockyard in Hong Kong. Just before the attack on Pearl Harbour, with the Japanese threat increasing hourly, the naval authorities decided that both Holt ships should be got away from Hong Kong as quickly as possible. *Tantalus* was in no condition to move, for much of her machinery had been removed for the refit and the only solution appeared to be for her to be towed to safer waters. This task fell to the tug *Keswick* which was to lead *Tantalus* for the journey, and by 5 December some generating capacity had been restored, the remaining machinery loaded, and coal for the *Keswick* taken aboard. The ships left at 7 a.m. on 5 December, and the message 'wishing the staff God speed' was relayed to Captain R. O. Morris as *Tantalus* passed Holt's Wharf.

A few days later, on 8 December, came the Japanese onslaught in Asia. All British ships in war zones were ordered immediately to seek safety where they could. *Tantalus*, being towed at 5 knots through hostile waters, was clearly in great danger, and Captain Morris decided to make for Manila, capital of the Philippines (then an American colony). He arrived there on 11 December, reporting to Holt's agents, and to the American authorities. At this stage Manila was bearing the brunt of savage Japanese attacks, and without her main machinery *Tantalus* was stranded helplessly as Japanese aircraft attacked Manila in waves. On Christmas Day, 1941, the day that Hong Kong fell, *Tantalus* was badly shaken by huge explosions from the attacks, and it was clearly only a matter of time before she suffered a direct hit. Captain Morris then took his men ashore, and on 26 December they watched the helpless vessel subjected to wave after wave of aerial attacks until she was finally ablaze and capsized. The crew returned to Manila, only to fall into the hands of the Japanese who entered the city on 3 January. The crew were all interned, and two were later executed by the Japanese for trying to escape.

The *Ulysses*, under Captain J. A. Russell, was more fortunate in her flight from Hong Kong. After strenuous efforts by the crew, dock workers, and Butterfield and Swire, the ship was able to leave the harbour at 2 p.m. on Sunday, 7 December, the day before Japan launched its attack on the colony. On board *Ulysses* were some of the staff of Butterfield and Swire, but for those left in Hong Kong there followed a period of great suffering as the colony quickly succumbed.

In another major war theatre, *Troilus* was one of an ill-fated convoy of six merchant ships which set out from the Clyde on 5 June 1942 to supply the beleaguered island of Malta. The convoy's escort consisted of one battleship, two aircraft carriers, four cruisers, 17 destroyers and four minesweepers. *Troilus* was Commodore's ship, and we may recall Captain Roskill's remark that 'Holt ships were very often chosen to embark one or other of the Commodores, and also to take station at the head of the convoy columns. This was without doubt a compliment to the skill and steadiness of the Masters and officers of the Company; but the positions of honour were also the positions of greatest danger.'[5] On this occasion the convoy reached and passed Gibraltar unscathed. But on 14 June some 40 aircraft attacked the convoy from various directions, and the A-A gunners in the merchant ships and escort vessels fought off waves of attacks with such resilience that only one merchantman was lost. *Troilus* herself accounted for two out of many enemy aircraft which were shot down. That same day the heavy ships in the escort returned to Gibraltar, and the five merchant ships continued with an escort of one cruiser, nine destroyers, and the four minesweepers. In the evening came another sudden air attack on the *Troilus*, but with bombs cascading all round, and missing, the ship's gunners were able to bring down the aircraft as it pulled out of its dive.

The next day, 15 June, as the convoy pressed on for Malta, an Italian squadron appeared. The British destroyers went to attack, and while they were thus engaged, German aircraft attacked once more. One merchant ship was sunk, and the only tanker in the convoy damaged and taken in tow. Meanwhile two of the escort destroyers had been damaged, and a succession of air attacks continued, crippling another merchant ship. By now it was decided to concentrate the remains of the escort fleet on protecting the two relatively undamaged ships, *Troilus* and *Orari*, a vessel of the New Zealand Shipping Company. Two badly disabled merchant ships and a destroyer were deliberately scuttled, and the small convoy moved on, aided by fighters from Malta. In the early hours of 16 June the convoy struck a minefield. *Troilus*, *Orari*, and four of the escort hit mines, and one of the destroyers was lost. As the merchant ships at last reached harbour with their precious supplies they found themselves in the midst of yet another heavy air raid. *Troilus* went on to survive a number of further war adventures before becoming the last Holt casualty of the war, when, as recounted already, she ended her days in the same Indian Ocean where an earlier *Troilus* had been the Great War's First Blue Funnel victim.

Finally, the loss of the *Centaur*, the saddest of all the many tragedies which struck the Holt fleet. This little 3222 ton ship was taken off the

31 HM Australian hospital ship, *Centaur*, sunk by Japanese submarine, 14 May 1943 (*photograph courtesy of Liverpool Museum*)

Australian-New Guinea run at the end of 1942 and converted at Melbourne into a hospital ship. In accordance with the Geneva Convention, the Japanese were informed of this in January, 1943. But the following 14 May, while on a voyage from Sydney to Cairns, the *Centaur*, commanded by Captain G. A. Murray, was torpedoed by a Japanese submarine off Brisbane. No warning was given, no time for any distress signals, and the fully illuminated hospital vessel with 333 on board sank in less than 3 minutes. No boats could be launched, but a few managed to survive on two rafts and some floating wreckage, and the following morning were able to right an overturned lifeboat. For 36 hours the dwindling number of survivors drifted with no oars or sail. Eventually the surviving 63 men and 1 nurse were spotted by a passing aircraft and taken by a destroyer to Brisbane. In all, 269, including Captain Murray, lost their lives in the single biggest disaster to befall the Blue Funnel fleet.

Not that the home front was without its perils. Liverpool was an early target for German bombing raids, and Winston Churchill, recalling the events of 1940, wrote that 'as November and December drew on, the entrances and estuaries of the Mersey and Clyde far surpassed in mortal significance all other factors in the War'. The intensity of air raids in

32 India Buildings after the blitz, May 1941: note the remains of safes and filing cabinets (*photograph courtesy of Liverpool Museum*)

Liverpool increased after July 1940, and reached a crescendo in the first eight nights of May 1941. These attacks devastated large areas of the city centre and docks, took a huge toll of innocent civilian life, damaged 91 vessels, excluding naval ships, and put one third of the port's berths out of action.

India Buildings did not escape. The new building, planned a quarter of a century before, and fully completed for only 9 years, became a victim of the blitz only weeks after Sir Richard Holt's death. On the night of Saturday 3 May the Holt headquarters was virtually destroyed by fire which followed the bombing of adjacent buildings. Hugh Wylie, then in the Passenger Department, later published his wartime recollections in the company's house journal *Ocean Mail*. He and Charles Storrs 'climbed the stairs to the 6th floor of India Buildings to find just a vast empty space – all the inner walls had gone'. Only the mezzanine floor remained untouched.

Head Office was moved immediately to Lawrence Holt's house at 52 Ullet Road, and others of the staff were assembled at No. 54, Richard Holt's house. Thereafter the necessary dispositions were made, many for the remainder of the war. In Lawrence Holt's house was established the Naval Architects department, while No. 54 housed the Managers, the Steamship, Accounts, Passengers, Freight, and Cash departments. Other departments were scattered further afield. Engineers, for example, went to Iliad House, Birkenhead; others went to the Canadian Pacific offices in the Royal Liver Buildings, while the Chinese crew department went to Neptune Street. After a while there was more reshuffling, as first the mezzanine floor at India Buildings was reoccupied, and then the floor above. When war clouds lifted at long last in 1945 the reconstruction of India Buildings was an urgent priority. But there were many equally urgent tasks facing the Managers as they had to shape the company to meet its peacetime tasks.

11

Picking up the Pieces

The first was the war to end all wars. The second was, as we all know,
the war which never stopped.

R. H. Thornton

Even before the final capitulation of Germany and Japan in 1945 the
Managers had turned their thoughts to the post-war future. The
decision they took, to restore 'normal' pre-war capacity and services as
soon as practicable, had no obvious alternative. From it followed the
consequences which were to dominate the rebuilding programme in the
first decade or so after the war: the purchase of US-built 'Victory' and
'Liberty' ships as a stop-gap measure to improve capacity; the speedy
reconditioning of the existing fleet to meet Blue Funnel standards; and a
programme of accelerated building to produce a fleet of some 60 or so
well-equipped modern vessels by 1960 which would be able to service
traditional Blue Funnel trades at their traditional levels.

Before we look at these building programmes and the other responses to
immediate post-war needs, however, we should step back a little and consider the post-war environment in which the Managers found themselves.

The extent to which the economic world after 1945 was to differ from

33 Victory ship, *Memnon* (ex-Phillips Victory) renamed *Glaucus*, 1957

that existing after 1918 could not have been forecast by the most perceptive observer. Let us recall a few salient features of that earlier post-war era. Then there had been a brief, hectic boom, followed by a sharp collapse, a period of hesitant stability, and then the traumas of world depression. After 1921 the prevailing trend of costs and prices had been mostly downwards, very sharply downwards between 1929 and 1933, when interest rates, too, reached low levels. High levels of unemployment helped to weaken the bargaining power of trade unions, with the result that various labour costs, such as marine crews and dock labour, tended to be stable.

After 1945 the post-war restocking boom, far from giving way to slump and unemployment, ushered in a period of sustained full employment and rising costs. Under the post-war Labour governments (1945–51) inflationary pressures were relatively subdued, but in the 1950s and 1960s rising costs, especially labour costs, became an increasingly prevalent feature. Inflation had a number of consequences for Ocean. Obviously rising costs put pressure on profits, but there also arose a growing disparity between the fleet's historic cost and its replacement value. In other words, normal levels of allowance for depreciation were inadequate to cover future shipbuilding needs. For a Company whose proud tradition was one of internally generated funds for building this was a critical issue. Inflation also called into question the traditional policy of self-insurance, huge sums of money having to be kept on hand to repair or replace ships at sharply rising current costs, an expensive luxury when interest rates were high.

If domestic circumstances were changing, the international environment was, if anything, even more uncertain and unfamiliar. Indeed, one of the most striking features of the minutes of the Manager's Meetings held fortnightly in India Buildings is the extent to which they show a growing preoccupation with political events in the world at large.

Such a preoccupation is wholly understandable when we recall that the whole basis of Blue Funnel's operations rested on the Europe–Far-East trade, the reliance on Asian raw materials and primary produce to fill the ships, and key role of the Far Eastern Freight Conference (as the Conference was renamed in 1941). Now, whatever may have been Alfred Holt's problems in the pioneering days before 1900, they did not include the dangers that important trading ports or trade routes would be closed to British ships, or entire countries removed from the orbit of world trade. Indeed, prior to 1914 the spread of the British Empire, the dominance of free trade, and the power of the British navy, ensured that opportunities for unfettered trade would be safeguarded. Even in the inter-war years the continued British presence in Asia and the strength of the European-dominated conferences, ensured a reasonable share for

British shipping lines even if the Japanese, Germans, and Scandinavians had to be accommodated.

But after 1945 question marks began to appear about both the existence of certain trades, and about the existence of the traditional conference system, question marks which had their origins in politics. There were, from the Blue Funnel Managers' point of view, two principal problems. First, unrest and upheaval in many areas which were traditional markets and suppliers for Blue Funnel. At their worst such upheavals might result in new anti-colonial communist regimes which could remove their countries from normal trade channels altogether. Or, since such upheavals were often connected closely with anti-colonial independence movements, new national governments might be formed which would demand shares of existing trade for ships of their own national lines. In other words, Blue Funnel found itself wedded to a system built on a European-dominated world where Europe's manufactures were exchanged for colonial primary products. Gradually, after 1945 there was an inexorable movement towards independent, sovereign, 'developing', nations, with their own aspirations and an increasing say in international affairs.

A glance at the political situation in territories of particular importance to Blue Funnel shows vividly the uncertainty which beset so many of them during the decade or so following the end of the Second World War. The war itself was a potent force of political change. On the one hand the humiliating defeats suffered by British forces at the hands of the Japanese greatly weakened British prestige throughout its colonial empire (and the French and the Dutch suffered similarly). On the other hand the Japanese occupation forces had often for their own purposes stimulated anti-colonial independence movements, while the collapse of Japanese power in 1945 left a power vacuum which new nationalistic groupings were ready to fill. Another factor was the breakdown of wartime cooperation between the Soviet Union and the Western allies, and the emergence, after 1946, of the 'Cold War'. Tension or civil war between Communist and anti-Communist groupings was endemic in many of Blue Funnel's traditional spheres of operations in the post-war world, in China, Malaya, Singapore, Indonesia, and elsewhere.

It would be beyond the scope of the present book to detail the tortuous political upheavals which affected many of the major areas of Blue Funnel's operations in the post-war period. Yet some summary comments are necessary, for these events exercised a powerful influence over managerial decisions in the 1950s and 1960s.

The collapse of Japanese power in September 1945 saw in all hitherto occupied territories scenes of destruction, devastation, and often hunger and deprivation. As late as 1951 a United Nations report, surveying the economies of South and South-east Asia, could write that 'millions of

34 *Menestheus* at Kure, Japan, March 1946, converted under Admiralty control as a Royal Navy Amenity Ship for service with the Pacific Fleet (*photograph courtesy of Captain Allan N. Cabot*)

people in the countries of the region stand dangerously near the border-line between hunger and famine'.[1] Familiar as we now are with the 'economic miracles' of so many Asian economies in recent years it is all too easy to forget just how desperate the situation seemed in the ten or twenty years following 1945. In China, a vicious civil war between Mao Tse-tung's Communist forces and Chiang Kai-shek's Nationalists ended in the collapse of the Nationalists in 1949. Peking and Tientsin surrendered in January, and in April the Communists surged across the Yangtze, entering Shanghai in May, Canton in October, and Chungking in November. The Nationalist Government and many of its followers fled to Formosa (Taiwan), where a separate, American-backed 'China' was established, leaving the mainland to the Communists. Communist victory was followed by the virtual withdrawal of that country as a major trading partner with Western countries, and the nationalisation of Western-held assets.

Communist victory in China in 1949 and Russia's decision in 1948 to foster destabilising pro-Communist and anti-colonial uprisings in South-east Asia threatened to engulf the entire region in upheaval. In Malaya, Communist uprisings led to a state of emergency there which lasted between 1948 and 1960. Even though the security forces gradually overcame the Communist threat from the mid-1950s, there remained considerable tension beween native Malays and the large Chinese community. In 1957 the Federation of Malaya achieved independence, but it was to be nearly a decade before both internal and external threats to its stability and very existence could be said to have disappeared. Singapore also suffered a period of great turbulence in its move towards full independence and autonomy. A strong pro-Communist element threatened for a time to take Singapore into the Chinese camp, and a Communist takeover of the main political party in 1957 was only narrowly averted. It was partly out of fear of an anti-Malayan pro-Chinese Communist Singapore that Malaya agreed to form a federation with Singapore in 1963, the new state (which included Sarawak and Sabah as well) being called Malaysia. The new grouping lasted only a short time, however, and Singapore was forced to leave the federation in 1965 to become a fully independent state.

A pattern of political upheaval and often a great deal of fighting and bloodshed can be found in many other parts of Blue Funnel's 'empire'. In Indonesia, for example, the Dutch vainly tried to re-establish their power after 1945. By 1949 the Dutch had been obliged to concede the establishment of an independent Indonesia, and during the 1950s remaining links with the Netherlands were gradually broken under President Sukarno's nationalist and left-wing regime. When in 1957–8 rebellions in Sumatra and elsewhere against Sukarno's centralist government broke out, Sukarno established a military dictatorship. In 1960–2 came conflict with the Dutch over West Irian (Dutch New Guinea), and between 1963 and 1965 came a tense period of 'confrontation' between Indonesia and the newly established Malaysia, when for a time it seemed as if Indonesia would attempt to destroy the new creation.

In addition to the regions we have mentioned, there was at one time or another unrest and turmoil in the Philippines, Korea, Ceylon (Sri Lanka), Burma and 'French Indochina' – the French forced to recognise the independence of a partitioned North and South Vietnam in 1954, the resulting creations festering and ultimately gathering into the full-scale Vietnam War of the 1960s. Hong Kong seemed in peril as the Communists advanced in China in 1949, while nowhere in the late 1940s was the future more uncertain than in defeated and devastated Japan, a shattered country effectively under United States military occupation between 1945 and 1950.

With such a catalogue of chaos it may seem surprising that Blue Funnel's services recovered after the war as effectively and profitably as they did, and that the Far Eastern services retained their traditional significance in overall levels of sailings and profitability. Several facts help explain this apparent paradox. Most significant was that some elements of instability actually made for buoyant trade conditions, at least in the short run. This was seen most clearly in the Korean War, which erupted in June 1950 and lasted until 1953. The war sparked a worldwide demand for raw materials and precipitated sharply rising commodity prices, prosperity for primary producers, and high freight rates. The war caused Japan's trade to rise markedly, as that country was used as a base for United States military operations and became a large source of supplies for the war. In a similar way, too, the Vietnam War of the 1960s brought a significant expansion of trade throughout South-east Asia, while other events, such as the Malayan Emergency, the Chinese offensive against Taiwan in 1957, and Indonesia's conflict with the Dutch in 1959–60 all boosted regional trade in one way or another.

Another element making for trade buoyancy in the post-war years was once again that convenient factor which we have seen working between the wars: losses in one region tended to be offset by gains elsewhere. Not all disturbances occurred together. Thus, at times when Indonesia's trade might be slack, that with Malaya or Singapore might be thriving. No sooner had the mainland China trade been lost in 1949 than the Korean War encouraged a surge of growth in Japan which led into the spectacular 'economic miracle' there after 1953. To be sure, this miracle brought its problems for British industrialists, including ship-owners and shipbuilders, but it also produced a constant expansion of overall trade and prosperity in Asia, especially in the 1960s and 1970s. Even in the immediate post-war years, when much of Asia lay devastated in the wake of Japanese occupation, the very symptoms of disruption (such as shortages of raw materials) were causes of prosperity else-where, for example in the Australian and American trades. And if wartime destruction meant, for some years, shortages of traditional cargoes homewards to Britain, there were enormous pent-up demands for many industrial products which Britain was able to supply. We should recall that in the post-war years, at any rate until the early 1950s, Britain was able to revert to normal levels of production much more quickly than some of her traditional competitors in Asian markets (Germany above all), while Britain's merchant fleet was also quickly operational again. Exports from Britain were helped by the government's post-war export drive, by various measures of relief and aid given by the United States and international agencies, and by the world shortage of dollars which boosted trade within the sterling area. Thus the shortfall in homeward cargoes was compensated for by fully-laden outward vessels, a pattern which lasted until around 1953 and which was in sharp contrast to the pattern which had prevailed between the wars.

I

Against these general background remarks it is useful to consider the various phases into which Blue Funnel's post-war history naturally falls. The most obvious break occurred in 1965, the year in which Ocean celebrated its centenary. In this single year the shape of the company altered dramatically as a result of three changes: the acquisition of Liner Holdings (which meant the complete takeover of the Elder Dempster Lines), the decision to seek a listing for the company on the Stock Exchange, and the decision to form, in conjunction with three other shipping lines, a major container consortium, Overseas Containers Ltd. These changes will be considered in detail subsequently, but here we may just note some of their far-reaching consequences. By becoming a publicly quoted company Ocean lost much of the rather cosy family business ethos which had hitherto dominated the relationship between Managers and stockholders. Now Ocean had to be far more accountable to its stockholders, and was, for the first time, at risk from unwanted takeover bids. This situation in turn obliged the Managers to consider safety and better performance through diversification: the century-old tradition of a single-product company, of 'doing one thing and doing it well', as Hobhouse had always insisted, was now to be broken. In fact, of course, the acquisition of Liner Holdings had significantly altered the company's nature anyway. Hitherto the whole focus and expertise of Blue Funnel had lain in the Far East, with Australia and America very much subsidiary interests. Now the fortunes of the company were also bound up with the West African trade, and the acquisition of Elder Dempster and its various subsidiaries turned Ocean into a 'group' rather than a 'firm'. Ultimately the decision in 1965 to move to a cooperative container organisation was most significant of all. For by the beginning of the 1970s the traditional Far East liner trades had passed away from conventional Blue Funnel vessels to the huge container ships of the new consortium. If 1965 and the passing of the Far East trades to containers in the early 1970s marked a major break with the past, another also came in 1972. The acquisition of Wm. Cory & Son added a substantial land base to marine interests, and pushed the company's centre of gravity increasingly away from Liverpool towards London. The change of name in 1973 to Ocean Transport and Trading Limited reflected the change.

In retrospect, therefore, we may view the first two decades after the Second World War as a distinct phase in Blue Funnel's history, a phase in which the company retained much of its character as a family-firm shipping enterprise, specialising in a high-quality cargo-liner service to the Far East. Broadly speaking, too, the firm's record and earnings of profits was one of relative stability. Nevertheless, we would do well to remind ourselves that from the perspective of, say, 1965, the previous twenty years seemed far from uniform. On the contrary, the Managers were aware of a distinct period lasting until around the early 1950s which might be termed the 'post-war revival', for not until that time had the immediate wartime losses been made good and the normal pattern of services fully re-established. By that time most of the numerous post-war 'A' class vessels, as well as the 'P' and 'H' classes had been brought into service, while in the wider shipping world the year 1953 saw the re-entry of German and Japanese shipping lines to the Far Eastern Conference. By that time, too, employment had been built up once more to pre-war levels, with some 600 men and 300 women on the shore staff, 82 masters and 1500 other seamen on regular sea pay, and with 250 midshipmen under indentures.

As far as the Blue Funnel management was concerned the year 1953 also marked the end of an era, for it saw the retirement of both Lawrence Holt and Roland Thornton. With Leonard Cripps already gone (in 1944), the company's affairs now passed steadily into the hands of younger men. Although John Hobhouse remained the senior partner until his retirement in 1957, the major force in the company was undoubtedly Nicholson, who formally became the senior Manager when Hobhouse retired. The ascendancy of Nicholson was significant not simply because it brought to the fore a powerful and energetic personality at a critical time, but because it marked for the first time leadership outside the Holt family. From 1953, indeed, the long Holt dominance over company affairs, not without heart-searching and some bitterness, slowly evaporated.

The various benchmark years around which Ocean's fortunes ebbed and flowed should be seen in evolutionary rather than revolutionary terms. Thus, although in one sense the early 1950s saw the end of a phase of 'post-war revival', the programme of accelerated building continued until 1960. As far as management structure was concerned, for all the significance of the declining Holt influence and the impact of the 1965 changes, not until 1967 was there a significant departure from the traditional style of management with its centralised control in the hands of the senior partners. And notwithstanding the far-reaching moves towards diversification made under the chairmanship of John Nicholson and his successor Lindsay Alexander, until the early 1970s conventional cargo-liners continued to provide the largest single component of Ocean's profits with Far Eastern trades maintaining their traditional dominance.

II

In 1945 the immediate task of the Managers was to reconstruct a company which had suffered greatly during the five long war years. Blue

Funnel now had to set its face towards the future, to plan a major programme of fleet reconstruction, to rebuild the eastern agencies, to restore the conference system, and to adapt the Holt management to the needs of peacetime.

In their Annual Report to shareholders in February 1944 the Managers showed that they were already considering the post-war future. The alternatives, they stated, were two: either to go into liquidation or to rebuild the fleet and services. The Managers thought it both 'the duty and the interest' of shareholders to follow the latter course. They already had their eyes on the future shape of fleet, for large financial reserves had been accumulated which 'must go in building tonnage which will retain the firm's leading position in the shipping world'. It was realised that the reserves might well be insufficient to replace losses and that a post-war building boom would mean inevitable delays in the construction of new tonnage. Accordingly the Managers had discussed buying 'cheaper standard ships from the Government', which was a course eventually adopted. Apart from reconstructing the fleet the main problems foreseen at this early stage were to find employment for the sea staff returning from war service to a depleted fleet, the restaffing of the eastern agencies – in view especially of so many in Japanese hands and needing to return home to recover fully, and the enormous strain under which the ships' Masters had been serving for so long.

These tentative plans for peacetime, drawn up in 1943 with the outcome of the war far from decided and with Japan still dominant throughout the east, is a striking comment on the confidence of the senior Managers in ultimate victory. Indeed, the new *Telemachus*, delivered from Caledon that year, was a prototype 'A' class, designed deliberately with post-war needs in mind. The emphasis on what might be termed 'human values': worries about finding jobs for young seafarer concern for the health of those interned in prison camps, and a determination to assist returning Masters with 'every opportunity of rehabilitation' also speaks well of the company and its traditions.

During 1944 the Managers further refined their plans for peacetime developments. These steps were determined and carried out very much at the instigation of Roland Thornton. The assumption was made that all Blue Funnel (and Glen) trades would recover their pre-war shape, and that the company should aim at a minimum to recapture its traditional shares of these trades. Holts deliberately eschewed expansion or diversification, although some desultory consideration was given to an interest in air transport. At the same time no change was thought necessary in the overall structure of the company, its agencies or its management. However, with a view to strengthening the post-war management team, which was depleted by Richard Holt's dealt in 1941 and the retirement of

35 Charles Douglas Storrs, Manager, 1944–60

Leonard Cripps in 1944, three new managers were appointed in 1944 – John Nicholson, Charles Storrs, and W. H. Dickie.

The immediate difficulties facing the Managers can readily be understood. A depleted fleet, reduced to some 38 vessels from the original fleet, many of them becoming obsolete and in any case needing expensive overhauls to reconvert them from wartime use, was the sorry basis from which managers confronted the task of rebuilding a service which would require some sixty efficient ships. The depressing antiquity of the fleet, one third of which were still coal-burners, can be seen from Table 11.1.

We should emphasise that the vessels recorded in Table 11.1 for 1945 are those which survived the war; the table does not include the various additions to the Holt fleet which came from managed vessels or the

Table 11.1   Age and size of Blue Funnel fleet, 1934–45

Date	Number of vessels	Tonnage	Average age (years)
1934	77	654 383	14.75
1938	75	591 330	18.43
1945	37	268 791	25.57

Source: Ocean Archives: Fleet Books.

government standard ships. But the problem facing the company in 1945 with either its own ageing vessels or newly acquired ships not built to Holts' specification needs little emphasis. Moreover throughout the East the Company's property had been destroyed or damaged, while the Straits Steamship and NSMO fleets, the latter having lost three of its six ships, would also need rebuilding. At home there were post-war shortages of men and materials, while the company's office staff were still scattered in various types of temporary accommodation.

The job of rebuilding the fleet was attacked with vigour, much of the task of negotiating prices and terms with the government devolving on Roland Thornton. In 1944 Holts was able to purchase the *Empire Splendour* from the government, a 12-knot 7300 ton vessel built in 1942 as a war transport. Renamed the *Medon*, the ship entered the Blue Funnel service in 1946 when no longer required for war purposes. Two more opportunistic acquisitions were the 15 knot, 10 000 ton, *Rhexenor* and *Stentor*. Both had originally been intended to transport locomotives and other equipment to the Far Eastern theatres of war, but in 1945 Holts were allowed to purchase them while still under construction. *Rhexenor* came into service the same year and *Stentor* the year following. However, the bulk of the immediate post-war fleet requirements were met in two ways: first by the purchase of fourteen American-built wartime vessels, and second by the immediate ordering in late 1945 of eight 'A'-class liners.

The decision to buy standard wartime ships was not an easy one, and was only reached, in the words of one of the Managers, 'after much agonising thought'. Such a policy flew in the face of the company's long tradition of designing and supervising closely the construction of its own vessels, and, of course, the opportunity to acquire standard wartime vessels in 1918 had been ignored – with much self-congratulation when costs subsequently collapsed and trades evaporated. The deciding factor seems to have been the need for short-term profits in order to have sufficient funds to mount a full-scale rebuilding programme. At the close of the war the cash reserves available for such building amounted to some £13 million (the balance sheet showing over £998,000 held in cash on 31 December, 1945). However, the cost of even the unspectacular 'A' class vessels was to be at least £600 000 each, with the prospect that costs would rise still further. The need for new vessels and the elderly nature of the existing fleet argued in favour of a cheaper alternative, especially since the Managers never contemplated borrowing to replenish the fleets.

The reluctance to borrow was, of course, fully in the self-financing traditions of the company, and the Managers were frightened of incurring once more the type of debt burden which had arisen during the 1930s depression. Many shipowners including the ill-fated Kylsant group, had run into deep trouble through borrowing for new tonnage, and the importance of large liquid reserves to deal with unexpected contingencies became an article of faith for Blue Funnel Managers until the 1970s. Thus the company's cash reserves at the bank stood at £4.4 million at the end of 1951 (a good trading year), £2.3 million in 1954, and £4.1 million in 1957 (after another prosperous year). With the advantage of hindsight we may question this policy. After 1945 interest rates were low, the shipping business was buoyant until around 1957, and costs of new ships were increasing. Liquid reserves earning little was a costly alternative to fleet development. Indeed, while there is no question that the purchase of the American-built war tonnage at cheap prices was a correct one, it is arguable that more should have been bought at a time when freight rates were high and shipping space in short supply. Some British companies, such as the Ben Line, were able to buy such vessels not only as a means of restoration but of expansion. Ben, for instance, was able to steal a march on both Glen and P&O by capturing trades to the Far East while the other two were awaiting the return of their own vessels after refitting.

Certainly in laying out some £4 million from their reserves in the purchase of six American-built 'Victory' ships from the United States Maritime Commission and eight 'Liberty' ships (or 'Sams' since the original names all began with the syllable 'Sam') from the War Ministry, the company made a brilliant purchase. Just how good were the terms can be seen in Table 11.2, which shows the cost of certain additions to the fleet.

In other words, for well under half of what newly-constructed ships might have cost (though admittedly hardly to be reckoned as equivalent) Blue Funnel had acquired recently-built ships – Liberties in 1943, Victories in 1945 – at a time of rising costs and long delays at Britain's shipyards due to post-war shortages of labour and materials. Of course, the vessels had to be adapted to meet Blue Funnel's requirements, and this was one of the first managerial tasks of W. H. Dickie.

Dickie's somewhat unexpected elevation as Manager – certainly a

Table 11.2   Cost of Blue Funnel ships, 1930–47

Ship	Year acquired	Gross tonnage	Speed (knots)	Builders	Cost (£)
Maron	1930	6701	14½	Caledon	260 000
Rhexenor	1945	10 100	15½	Caledon	673 000
Myrmidon	1947	7688	16	Permanente Metals (Victory)	226 896
Eurymedon	1947	7314	11	Bethlehem (Sam)	135 197
Agapenor	1947	7664	15½	Scotts ('A' Class)	675 000

Source: Ocean Archives: Fleet Books.

post 'through the ranks' and out of character with Holt traditions – was expressly to take responsibility for the post-war rebuilding programme. Largely due to his efforts the new vessels were modified with a maximum of speed, a minimum of alterations, yet becoming most economical and serviceable vessels. The Victories were all in service for Blue Funnel during 1946 or early 1947, the Sams in 1947. Two of the Victories were placed with the NSMO fleet as part of the revival of the Dutch-based operations. And in 1947, too, came the first ships of the post-war programme with the delivery of four 'A' class vessels.

The 'A' class ships were, in John Nicholson's words, 'merely work-horses wrapped round the maximum single-shaft diesel power obtainable in the late '40's and early '50's . . . we built so many not because of their excellence but to make up for war losses'. The 'A's' were 15 knot motorships of traditional and rather unoriginal design, each carrying berths for 12 first-class passengers. By 1954 no fewer than 21 'A' class ships of various marks had been delivered, four of which had by that date been transferred to the Glen Line.

The Managers had in 1945 determined to use the 'A' class ships (as many as could be afforded and as circumstances demanded) to supplement eight fast 18 knot ships. These latter were to be capable of a four month round trip to Japan and Australia, and were to spearhead Blue Funnel's competitive drive. They were the more interesting 'P' and 'H' class vessels which made their appearance between 1949 and 1951. Driven by powerful steam turbines they were thus the first steamships built since the *Antenor* and *Polydorus*, acquired in 1925. They had a service speed of around 18 knots, were of some 10 000 gross tonnage, and were equipped for 35 first class passengers.

The *Peleus* class vessels, the flagships of the post-war fleet, entered the Far Eastern trade in 1949 and 1950, while the *Helenus* class ships, with more refrigerated space and adapted for the Australian trade, appeared

36 Passenger lounge, Peleus class, 1949 (*photograph courtesy of Liverpool Museum*)

between 1949 and 1951. The four 'P' class ships were able to give a monthly service to the Far East, and once more allowed the Managers to make the traditional claim of having the fastest cargo ships in service. This, though, was before the re-emergence of Japanese and German competition after 1953, before the introduction of new Scandinavian vessels, and before competition from large fast Ben ships became a matter for lively concern in the late 1950s. The 'P' class ships were the fastest yet put into service by Alfred Holt and Co. From Birkenhead the ships normally called at Rotterdam for further loading and then journeyed non-stop to Singapore in only 20 days. This compared with 25 days for the fastest P&O ships in the early 1950s, and with 30 days for Blue Funnel's pre-war *Antenor* class.

The combination of the new fast ships and the various 'A' class and other Blue Funnel ships allowed the company to re-establish its familiar weekly service between Europe and the Far East during 1950. This

Table 11.3   Far East weekly service, 1939 and 1950

Year	Ships required	Average speed (knots)	Average round voyage	Annual carrying capacity per ship*
1939	21	14½	150 days	345000 tons
1950	17	16½	130 days	397000 tons

*Figures are per ship each way, tons of 40 cu.ft.
Source: Ocean Archives: Managers' Minute Books.

service had operated between the wars, indeed since 1883, but by 1950 the sailings were considerably more efficient as Table 11.3 shows.

The massive accelerated building programme was further supplemented early in 1949 by the transfer from the Glen Line of four elderly 9500 ton, 12 knot, motorships, built just after the First World War. The fleet also acquired in 1949–50 three 8900 turbine ships ordered originally by S. and J. Thompson's Silver Line, but taken over and modified by Blue Funnel while still under construction.

As a result of these various acquisitions the Blue Funnel fleet had more or less recovered from its wartime depletions by the early 1950s, although the programme of accelerated building was not finally completed until around 1960. The numbers of new ships delivered makes impressive reading, no fewer than 45 between 1947 and 1959, or around 3½-ships a year. The year by year totals are given in Table 11.4.

At the same time the fleet's total tonnage was rapidly repaired to prewar levels as Tables 11.5 and 11.6 demonstrate.

One consequence of these energetic building efforts was a much healthier age structure of the fleet. Indeed, the average age of the fleet was nearly halved by 1950, as Table 11.6 shows, while even if we were to omit the war-built Victories and Sams, the average age in 1950 was only 14.48 years – considerably younger than on the eve of the war.

There can be no doubt that by the beginning of the 1960s Holts had rebuilt a splendid fleet, with all the attention to structural and technical detail which had traditionally been the hallmark of Blue Funnel ships.

Table 11.4   New ships delivered, 1947–59

1947	4	1951	2	1955	2
1948	4	1952	2	1956	2
1949	8	1953	4	1957	4
1950	8	1954	2	1958	1
				1959	2

Source: Ocean Archives: Fleet Books.

37 Discharging cargo from the *Gorgon* at Singapore in the 1950s

38 Blue Funnel ships at Singapore, May, 1953: *Laomedon, Anchises, Ascanius, Gorgon, Pyrrhus*)

Table 11.5 Blue Funnel Fleet, 1945–50
(gross tonnage, end of year)

1945	268791
1946	334727
1947	443100
1948	434866
1949	526861
1950	540111

Source: Ocean Archives: Fleet Books.

After the wartime devastation and in the teeth of post-war difficulties this was a notable achievement, and *The Economist* rightly reflected that 'the management has achieved its dominant position in the Far Eastern Conference by being far-sighted enough to build consistently faster ships with better cargo handling equipment than most of their British competitors (including P&O) and many of their foreign ones'.[2] But *The Economist* also wondered whether the traditional policy of self-insurance, and 'building stronger and more expensive ships than anyone else, and running them more carefully' (the journal noted that all charts for Blue

39 Seamen's Mess 'A' class (*photograph courtesy of Liverpool Museum*)

Table 11.6   Blue Funnel fleet, size and age, 1934–61

	Number of ships	Gross tonnage	Average age
1934	77	654000	14.75
1938	75	591000	18.43
1945	37	269000	25.57
1950	64	540000	12.66
1961	59	489000	12.39

Source: Ocean Archives: Managers' Minute Books and Fleet Books.

Funnel ships had courses inked in by the office staff), did not tie up resources better used elsewhere, and asked 'is it wise to build heavy, expensive ships to last 25 years, when the period of writing off a ship is decreasing to 15 or even 10 years or less?'

It is appropriate at this stage to say something of the work of W. H. Dickie, for he was the Manager most directly concerned with supervising Holts' post-war building programme. Dickie was a most untypical Holt Manager, elevated in response to unusual circumstances. He held no university degree, spoke with a north-east accent, and had served no time as Assistant Manager. He was, though, a trained and experienced shipbuilder. He had joined the company from Hawthorn, Leslie and Co., at the age of 25 in 1920, first as Assistant Superintendent Shipwright, and then after 1933 as Superintendent Shipwright. His appointment as Manager in 1944 owed much to Lawrence Holt, with whom Dickie had

40 *Sarpedon* at Holts' Wharf, Hong Kong, October 1962

41 William Hugh Dickie, Manager, 1944–54

worked closely in the 1920s on various aspects of ship maintenance. Doubtless the Managers recalled the absence of shipbuilding and technical expertise at managerial level which had affected the company for the worse after 1918, and were determined to have an expert voice in the many important post-war decisions which would have to be taken. Dickie, in a brief term of years as Manager, served the company well. He assumed almost single-handed the task of day-to-day supervision of the large rebuilding programme, paying countless visits to the various shipyards where construction was under way, and attending to the many detailed and complex problems which arose in yards stretched to full capacity and with innumerable delays and obstructions. As mentioned already, the speedy and efficient modification of the fourteen standard war vessels was largely a product of Dickie's energy and competence. It should be added, however, that Dickie had neither the training nor experience to hold much sway with Flett and the Naval Architects department. As in the inter-war years the dominant influence on ship design tended to favour sturdy, eminently dependable, but also rather costly vessels in terms of construction and manning.

Dickie's role in the post-war revival of Blue Funnel was not confined to the details of the building programme. He was able to play a significant role in the development of the engineering departments, and in the early 1950s created a special personnel department to handle the recruitment and shore management of seagoing engineers, a task undertaken hitherto by junior superintendents. He was involved too in the pioneering decision to put the powerful 14 000 horsepower engines in the 'P' and 'H' classes on a single shaft, and was also instrumental in the successful adoption of boiler oil in diesel engines – a significant cost saving for the company. Dickie also improved Holts' methods of firefighting aboard ship, playing an active role in developing a new system following the loss of the *Menestheus* by fire in 1953. Another of Dickie's legacies was the Odyssey Works, initiated by him in the immediate post-war years. Some Managers at the time felt that ship replacement should have first call on the company's resources, but Dickie was able to convince them of the importance of the Works. The Odyssey Works brought together the various scattered departments which looked after repairs and overhauls to Blue Funnel ships in Liverpool, and, when the premises were completed in 1952 at Birkenhead, housed the technical superintendents and their staff, the Marine, Medical and Stewards Departments, and the newly-founded Engineering Apprentices School.

This latter facility reminds us of the continued thought and initiative which Alfred Holt and Company have given to the training of both engine and deck staff for service on Blue Funnel ships. From an early date Holts paid considerable attention to the training and education of sea staff. At first Blue Funnel ships were manned by crews experienced in the rigours of sailing vessels, many of the most redoubtable Masters and senior officers coming as former sailing men from Wales, the north-west of England, and Scotland. As time went by the supply of those trained in sailing ships necessarily dwindled, and the company became involved in training its own future officers. In 1916, on the initiative of Lawrence Holt, a Midshipmen's Department was set up, and a hostel opened in Liverpool at Riversdale Road for periods of shore leave. The term 'midshipman', incidentally, echoed the traditions of Royal Navy and East India Company nomenclature, and not until 1967, on the merger of the Blue Funnel and Elder Dempster fleets, was the more usual term 'cadet' adopted. During the inter-war years young midshipmen received their sea-training on board various Blue Funnel ships which carried a number – usually four – supernumerary to the crew. These midshipmen then underwent a period of intensive training in all aspects of seamanship, and from the ranks of such carefully selected recruits sprang many long-serving officers in the Blue Funnel fleet.

Following the Second World War the company extended its training facilities, designating the new *Calchas*, delivered in 1947, as a full training ship. The deck crew of the *Calchas*, except for the Bosun and Carpenter, consisted entirely of midshipmen. There was a standard deck complement of 18, consisting of one senior midshipman holding the rank of Leading Seaman, ten others who had received their Efficient

42 Launch of *Calchas*, 27 August 1946, by Mrs L. D. Holt, accompanied by
Lawrence Holt (*photography courtesy of Liverpool Museum*)

Deck Hand certificates, and a further seven too early in their careers to
have qualified and who took the place of Ordinary Seamen and Deck
Boys. Additionally, as on other Blue Funnel ships, four more midshipmen were carried. Later, in 1956, the *Diomed* replaced the *Calchas* as a
midshipmen's training ship. This same *Diomed* was also used to give
promising India Buildings staff the opportunity of a voyage to the East,
and in this way many such individuals while in their twenties obtained
experience of immense value both to themselves and the company.

By the early 1960s around 280 midshipmen were in training at any one
time, with an annual intake of around 80, some of whom were drawn
from Asian countries. Boys joined the School at 16 or 17 years of age,
and spent 4 years taking the Second Mates' Certificate, then normally
sailing as Extra Third Mate. Normal progression would lead to a First
Mates' Certificate after twelve months sailing, then an additional 2½
years at sea as Third Mate when the Masters' Certificate would be taken.
The average recruit would be sailing as Second Mate around the age of
26, as First Mate around 30, and would probably become a Master in his
early forties.

The significance of the Holt training programmes can be seen from the
fact that the overwhelming majority of deck officers started as midshipmen

with the company. Those few recruited from outside had to match exacting standards; none were taken under the age of 23 and they had to possess at least a First Mate's Certificate and, in practice, often a Master's Certificate.

As far as the engine room was concerned, the basic channel of recruitment long remained that of a four-year apprenticeship in engineering workshops, as instituted as far back as 1862. Aspiring recruits to the merchant navy took an Engineering Certificate of Competency issued by the Board of Trade. In 1952, in a world of rapidly evolving technical change, the government allowed that initial training be organised by shipping concerns. Holts accordingly started a scheme for training engineering cadets, annually taking in 24 young recruits aged 16–18 for a programme which included sea training, courses taken at local Technical Colleges, and a year in the company's engineering workshops at Odyssey Works.

In 1963 a new training and residential school, *Aulis*, was opened for both engineering cadets and midshipmen at Riversdale Road on the site of the old Midshipmen's Hostel. The new building could cater for over 100 recruits, midshipmen staying for 6 months and engineering cadets for two years. With better facilities the numbers undergoing training increased, and by the mid-1960s there was an annual intake of 50–60 engineering cadets. In addition, three Cadet Catering Officers were also taken from grammar schools after Ordinary level examinations for training towards a career as Catering Officer.

Deck ratings on Blue Funnel ships before the Second World War were usually recruited as boys, and gained their experience and training at sea. At the end of the Second World War the government made compulsory a twelve-week period of pre-sea training, such training being undertaken for the merchant navy as a whole by the two National Sea Training Schools at Sharpness and Gravesend. An alternative route for suitable youngsters was provided by Holts. Prior to 1956 Blue Funnel recruits were sent to the Outward Bound Sea School at Aberdovey for their initial training, but in that year the company opened its own Deck Boy Training School. In 1958 the School was transferred to new premises at Odyssey Works, Birkenhead, while the following year a Stewards' Training School was incorporated there. By the mid-1960s around 65 Deck Boy recruits were enrolled each year and a similar number of Catering ratings. Also at Vittoria Dock a new Lifeboat Training School was established, the first having been started by Lawrence Holt during the war.

Asian crews, who numbered around 3000 at any one time in the hey-day of Blue Funnel sailings, were recruited differently. Mostly they were recruited in Hong Kong (though many taken on there came from the Chinese mainland, especially from the Canton region), but some were recruited in Singapore and some from among the Liverpool Chinese community.

Sadly, the varied additions to Odyssey Works, and a new laboratory added in 1955, were not seen by the Works' founder. Dickie died suddenly in 1954 at the age of 59, his health impaired, it was generally believed, by the inordinate strain of the work he had undertaken in overseeing the reconstruction of the fleet.

This fleet, we have observed earlier, was becoming more costly. Shipbuilding costs had inevitably risen in the Second World War, although the virtual impossibility of new construction for merchant shipping operations made this somewhat irrelevant. After the war the trend continued upwards, and rising costs were joined by considerable delays at Britain's overstretched yards, at a time of acute labour shortage. This was much to the consternation of the Managers. At the annual meeting in 1948 they made a rather striking comparison between the 1906 *Bellerophon*, sold that year for scrap, and the *Achilles*. The Bellerophon had cost only £105 000. This represented £11 14s 8d per ton gross. The *Achilles*, delivered in 1948, cost £675 000 or £82 7s 6d per ton. This was a rate of increase around double that of the general rate of price rises between the two dates. Moreover shipowners in the immediate post-war years were unable to make fixed-price contracts with the yards, and so were compelled to pay for additional costs arising from wage increases and higher raw material prices. In 1949 the situation became such that the company suspended its replacement programme for a while, although early the following year building was commenced once more. The general upward drift of prices continued throughout the 1950s, and although each successive generation of ships incorporated improvements in design and efficiency, the inflationary trend was only too evident (Table 11.7).

A nearly threefold increase between the cost of the 'A' class *Agapenor* and the 'M' class *Maron* was a hefty additional burden and, superior as the latter vessel was, neither the speed nor the cargo-carrying capacity kept pace with this rate of inflation.

Table 11.7   Cost of certain vessels, 1947–60

Ship	Builders	Gross tonnage	Year	Cost (£)
Agapenor	Scotts	7664	1947	675 000
Autolycus	Vickers	8236	1949	770 500
Lycaon	Vickers	7859	1954	1 247 000
Diomed	Caledon	7980	1956	1 350 000
Ajax	Vickers	7969	1958	1 450 000
Maron	Caledon	8529	1960	1 879 000

Source: Ocean Archives: Fleet Books.

III

Alongside the problems of fleet reconstruction an urgent priority of the company after 1945 was the reconstruction of India Buildings and the reassembly of an effective administrative headquarters in Liverpool. It will be recalled that following the blitz of May 1941, the main departments were perforce housed in makeshift accommodation in a variety of scattered locations. The hub of operations became the late Richard Holt's home at 54, Ullet Road, where the Managers took over the fine library for their headquarters, and the steamship, accounts and other departments were also found a home there. Naval architects were next door, while other departments were dispersed to Iliad House in Birkenhead.

At India Buildings itself the mezzanine floor had been undamaged, and a City Office was quickly established there. After some patchwork repairs it once more became possible to use the first floor which was then able to house the steamship, freight, and a few other departments. These arrangements were naturally inconvenient and made for much futile journeyings, mislaid paperwork and patchy communications. A dismembered India Buildings was also a financial burden. Ocean stood as guarantor to India Buildings' debenture holders, and interest payments on this account alone cost £35 745 in 1944, a full quarter of the total sum distributed by the company as dividends. At the war's end, therefore, the reconstruction of India Buildings became imperative, with the prospect once more of letting to tenants and to relocate in one headquarters as many key departments as possible.

Rebuilding anything in the immediate post-war years was no easy task. In those years of shortages and austerity building licences were granted by the government only for what were deemed essential purposes, and with the huge toll of civilian housing and so many key industrial sites destroyed there were any number of candidates for building permits. At the beginning of 1945 (with the war still in progress) the Managers reported to shareholders that the present dispersal of offices 'did not make for the best results' but they had not been able to get the necessary building permits 'as all available labour is being allotted to the London area to set right the guided missile and bomb damage'.

Somehow or other, and most fortunately, the War Damage Commission was persuaded that Ocean's needs were essential. Although no improper pressure was brought to bear Ocean was certainly helped by the good offices of Sir Stafford Cripps, then a leading member of the new Labour Government. Stafford Cripps had, of course, a long-standing if indirect connection with Holts through his elder brother Leonard, and John Hobhouse was a cousin. It was Stafford Cripps, moreover, who had directed the young Michael Foot to a short-lived post in India Buildings in 1938. Few shipping companies can boast that a future Opposition Leader sprang from their ranks. In any event Hobhouse made a direct approach to Stafford Cripps, and permission was received in October 1945 for a partial renovation of India Buildings. The Company had to be satisfied with a more or less exact replica of the old buildings; only a very few improvements could be incorporated, such as larger lifts and a better heating system.

As a prelude to the larger task of reconstruction, utility offices were soon established on the first and second floors, and in May 1946, exactly a year after the end of war with Germany, the company was able to relinquish its emergency quarters in Ullet Road, and operate once more from its traditional base. For the next few years business was conducted in an environment of the banging and crashing which necessarily accompanied construction. Those departments like the Chinese Crew section and a few technical departments which remained away from India Buildings may have felt relieved to have done so. Not surprisingly, in view of the enormous damage to be rectified in a Britain acutely short of labour and many essential materials, there were many aggravating delays in the rebuilding programme. The new India Buildings, the third incarnation since 1834, was only finally completed in 1953.

IV

The war and Japanese occupation had left throughout the East scenes of destruction and devastation, often the breakdown of law and order, the disappearance of many of the trappings of normal civilised life and the infrastructure of commercial activity, and often political chaos. As the early Blue Funnel ships nosed their way into war-torn ports after the war, usually bringing relief supplies and goods for the allied troops who were administering the stricken territories, they entered a world vastly different from that they had left five years before.

The *Anchises* voyaged in 1947 to Singapore, Shanghai, Chingwantao, Taku Bar, Yokohama, Shimidzu, and Kobe. The Master, J. E. Watson, recorded his impressions in letters to a school which had 'adopted' his ship (the Ship Adoption scheme had started in 1936 and several ships of the Blue Funnel line were among the earliest to participate). 'Shanghai', he wrote:

> appears to be in a chaotic state. The city is now entirely in Chinese hands and anybody knowing China knows what that means. The Chinese monetary system has depreciated to an abysmal level and people there now talk of millions of dollars. The rate of exchange

whilst we were there stood at 480 000 dollars to the £ sterling. The essentials of life are very meagre and expensive, consequently destitution is rife and widespread. The export market in Shanghai has dwindled to a mere shadow of pre-war days and imports mainly consist of materials consigned to relief organisations, as in our case.

Chingwantao in North China was a coal exporting port, but *Anchises* found no coal, the mines being then in the hands of Communist rebels fighting the Nationalist government. A little cargo came from Taku Bar (bristles, goat and camel wool and strawbraid), and then on to Japan for the ship's first post-war visit. In Yokohama the American occupation forces 'control everything – labour, ships' movements, quarantine, sentries, policy – in fact, soldiers everywhere'. Much of the city was destroyed by bombing, and there was 'little or nothing offered for sale'. At Shimidzu some tea was loaded for Tangiers, while at Kobe, where the approaches were still dangerous because of mines left from the war, 'the place appeared to be a shell of its former self. Kobe in pre-war days was a hive of industry, but today it looks as if all the bees have been "smoked out" '.

Resuscitation of the Eastern trades after 1945 was eased for Holts by the surprisingly limited extent of war damage to Eastern properties. In Shanghai, Holt's Wharf at Pootung was virtually unharmed, though badly needing maintenance. In Hong Kong the company's warehouses were all intact, although some damage had been done to piers in air raids. In Singapore the Ocean building had been struck by a small bomb in 1942, but damage had been repaired by the Japanese. Here, other than general maintenance, there was little delay or expense involved in restarting Holt operations. Already by February 1946 George Palmer Holt, on a visit to the area, could report to Hobhouse that 'Singapore gives the impression of being well on the way to recovery'. Only in Indonesia was there significant damage, with two warehouses, one at Surabaya and one at Djakarta, destroyed, while most of the hoists, cranes and other equipment had disappeared. In the Indonesian ports the Holt wharves and equipment were quickly brought back into working order, thanks very largely to the ingenuity of Captain H. Brouwer (who was sent out after the war to look after the company's interest) and to K. Berkhoff. The latter was Senior Wharf Manager in Indonesia, had worked for Holts there since 1921, and in making arrangements to restart Holt operations after the war was described as combining 'the enterprise of a buccaneer with the guile of a serpent'.

The relative immunity of Eastern property is a testament both to the suddenness and success of Japanese attacks and to the equally sudden collapse of Japan after August 1945. But if property was relatively unscathed the toll of war on Holts' Eastern staff was grievous. Many of

43 Holt's Wharf, Kowloon, August 1945 (*photograph courtesy of Liverpool Museum*)

the company's employees had been interned by the Japanese, and many had suffered and died under unspeakable conditions. One episode must stand for many. The Japanese invasion of Singapore began on 8 February 1942, and fierce fighting lasted for a week. With the inevitable end in sight some of the Mansfield staff made an attempt to reach Sumatra in two Straits Steamship vessels and an Imperial Airways launch. The Straits Steamship boats were captured by the Japanese, and of the seven Mansfield staff interned, only two survived the war. The group on the launch, however, who included F. L. Lane, later Chairman of Mansfields and the Straits Steamship Company and Director of Ocean, managed to reach safety at Palembang.

The major and urgent task of rebuilding the Mansfield organisation was put in hand before the end of the war. Planning was undertaken by John Hobhouse and C. E. Wurtzburg, both of whom had an unrivalled knowledge and experience of Mansfields and Straits Steamship. Charles Wurtzburg had joined Alfred Holt and Co. in 1913 at the age of 22, having taking a brilliant first-class honours in Classics from Emmanuel College, Cambridge. He joined Mansfields in 1921 and succeeded H. E. Somerville as Chairman in 1932 before returning to London to take over Glen and McGregors in 1937. His career in various spheres was a most distinguished one. In the Straits Settlements he was a respected member of the Legislative Council and Executive Council, while in 1951 he became President of the Chamber of Shipping. He retained a close involvement with Eastern affairs and made a life-long study of Sir Stamford Raffles, the founder of Singapore. His classic biography, *Raffles of the Eastern Isles*, was published after his death in 1954.

John Hobhouse, too, had maintained the closest connections with Straits affairs since his initial spell in the Mansfield office in 1919–20. Later, in 1953, the then Chairman of Mansfields, A. McLellan was to write of Wurtzburg and Hobhouse, 'Mansfields will always owe a debt of gratitude to these gentlemen, for the success of the firm today is due in no small measure to their successful initial planning and later the careful selection of post-war recruits.'[3] No small part in Mansfield's revival too, was due to the short-term appointment at the end of the war of R. S. MacTier. MacTier, who was knighted in 1961 for his services to shipping, had served with Mansfields in Singapore for most of the 1930s, returning, with Wurtzburg, to Glen in 1937. MacTier, incidentally, was one of the few of the European agency staff to speak fluent Chinese. Immediately following the end of the war MacTier returned to Mansfields as Chairman to guide the company through its early difficult months, and he also at that time acted as shipping adviser to Lord Killearn, British Special Commissioner in South-East Asia. In this short period MacTier also played a significant role in the formation of Malayan Airways, which was the forerunner of Singapore Airlines and Malaysian Airways System. The enduring interest and involvement of Blue Funnel men like Wurtzburg, Hobhouse, MacTier, and George Palmer Holt in the affairs of South-east Asia was characteristic of Holt Managers. Thus some years later, in 1952, it was wholly within Company tradition that Hobhouse and MacTier should have become Governors of a newly founded Malayan Teacher's Training College near Liverpool and that the company should have supported the institution (opened by the Malayan government to train Malayan students to teach English as a second language) from its inception.

Mansfield's first post-war Board Meeting was held, under MacTier's chairmanship, in March 1946, the same month that the British Military Administration in Singapore ended and civilian government of the colony resumed. MacTier left Singapore to return to the United Kingdom in September 1946, and A. McLellan succeeded him as Chairman of Mansfields. During McLellan's period of leadership Mansfields was able slowly to recover from the wartime upheavals and by the time he handed over to F. L. Lane in 1953 the agency was once more playing a full part in Blue Funnel's business. Among post-war developments may be mentioned a continuation in pre-war interests in airline business. When Malayan Airways was formed in 1947, with Straits Steamship as a major shareholder, Mansfields became managers. Although the QANTAS general agency acquired in 1934 was lost in 1947, when the

Australian airline opened an office in Singapore, Mansfields remained as sales agents for QANTAS. At the same time the company continued as agents both for BOAC and Pan American Airways. The former agency had grown from the start in 1933 of an air service, by what was then called Imperial Airways, between the United Kingdom and Singapore. And when Pan American extended its trans-Pacific service from Manila to Singapore at the end of 1940 Mansfields were appointed agents for that company too.

Another significant step was the opening of a Mansfield office in Kuala Lumpur on 1 January 1947, with E. J. C. Gardner the first manager. By this time, despite all the difficulties and deprivations of the early post-war years, there were growing signs of a return to normal activities. A senior government official, on a visit to the Singapore Harbour board, was reported by McLellan to have remarked, 'it is a most cheering sight to see so many Blue Funnel vessels alongside the wharves – it is just like pre-war days and extremely encouraging to realise how rapidly the Blue Funnel Line and British shipping generally is again getting into its stride.'[4]

It was one thing to renovate buildings, recruit staff, and build up once again the infrastructure of shipping agency business. It was quite another to organise a shipping line as many-sided as Straits Steamship. Mansfield's business was, of course, integrally bound up with Straits Steamship, and the shipping line in turn provided significant feeder services for Blue Funnel ships. Rebuilding Straits Steamship was therefore a major concern of Ocean in the immediate post-war years.

On the eve of the war, it will be recalled, the Straits Steamship Company was a busy, thriving concern, having recovered quickly from the world slump of the 1930s under the effective leadership of Charles Wurtzburg. By 1940 the Company served no fewer than 84 ports throughout the region, with a network of interconnected services linking Singapore and Penang with ports throughout the Malayan peninsula, southern Burma, southern Thailand, eastern Sumatra, Sarawak, and North Borneo, and, through the Ho Hong deep-sea routes, between China and Rangoon. A flourishing trade between Singapore and Bangkok was also maintained, in close cooperation with the Siam Steam Navigation Company, and in Bangkok itself the company had maintained the Bangkok Wharf Syndicate in conjunction with Swire's China Navigation Company and the Borneo Company.

Following the Japanese invasions of South-east Asia at the end of 1941 and early months of 1942 the Straits Steamship Company abruptly ceased operating. Most of its fleet – some 31 out of a total of 52 vessels – was requisitioned by the naval authorities to serve under the White Ensign, and of these ships only six survived the war. Of the remainder, only fourteen managed to escape to safety, making the perilous journey of over 1000 miles to Colombo or Western Australia. Meanwhile, after the collapse of Singapore in February 1942, the Straits Steamship Company was re-formed in London where, for the rest of 1942, Ocean looked after its affairs. On 31 December, Wurtzburg was appointed Administrator of the Straits Steamship Company. Curiously there was already an enterprise registered in the United Kingdom called the Straits Steamship Company operating in the Menai Straits, making a formal change of name necessary. So, in 1943, the Singapore Straits Steamship Company was registered, with J. R. Hobhouse as Chairman. In addition to Wurtzburg, the other directors were Lawrence Holt, A. Jackson, and H. E. Somerville. Later, in May 1944, Wurtzburg replaced Hobhouse as Chairman.

The early tasks of the newly formed company were largely administrative and legal, there being no ships to run and little business to transact. But at the end of 1944 the company was charged with the management of what was planned to be a fleet of 86 new cargo ships to take part in operations against the Japanese under South-east Asia Command (SEAC). Managing meant supervising, manning, and operating the ships, a tremendous feather in the company's cap which thus was given the responsibility of running all the small cargo ships within the area of SEAC's operations.

In the event the sudden capitulation of Japan in September 1945 meant that the new vessels never served their intended purpose, and many, indeed, were never built. However the unfinished armada was able to play a useful part in the Straits' fleet reconstruction after the war, contributing no fewer than 29 vessels which were acquired in July 1946. Most of the ships were small vessels, equivalent to the pre-war 75 tonners, while of six larger vessels one went to the Sarawak Steamship Company. In 1947 came further additions, three other ex-Ministry ships, two new 75 tonners from England, and four Australian-built ships. Due mainly to these purchases the Straits Steamship Company had rebuilt its fleet by 1947 to 53 vessels, totalling 38 386 tons compared with 51 vessels, 38 103 tons pre-war. The Chairman, H. J. Toms, was able to state, 'we have regained, with extraordinary rapidity, a position closely approaching our pre-war standing'. In another way, too, the pre-war situation had returned, for on 1 January 1947 the London-based board resigned, and the company, now under its old name, was re-registered in Singapore. Unfortunately for a variety of reasons many of those regional trades which had been the mainstay of the company's pre-war operations were either severely curtailed or even lost altogether. In the post-war world too, Straits found itself in the same predicament as its parent, Ocean, faced with rising costs and intensified competition.

Finally, we will look briefly at the situation further east, where Swires were also faced with the formidable tasks of rebuilding a devastated enterprise. Butterfield and Swire had suffered more substantially than Ocean or Mansfields through wartime property damage, and the task of reconstruction after 1945 was more arduous. In the waves of Japanese attacks which accompanied the bombing of Pearl Harbour on 8 December 1941, most of the river and coastal fleet of the China Navigation Company had been captured or destroyed. Others were deliberately scuttled in Hong Kong harbour before the Japanese invaded. In all, 20 out of 38 ships were lost, while in Shanghai and elsewhere on the Chinese mainland, Swire's wharves, warehouses, and other properties were destroyed or damaged. In Hong Kong itself the damage was formidable, with buildings and godowns destroyed, and the Taikoo Sugar Refinery and Taikoo Dockyard in ruins. Only a few of Swire's eastern staff escaped internment, and many died in the hands of their captors.

Details of the Swire revival after 1945, surely one of the great success stories of post-war British enterprise, would carry us beyond the scope of this history, but in view of the very close and continuing connections between Ocean and Swires it is appropriate to mention a few highlights. During the war itself the affairs of the dismembered company were largely directed by Warren Swire in London and from the wartime base of the China Navigation Company established in Bombay. Warren Swire died shortly after the war, and the task of re-creating an efficient organisation then fell very largely on the capable shoulders of John Kidston Swire who became Chairman in 1946. The post-war situation was bleak. In mainland China there was no chance of rebuilding what had hitherto been one of the cores of the company's operations. The trade on the Yangtze river and the river properties were gone permanently and the China Navigation Company was even barred from the inter-port coastal trade under new cabotage laws passed in 1943. With the Communist takeover in mainland China in 1949 Swire's presence in China, dating back to 1866, was doomed, and the final closure of the last Butterfield and Swire office came in 1954. Gone, too, was the lucrative pre-war emigrant trade from Swatow to Bangkok and the Straits, which Butterfield and Swire had managed for the China Navigation Company in the inter-war years.

Swires' revival was built largely on successful diversification and the development of new trade routes for the China Navigation fleet now driven from its traditional spheres. In 1948 Butterfield and Swire bought a controlling interest in a small two-year old Hong Kong airline, Cathay Pacific. Under Swires' guidance Cathay Pacific was destined to become one of the world's finest and most successful airlines. Also in 1948 Butterfield and Swire opened an office in Tokyo, and this move heralded a major increase in the company's presence in Japan, which grew at a rapid rate both during the Korean War and in the subsequent years of the 'economic miracle'. In 1952 Butterfield and Swire set up a branch in Australia and later expanded in New Zealand and Papua New Guinea. In place of the diminished river and coastal trades of the China Navigation Company, new deep-sea trades were sought, especially in the south-west Pacific. With most routes passing through the 'home port' of Hong Kong, trades such as Japan and China to Indonesia, and Japan to Australia and Thailand were developed.

Despite the massive extension of diverse interests, the parent company of the Swire group, John Swire and Sons, remained a family firm, retaining much of the personal and individual flavour which was once so characteristic of the best elements of British business enterprise. Family leadership continued, the great traditions of the firm being maintained by John Anthony Swire as Chairman of the company from 1966 until 1987 in succession to John Kidstone Swire. A comparison with Ocean is tempting. Both firms stemmed from great Victorian, Liverpool-nurtured, founding fathers, close friends and mutually supportive in business. Lest we conclude from the Blue Funnel story that there is anything 'inevitable' about the gradual decline of family entrepreneurial drive and zest as generations unfold, or that organising big business necessarily involves public quotation, the history of Swires provides a salutary alternative.

12

Into the 1960s: Calm Before the Storm

One of the greatest – perhaps *the* greatest – event in my lifetime has
been the awakening of East Asia.
(G. P. Holt)

The retirement of Lawrence Holt and Roland Thornton in 1953, the
sudden death of W. H. Dickie in 1954, and the approaching retirement
of Hobhouse in 1956 (delayed until 1957 by the Suez crisis) made
imperative the recasting of the managerial team. These were not the
only losses. In 1955 Brian Heathcote, an Assistant Manager who had
taken a wide measure of responsibility for staff and training matters,
retired after 36 years with the company. Storrs, due to retire in 1960,
was also approaching the end of his career with Blue Funnel.

The period of Hobhouse's leadership between 1953 and 1957 was in
some respects, and unfortunately, an 'interregnum'. John Hobhouse,
who was knighted for the services to shipping in 1946, was in many
ways typical of the Alfred Holt and Company as it existed between the
wars. In outlook conservative, he saw his role with Ocean as a way of
combining both commerce and public service, and in the latter he was
active both within his adopted Liverpool and nationally. Within the
company his main spheres of interest and responsibility lay in the Straits
and Indonesian trades. His eighteen month sojourn in Singapore in 1919
gave him a close working relationship with Mansfields and with the
Straits Steamship Company, remaining 'a staunch friend' for 37 years in
the words of Professor Tregonning, historian of the Straits Company.
The Dutch component, Meyers and NSMO, were further interests, and
after the Second World War Hobhouse played a leading role in revital-
ising trade in the Singapore and Dutch branches of Ocean activity.
Through his work with the Straits trades, Hobhouse was also instrumental in introducing the bulk carriage of palm oil and latex in Blue
Funnel ships in the 1930s, both important developments which gave
Blue Funnel a pioneering role in these trades and produced major
cargoes at a time of general world depression. The pilgrim traffic was
also within Hobhouse's sphere of responsibility, and characteristically
he became involved in devising improved international regulations to

282

govern this trade. The introduction of company pensions in 1936 was another valuable achievement, especially notable, perhaps, because the individualist Richard Holt was philosophically opposed to pensions. None the less, Hobhouse persuaded Holt otherwise, and brought in a scheme which remained the basis of the company pension scheme until the 1960s.

On a wider front Hobhouse was active in many Liverpool and Liberal organisations, many of the former representing social issues in which he was deeply involved. For example, before the Second World War he was a member of the City Council, of the Liverpool Improved Houses Movement, the Central Housing Advisory Committee, the Blue Coat Hospital Board of Trustees, the School of Tropical Medicine, and numerous other Liverpool-based bodies. In the war he worked for the Ministry of War Transport as Regional Shipping Representative, was Chairman of the Liverpool Steamship Owners' Association (1941–3), and Chairman of the General Council of British Shipping (1942–3). After the war his numerous activities included Chairman of the British Chamber of Commerce for the Netherlands East Indies (1945–51), and Chairman of both the Liverpool and national associations of port employers (the former 1947–54, the latter 1948–57).

It was unfortunate for Ocean that Hobhouse did not play the effective part in post-war reconstruction which might have been expected. This may partly be explained by the toll of exceptionally heavy war duties, but it was due also to the tragic loss of one of his sons in a school accident just after the war, a blow from which he never recovered, and as senior partner between 1953 and 1957 he showed little dynamism.

There was no doubt, of course, that the mantle of senior partner and effectively of Chairman of the Company would fall on the shoulders of Sir John Nicholson, once Hobhouse retired. Nicholson's 'pair' when he joined the company in 1932 had been George Palmer Holt, an exact contemporary. But Nicholson's outstanding leadership qualities had led to his becoming an Assistant Manager in 1938 and a full Manager in 1944. George Holt, notwithstanding his family claims, did not become an Assistant Manager until 1947, and a Manager until September 1949.

During the mid-1950s a series of important appointments were made which in effect determined the leadership of the company into the container age of the 1970s. In 1953 two post-war 'crown princes', R. O. C. Swayne and J. L. Alexander were made Assistant Managers, and in November 1955 became full Managers. Both epitomised the Holt combination of academic attainment and social acceptability. Both had joined Holts soon after the war, Swayne in 1946 and Alexander, straight from University, the following year. Alexander especially came with a brilliant academic record, with a First Class degree in Modern History and the choice before him of the diplomatic service, a university career, or Holts. Perhaps he saw in Holts a bit of each of the others. They were joined as Managers in November 1955 by a third less traditional type of appointment. He was E. G. Price, who had joined Butterfield and Swire in 1925 and spent all his career, outside war service, in the east. After the war he became manager of Butterfield and Swire's Japan branch, and worked closely with Nicholson during the 'Mitsui Fight' (see p. 302). He spent the last eight years of his career in India Buildings, and with his unrivalled knowledge of the eastern trades was able to play a valuable part in the homeward trades from this region.

A fourth managerial appointment in 1955 took place at the beginning of the year, when R. S. MacTier was brought in to replace the loss of technical expertise suffered through the death of W. H. Dickie. MacTier had been educated at Eton and Magdalen College, Cambridge, where he read mechanical sciences before joining Holts in 1928. The early part of his career was spent mainly with Mansfields and the Straits Steamship Company in Singapore, before returning with Wurtzburg to the newly acquired Glen Line. It was from Glen that he was brought to India Buildings, and with Ocean he became largely responsible for the technical operation and manning of the Blue Funnel and Glen fleets. It is interesting, and a comment on the calibre and contribution of the Holt management, that three of the 1955 Managers, MacTier, Alexander, and Swayne, were later knighted for their services to shipping. MacTier, incidentally, was distantly related to the Holt family, for his mother was a Hobhouse. The family tradition was maintained at around this time by the appointment of Richard Hobhouse, Sir John's son, as Assistant Manager in 1957. Richard Hobhouse became a Manager in 1963 and his retirement in 1976 brought to an end the long record of Holt 'clan' Managers.

To detail the varied contributions of each and every one of a growing list of Managers and directors would be both tedious and invidious. As we will see later, both Alexander and Swayne had fundamental roles in the new developments Ocean was to take in the 1960s and 1970s, Alexander especially as architect of the Cory takeover and Swayne through his work with OCL. But others, too, left imprints on their own territory. MacTier and Taylor on the technical and engineering side, Chrimes through his modernisation of the pension fund and improvement of office management, George Holt in the handling of Indonesian business, St Johnston through his revitalisation of the American trades, Glasier and Hobhouse as ships' husbands, and so on. What, indeed, is striking is the way the Ocean management structure allowed a small number of able individuals to focus their talents in particular areas while still retaining some form of overall collective responsibility. Table 12.1

Table 12.1   Ocean managers, main areas of responsibility, June 1963

	First string	Second string
John Nicholson	Outward freight	Cash
	European agencies	Engineers
	Accounts	Roxburgh/Scott
	Berthing outwards	Caledon Kincaid
	Far East Conference	
	Glen, McGregor	Meyer, NSMO
	Straits Steamship	*Politics*: Japan, Hong Kong
	Liner Holdings	
Lindsay Alexander	Inward freight	Holt's Warehouse
	Berthing homewards	American department
	Stevedoring	Far East Conferences
	Hong Kong property	New York Conferences
	Stapledons	Trans-Pacific Conferences
	Odyssey Trading	Straits Steamship
Areas:	Malaysia, China, Japan	Philippines
	Hong Kong, Suez Canal	Ceylon
Ports	Malaya, Borneo, Bangkok	Japan, Hong Kong, Philippines
Politics	Japan, Hong Kong	Philippines
	Malaya, Bangkok, Borneo	
R.S. MacTier	Engineers and	Glen/McGregor
	shipwrights	Mansfields
	Naval architects	Liner Holdings
	Laboratory	
	Technical	
	Engineers	
	Cadets	
	Birkenhead property	
	Rea Towing	
	Caledon	
R.O.C. Swayne	Legal	Passengers
	Australian conferences	Steamship
	Chinese ratings	Victualling
	Roxburgh, Scott	Masters and Officers
	Mansfields	
	Graduate recruiting	
Areas:	Australia	European ratings
Ports	Australia	India Buildings
Politics	Australia	Philippines
G.P. Holt	Java Conference	Australian conferences
	Steamship	W. Australia trade
	Ships' Husband	Chinese ratings
	Masters and Officers	

Table 12.1   (*continued*)

	First string	Second string
G. P. Holt	European ratings	*Areas:*
	Philip Holt Trusts	Indonesia, Australia, Jeddah
	Meyer, NSMO	*Ports:* Indonesia, Australia
	Apprentices	
Politics	Indonesia	
K. St Johnston	American department	Outward freight
	Passenger department	Legal
	New York Conferences	Berthing outwards
	Trans-Pacific Conferences	Stevedoring
		Hong Kong property
Areas:	Philippines, Jeddah	China, Japan, Hong Kong
Ports:	Japan, Hong Kong,	
	Philippines	
Politics	Philippines	
H.B. Chrimes	Cash	Accounts
	Victualling	Rea towing
	Office staff	
	Stewards	
	India Buildings	
	Rea paint	
R.H. Hobhouse	Vessels accounts	Inward freight
	Indonesia commercial	Berthing homewards
	Medical	Java Conference
	Holt's warehouse	Stapledons
	Indonesia property	*Areas:* Malaya, Suez Canal
	Areas: Indonesia, Ceylon,	*Ports:* Malaya, Borneo, Bangkok
	Aden	*Politics:* Indonesia
R.J.F. Taylor (Assistant Manager)		Engineers and shipwrights
		Naval architects
		Laboratory
		Technical estimate
		Cadets
		Odyssey works

Source: Ocean Archives.

illustrates the division of responsibilities among eight Managers and one Assistant Manager in mid-1963.

The table is incomplete, for a full listing of all areas covered by the Managers would be too long and too congested. But we can, nonetheless, learn much from the table about the functioning and structure of Ocean's management. For example, the distribution of 'first' and 'second' strings was a long-standing arrangement to preserve continuity both when a Manager retired and when one went on a not infrequent journey overseas. This thoroughness is reflected in the range of subjects covered, which (along with the omissions mentioned) embraced every aspect of Ocean operations, from midshipmen's hostels to local charities. We may note too, that the Managers themselves took on special areas of expertise and responsibility. Thus MacTier was very much on the technical side, Alexander and St Johnston on the commercial, Chrimes was involved especially with staff matters, Holt with seafarers and Indonesia, and so on. Another point worth noting is the duties given to R. J. F. Taylor, who had become an Assistant Manager at the beginning of 1962. His duties dovetailed with those of MacTier, whom he was being groomed to succeed in overall responsibility for technical matters. Thus did the Ocean Managers ensure continuity and a painless succession.

The number of Managers, or Managing Directors as they were called after 1965, grew with time. In 1953, after the departure of Thornton and Lawrence Holt, there were five Managers. So there had been in 1913. By 1965, when Ocean became a publicly quoted company, there were nine 'directors', and by the end of 1967, twelve. Although we cannot dwell upon the individual roles of each and every member of this growing list, it is appropriate to consider the many-sided contributions of Sir John Norris Nicholson, who was senior Manager and Chairman between 1957 and 1971.

As historians come to deal with events bordering on the present day, so their two great allies, hindsight and the inability of the dead to answer back, become less helpful. But with whatever circumspection we should review recent changes, the very virulence of upheaval in the 1960s and 1970s makes it essential that we should reflect upon the managerial decision-makers who took Ocean into the modern era.

The late 1950s, when Nicholson's leadership began, marked something of a turning-point in Ocean's fortunes, as we have seen. In commercial, managerial, and family terms, and even in the nature of stockholding, the former era was passing rapidly. It was John Nicholson's task to preside over the transformation, which saw a restructuring of capital, stockmarket quotation, containerisation, diversification, management reorganisation, and takeover battles. When he handed over to Lindsay Alexander in 1971 the transformation was far from complete, but the course was set.

Lying beneath the surface of this catalogue of achievements exist some large and largely unanswerable questions. Are we to view the emergence of a large conglomerate group from a single-product family enterprise as a 'success'? Was the transition achieved efficiently? Were

44 Sir John Norris Nicholson, Manager and Director, 1944–76; Chairman, 1957–71

the group's subsequent difficulties in the early 1980s in some way due to the decisions of a previous age? Did the run-down of traditional operations, with the unhappy consequences for fleet and seafarers, have to take place so speedily? Was it necessary for Ocean, if it was to develop container services, to sink its identity so completely in a London-based consortium? The questions can be raised, but even to attempt coherent answers would take us far beyond the chronological limits of this history. But what is relevant here is that the enormous authority wielded by Nicholson within the company made him personally very influential in the path the company took. His influence, moreover, was felt both in the composition of the managerial team and in the way the Managers could operate within their own spheres.

Nicholson's credentials to lead the company were impeccable. He had been in 1932, arguably, the first 'crown prince' to enter the company, in the sense that he was the first from outside the Holt family clan to be recruited explicitly to progress towards full partnership. He was a wealthy and well-connected young man, born in 1911 and succeeding at an early age to his grandfather's baronetcy (his father was killed in the war). Like Richard and Lawrence Holt, he was a product of Winchester, and he went from there to Trinity College, Cambridge before joining Holts. His path to India Buildings was somewhat unusual, a career in shipping being the idea of Nicholson's stepfather, Lord Mottistone, who approached Richard Holt, a parliamentary and personal friend.

Nicholson came at the grimmest period of depression, in 1932. He joined at the same time as George Palmer Holt, and both lived for a time with Richard and Eliza Holt at 54 Ullet Road. Without wishing to invade the realms of the psychologist we can guess that this sojourn with the leader of the company, then at the height of his powers, made a deep and lasting impression on Nicholson. On the one hand he came to see in Richard Holt an ideal business leader, and the Richard Holt style of management – paternal, critical, and high-principled with a propensity to keep all aspects of company affairs within his own jurisdiction – as a mould for his own future conduct of affairs. On the other hand Nicholson came to see himself as an upholder of Holt traditions, and in some ways as Richard Holt's heir. He often referred quite unaffectedly to 'Uncle Dick' and 'Aunt Eliza', and harked back always to his days at Ullet Road. Nicholson's position in these early years must have been a somewhat odd one. That he was heir was soon apparent with his promotion to Assistant Manager at the end of 1938 at the age of only 27, and with a desk on the 'quarter-deck' unlike the other Assistant Managers. His contemporary in age and arrival, George Holt, was given no such preferment and other members of the Holt family were also denied. The other new Assistant Managers in 1938, C. D. Storrs and Brian Heathcote, were both in their forties, which left only Nicholson in his own age-group as an obvious candidate for future partnership. Yet Nicholson was an 'outsider', not a cousin, or even a cousin of a cousin. His ascent of the partnership ladder, and knowledge of Richard Holt's unyielding judgements on the claims of his own family, may well have coloured his own later rather assertive dealings with aspirants from the Holt family.

Sir John Nicholson's career with Holts was predictably brilliant. He was made a full Manager in 1944, alongside W. H. Dickie and C. D. Storrs, as part of the company's plans for post-war reconstruction. He became senior partner in 1957 and first Chairman in 1965, and remained at the head of the company until 1971. Among a plethora of outside appointments he became Chairman of the Liverpool Port Employers Association between 1957 and 1961, Chairman of Martins Bank, 1962–4, and President of the Chamber of Shipping, 1970–1, when he was awarded a KBE.

Nicholson was a natural leader. He combined a forceful personality with immense energy and an agile mind, not afraid to shoulder responsibility. Although naturally conservative in outlook, he was capable of facing the necessity for change and, when once convinced of a particular course of action, would throw himself wholeheartedly into it. There can be no doubt that many of the major developments of the 1960s owed a great deal to Nicholson's vision and forceful leadership, and he did much to modernise the company. To give just a few examples, under his guidance the capital of the company was written up in 1959 for the first time since 1902, the annual reports to shareholders became more informative, tax advantages from Bermudan investment sought, Japanese yards engaged for shipbuilding, and the 'family business' character was diluted both by the introduction of outside recruits and the exclusion of a number of Holt family contenders. Fundamentally Nicholson, unlike John Hobhouse, was quite prepared to see Ocean expand and diversify. He promoted the purchase of Liner Holdings, and later encouraged a broadly based programme of diversification. Above all, once convinced of the need for containerisation, he became one of the key figures in promoting Overseas Containers Ltd. Sir Andrew Crichton, first Chairman of OCL, later recalled Sir John Nicholson's personality and great contributions in the formative period of OCL. Nicholson

> was a man of unbounding and restless energy with a quick and highly intelligent mind. He was excellent in debate, always with a full grasp of facts which meant that his contribution was consistently valuable . . . Sir John Nicholson's approach was one of concern with the all-important aspect of OCL's financial performance and the earliest achievement of profitability. This was the anxiety of us all, but surrounded as we were by the clouds of problems in the shape of opposition by dockers, the difficulties of building ships, depots and equipment, organising and recruiting staff, it was healthy that one of our number should directly and objectively be concentrating our minds on the ultimate results.

A less positive side to Nicholson's chairmanship was a certain lack of confidence in his colleagues and an unwillingness to delegate responsibility. He followed in the footsteps of Richard Holt, keeping both overall direction of policy firmly in his own hands, and, before 1965, such traditional responsibilities as investment policy, allocation of partners' remuneration, and allocations of stock. This was understandable, for Nicholson was guiding a relatively inexperienced team and was easily the most qualified to control financial and investment issues. Moreover under Nicholson the practice of full debate and collective responsibility for decisions was maintained, and as Chairman and senior partner he allowed full rein to the other Managers in their own respective spheres. Nicholson could be formidable in debate, but his impatience sometimes inhibited younger colleagues. He did not suffer fools gladly and preferred not to suffer them at all.

Nicholson's attitude to financial policy was shaped by his pre-war experience in the depression, when the company became deeply and dangerously indebted to the bank. Nicholson was determined, therefore, to maintain a high ratio of liquid reserves which in turn dictated caution in shipbuilding and also left Ocean exposed to possible takeover attention once the company became publicly quoted. A second attitude concerned management. Nicholson preserved Richard Holt's managerial structure and style until well into the 1960s, which meant heavy executive responsibilities for all Managers, and overall control by the Senior Manager from whom initiatives flowed. Younger Managers were not encouraged to take significant initiatives, nor, with their heavy work loads, was it easy for them to do so. We may see in these attitudes three unfortunate consequences in the 1960s. One was excessive 'tinkering' with the details of ship design and operating methods, when a more fundamental approach to the problems of cost reduction and structure of the business was required. Second, the early stages of diversification were carried out in an unplanned and even haphazard way, owing much to the initiatives of Nicholson himself, and leading sometimes to disappointing results. Finally the rigidity of the management structure left little room for badly needed financial and accounting expertise, (a gap which Nicholson was slow to perceive), which in turn made the implementation of a coherent diversification plan well-nigh impossible.

We should emphasise, too, that Blue Funnel's standards and traditions were borne by many besides the Managers, and long-service and loyalty were characteristic throughout the company's operations, on sea and shore. To catalogue the long list of prominent individuals, powerful personalities, and legendary Masters would be as tedious as a more selective list would be invidious. Nor should we forget that the corpus of the company's history not only embraces a Ben Dawson, who entered service with Holts in 1892 and retired in 1946 after a lifetime with the company which included 40 years as Chief Steward on cargo, passenger, and troop ships; but it embraces also, as the obituary note recorded, 'G. A. Smith, Third Mate: murdered at Menado, 13 June, 1952, 1 year's service'.

Key figures in the Ocean hierarchy were the heads of the various sections and departments, the numbers of which grew with the expansion and complexity of the company's operations. It is something of a curiosity that a company which gave such detailed and meticulous attention to the naming of ships should have taken until 1957 to give official titles to these sectional heads, for only then were such titles as Departmental Manager created 'to mark the status and responsibility' of the job. Yet this lack of rank and flamboyance was very much in keeping with company traditions, and was reflected, too, in the absence of other big business trappings: no Rolls Royce dignified the company fleet of cars, and only senior executives travelled first class. Working hours were long, often prior to 1939 until 9 p.m. at night and with Saturdays included. Not until 1965 was Saturday work ended. With Greek-named ships went somewhat Spartan conditions ashore.

The main departments in the organisational hierarchy were the Steamship Department, Inward and Outward Freight, Conference, and Passengers, coupled with the Accounting and Engineering sections. Key roles in their respective spheres were played by the Marine Superintendents and Naval Architects. Beyond these were a multiplicity of departments whose numbers grew as time went by, dealing with such matters as Asian crews, stores and victualling, stevedores and warehousing, training, agents, the American trades, publicity and advertising, and the legal and secretarial business of the company.

Although a complete list of departmental and other sectional heads would be impossible, mention should be made of a few. Major figures in the old India Buildings were certainly F. P. Jackson, head of the Outward Freight department, and 'Sam' Townley, head of Steamship. The Steamship department looked after the operational maintenance of the ship schedules between Port Said and the United Kingdom homewards until loading for the outward journey had been completed, and the department also supervised the appointment and administration of the ships' crews (for Glen ships as well as Blue Funnel). Townley had joined Holts in 1881, at a time when Steamship and Inward Freight were run as a single department with a staff of three, and retired in 1935. Both Townley and Jackson were long-serving and powerful figures in the company, but there were many others whose contributions were outstanding. These included W. Nelson in the Conference Department, Nelson joining the Company in 1888 in the post section (like all clerical recruits until well after the Second World War) and retiring in 1943; J. T. Miller, a successor to Townley as head of Steamship; A. Toone in Outward Freight and J. Matchett in Inward Freight, who retired in 1947 and 1948 respectively; and A. H. Colenso, who had arrived originally in

45 India Buildings after rebuilding, 1950s

1915 as a result of the Indra Line purchase, in the American department. Post-war stalwarts included H. E. Price, who retired as head of Steamship in 1955; E. Storey, who guided the same department after 1958; C. Cameron Taylor in Outward Freight; A. B. Tytler in Inward Freight; A. F. Stoker in the Conference department, and D. Evans in the American department. In the early 1960s came a new generation of departmental heads, among whom may be mentioned H. W. Garton (Inward Freight), H. R. Disley (Outward Freight), and J. T. Utley (Conference). At this time, too, the company placed increasing emphasis on publicity and advertising, and this department, under H. Wylie, was much involved in the extensive public relations work surrounding Ocean Steam Ships' centenary celebrations in 1965.

In other spheres the place of W. T. Threlfall certainly deserves notice. He joined Holts in 1914, returned after war service to the Accounts department in 1919, and had become the principal force there by around 1930. In 1936 he became Chief Accountant and Secretary of the Company and exercised a strong influence over many aspects of Company affairs until his death, six weeks before he was due to retire in 1957. Himself unqualified, it was said that he was reluctant to encourage those with formal professional qualifications within his section. On his retirement he was succeeded by H. O. Davies as Chief Accountant and J. Greenwood as Secretary, the latter appointed as Secretary and Chief Accountant in 1965 and to the Board in 1970.

Two further individuals who made extensive and wide-ranging contributions were Captain M. B. Glasier and Brian Heathcote. Glasier joined Blue Funnel in 1924 and during the war years served the company with distinction as Marine Superintendent. Resigning to join the Elder Demster board in 1951 he returned as Ocean director in 1966 when the two companies amalgamated. He was deeply involved in matters affecting stevedoring and dock labour in London and Liverpool, and his forward-looking views earned much respect. Heathcote, like Glasier, worked closely in a number of areas with Lawrence Holt, and both shared his involvement in the Outward Bound movement. Heathcote, who became Assistant Manager in 1938, was an Instructor Lieutenant with the Royal Navy during the First World War, and in 1918 was engaged by Lawrence Holt to take charge of the new Midshipmens' department. He also helped in the running of the company's Laboratory, taking control of its management in 1933. He was prominent in many areas connected with the welfare of seamen, being at various times Chairman of the Liverpool Port Welfare Committee and the Liverpool Seamens' Welfare Committee, and two years before his retirement from the Company in 1955 he succeeded Lawrence Holt as Chairman of HMS *Conway*.

I

The immediate post-war years saw buoyant trading conditions. Indeed, in sharp contrast to the years following the First World War, there was a lengthy period, almost uninterrupted, of high demand for shipping services which lasted until the late 1950s. Under these circumstances the rebuilding of commercial operations after 1945 was largely a question of re-establishing services so that the freight coming forward could be handled. In 1950, though, there was a modest extension of operations when Holts became partners with the De La Rama Steamship Company (a Philippine line) and the Swedish East Asia Company in a service called De La Rama Lines. The service had been started before the war by De La Rama and Swedish East Asia to run between the Philippines and Hong Kong, across the Pacific to the west coast of the United States. After the war the service was extended to New York and other east-coast US ports. Butterfield and Swire acted as agents in Hong Kong, China, and Japan; De La Rama Steamship in the Philippines; and Funch, Edye in the United States.

Until 1950 the company's trading operations were conducted very much in the lingering shadows of the war. In almost every area of traditional Blue Funnel enterprise in Asiatic waters, in Malaya and Singapore, in Indonesia, China, Japan, and the Philippines, the extent of wartime destruction and dislocation was such that the speedy revival of homeward cargoes was impossible. As a result, outward cargoes for some time exceeded homeward, a reversal of the inter-war trend, when the much bulkier staples of the homeward trades had meant that roughly one-quarter of the outward carrying capacity had been empty. After 1945, on the contrary, there was a lengthy period of what might be viewed as post-war restocking in Far Eastern territories, a trend which was coming to an end in 1949 but was then masked by the influence of the Korean War which broke out in 1950 and lasted until 1953. Thereafter the traditional tendency for homeward trades to exceed in bulk those outwards was restored, except in the Australian trade. This trend was indeed reinforced by economic developments which tended to mean a slow rate of growth of British exports after 1953. British exports suffered a significant decline in such traditional staple trades as textile sales to the Far East. Already long before 1939 these trades had faced growing industrialisation in Japan, India, and elsewhere and in the 1950s and 1960s a revival of such competition, the further rapid industrialisation of Japan, and the industrial advance of such countries as Hong Kong, Singapore and Taiwan all meant the further erosion of Britain's markets.

Table 12.2 shows the early post-war strength of Blue Funnel's outward trades, a strength reinforced by government policy which required shipowners to give priority to British export trades.

The traditional pilgrim trade was quickly revived, although there were some further significant changes in the way pilgrims were carried. The changes were in part induced by continual pressures from the authorities to improve the conditions under which the pilgrims were carried,

Table 12.2   UK–Far East trades, 1939–53

	Outward cargoes from Birkenhead (tons)	Homeward to Liverpool (tons)
1939	286 000	345 000
1946	359 000	216 000
1949	540 000	440 000
1950	625 000	453 000
1951	701 000	497 000
1952	732 000	436 000
1953	600 000	414 000

Source: Ocean Archives: Voyage Accounts.

the government of newly independent Indonesia understandably wishing to be closely involved. A solution to the problem was found in 1949 with the conversion of the *Tyndareus* as a full-time pilgrim ship which supplemented and eventually replaced the scheduled liners. The *Tyndareus* made two or three return journeys annually between South-east Asia and Jeddah, originally carrying nearly 2000 pilgrims on each trip, though subsequently, after negotiations with the Indonesian authorities, the amount of space allocated to each pilgrim was increased. For Ocean, George Palmer Holt proved a sympathetic and skilled negotiator with the Indonesians, and became a considerable authority on the pilgrim business. Certainly the voyages of the *Tyndareus* meant a considerable improvement in conditions under which pilgrims were carried. The ship had hospital facilities, a cinema, a public address system, and other features which meant that pilgrims could feel their own needs now came first, rather than being considered simply another variety of cargo. George Holt later wrote of the *Tyndareus*:

> Of course, the pilgrim trip was no luxury cruise and this elderly freighter was not an ideal vehicle for the road to Mecca. Undoubtedly there was overcrowding at the beginning and considerable hardship from the tropical heat had to be endured by those simple, kindly folk from the villages of Indonesia and Malaya, for whom the sea of faith is still at the full and the pilgrimage a goal for which their life savings are gladly sacrificed. But at the start of every pilgrim voyage she looked a picture, bright with new paint and spotlessly clean. Her victualling was always of high quality, and her menus carefully attuned to the varying tastes of her passengers. Not once during her pilgrim career did she suffer a serious engine breakdown. High praise must be rendered to the Masters, headed by Captain Tom Phillips (who took her on her first pilgrim season in 1949), deck officers, engineers, pursers and stewards, who put their heart into this notable venture. Nor should her faithful Chinese crew be forgotten, nor the Malays and Indonesians who, starting in the humble capacity of casual cooks and sweepers, were gradually built up into a smart, uniformed body of regular attendants. It is a remarkable fact and a tribute not only to the devotion of her crew, but also to the enthusiastic support and hard work contributed by superintendents and agents at Djakarta, Singapore and Jeddah that, in spite of the *prime facie* handicaps of age and modest speed, there have never been any serious complaints either from pilgrims or governmental authorities responsible for the pilgrimage, about *Tyndareus*. Indeed, she has received repeated expressions of praise.

Initially the *Tyndareus* carried pilgrims both from Malaya and Indonesia, with Mansfields in Singapore and Maclaine Watson in Java continuing

46 *Gunung Djati*, dining room, 1959 (*photograph courtesy of Liverpool Museum*)

to act as agents for the trade as they had done from its inception. However, in 1953 the trades were separated, as the new political circumstances in South-east Asia meant that conditions of the service were increasingly becoming a matter of individual arrangements with the authorities concerned. The Malaya section was inherited by Swire's China Navigation Company, while the Holt interest was henceforth confined to Indonesia. So close had been the long association of the *Serompong Biru* (Blue Funnel Line in Malayan) with the Straits pilgrim trade that the first two ships of the China Navigation Company were obliged to bow to tradition and make their initial voyages in Blue Funnel colours. Despite early doubts the *Tyndareus* pilgrim voyages proved very successful and profitable, and the 44-year-old ship was only taken out of service in 1960 when it became evident that, without substantial renovations, she would not pass her 1961 survey. The *Tyndareus* was now replaced by the specially prepared *Gunung Djati*, a project in which George Holt again played a considerable part, and a vessel which, he wrote later, was 'generally acknowledged to be the finest pilgrim ship which has ever called at Jeddah'.

The *Gunung Djati* deserves a special mention. She had started life as

47 *Gunung Djat,* in 1960

the *Pretoria,* built in Hamburg in 1936 as a passenger ship belonging to the Deutsche Ost Afrika Line for the Hamburg to Capetown run. Taken as a prize by the British government in 1945, she served as a troop-carrier, eventually taking the name *Empire Orwell,* and was purchased by Holts from the Ministry of Transport in November 1958. The conversion of the ship as a pilgrim-carrier was carried out in close consultation with the Indonesian authorities, G. P. Holt playing a leading and sympathetic part in the arrangements. The *Gunung Djati* retained accommodation for 106 First Class passengers, but remaining cabins were stripped to provide large open spaces for unberthed pilgrims, and beds (or *Baleh balehs*) were fitted for 200 pilgrims. She was crewed by Indonesian ratings and made two journeys a year in each direction between Java and Jeddah. Features of the refitted ship included the decoration of a large area on the 'B' Deck as a mosque, and a movable arrow to indicate to pilgrims the direction of Mecca throughout the voyage. The ship was unique in the Blue Funnel fleet not only on account of its German origins and specialised purpose, but it was also the only vessel with two funnels and a non-classical name. The name was suggested by the Indonesian organisation running the pilgrimage, Sunan Gunung Djati being one of the earliest Islamic missionaries from Java who rose from a humble background to become a Sultan, and was buried on the slopes of Gunung Djati ('the hill covered by a teak forest').

Unfortunately changes in the methods of presenting the company's balance sheet and profit and loss accounts after 1948 make direct comparisons of the overall prosperity compared with the inter-war years impossible. From 1948 Holts, edging perceptibly towards a 'group', began to publish consolidated accounts. At first the consolidated figures covered the traditional Blue Funnel shipping lines (Ocean, China Mutual, and NSMO), the Glen Line and its subsidiaries, together with India Buildings Ltd., Mansfield and Co., and Aitken, Lilburn, and Co. The latter company, Blue Funnel's outward agents in the Glasgow–Australia trade from its inception in 1901 was bought by Ocean, and from 1 January, 1947 the business was run jointly with that of Roxburgh, Colin Scott and Co., so that the Australian and Far Eastern agencies were under one control. The accounts of Roxburgh, Scott were consolidated from 1953 and of Odyssey Trading in 1957. Nevertheless, the item 'trading profit' in the consolidated accounts remained composed substantially of the net earnings of the shipping lines and consequently can be used as an appropriate guide to the fortunes of Ocean's shipping business. Table 12.3 shows the figures for trading profit between 1947 and 1961.

Bearing in mind that the figures shown in this Table include not only the Glen Line but various other subsidiaries, we may note four points of

Table 12.3 Consolidated gross trading profit, Ocean Group, 1947–61

	(£000)
1947	2739
1948	3461
1949	5120
1950	5061
1951	6449
1952	8146
1953	5591
1954	5017
1955	4564
1956	5226
1957	8657
1958	6185
1959	5190
1960	6958
1961	5811

Source: Ocean Archives: Reports to Shareholders.

immediate interest: the very high returns which occurred in the early 1950s and again in 1957; the rapid increase in earnings in the late 1940s; depressed earnings for several years after 1953; and, finally, a period of generally stagnating earnings after 1957.

The recovery of earnings after 1947 calls for little comment, and indeed, has already been reviewed earlier. Prosperity was based on the general expansion of trade after the war, coupled with the arrival of the company's new ships, at a time when competition from other shipping lines had yet to make a marked impact on Ocean's trades. Two points may be underlined. One is that Ocean was excellently placed to acquire a strong position in the Japanese trades before Japan herself was able to mount a renewed incursion into world shipping operations in 1953. The second was the boost given by government policy to trade within the sterling area, which also benefited Ocean. The British government in 1946 directed shipping companies to develop where possible trade with sterling area countries, which meant in practice countries within the British Commonwealth. Many of Ocean's traditional trades were firmly based within the sterling area, in Malaya, Singapore, Hong Kong, and Australia, and the general redirection of trade after the war towards the sterling area was singularly helpful to Holts.

The upsurge in earnings at the beginning of the 1950s shown in the Table can be attributed largely to the impact of the Korean War. In June 1950 North Korea mounted a surprise attack on South Korea, and a long and bloody struggle ensued, with Communist China giving aid to the North and the United States and other nations, on behalf of the United Nations, supporting the South. An armistice was finally signed on 17 July, 1953. The war produced almost inevitable prosperity for shipping, for the large volume of war supplies needed gave a boost to trade throughout the Far Eastern regions, while fear of a more widespread conflict also encouraged the stockpiling of basic raw materials and other strategic goods at inflated prices. Since neither the German nor Japanese fleets had yet re-entered world trade the war presented windfall gains for both the Blue Funnel and Glen fleets. We have seen already how the volume of exports from Birkenhead to the Far East rose by about one-third between 1949 and 1952, and a few instances of the enhanced earnings of individual voyages will show the pattern even more vividly. Thus in mid-1950 the *Autolycus* completed her fourth voyage to Japan having earned a gross sum in freights of £150 832. The sixth voyage, also to Japan, but made in the war months of 1951, brought in gross earnings of £191 247. Not only was this an increase of one-quarter of the average for the first four voyages, but was a sum not achieved again until 1957, with earnings again exceptionally high due to the closure of the Suez Canal. Or the *Agapenor*, delivered in 1947, whose first eight voyages earned average freights of £148 000. Then came a series of much higher earnings, with nearly £200 000 received from two successive voyages departing in 1951. As in the case of the *Autolycus* these figures were not reached again until the period of the Suez closure.

With the eruption of the Suez hostilities and subsequent closure of the Canal, Stapledon's agency in Port Said was forced to close and the British and Maltese staff employed there obliged to leave. The Canal opened once more in mid-1957 and in 1958 Stapledon's were re-activated under British management, largely through the efforts of N. B. Craig, senior partner from 1955 until 1963. The long connection between Holts and Stapledons, which had existed since 1870, came to an end on February 1, 1964, when William Stapledon and Sons ceased to exist as a separate entity. Stapledons was then absorbed into the new-formed Assiut Shipping Agency, an Egyptian-managed concern, which became Blue Funnel agents at Port Said.

As in the Korean War years, the high freights in the aftermath of the Suez crisis were also exceptional and even more short-lived. In July 1956 came news from Egypt that Colonel Nasser had decreed the nationalisation of the Canal. After a few months of diplomatic crisis Israel attacked Egypt in late October and British and French forces moved to protect the Canal. After only a few weeks the British and French were obliged to call off their action, while the Canal itself, blocked and closed during the hostilities in October, remained shut until the middle of the following year. No Blue Funnel ships, fortunately, were caught up in the hostilities, although a worried telegram shot out from India Buildings on 30 October to Captains Sparks, Gould, and Phillips on the *Gleniffer*, *Ascanius*, and *Theseus* respectively: 'WHERE ARE YOU REPORT SOONEST GOOD LUCK ODYSSEY.' Liners traditionally using the Suez route could readily be directed to the Cape or across the Pacific. Indeed the Australian service had been re-routed to the Cape from July, while after October a profitable service to Japan via the Panama Canal was run. But since these journeys took longer for the many ships forced to use alternative routes, this naturally meant that additional tonnage would be needed to carry the same volume of goods. Inevitably freight rates, and profits, rose substantially. The impact of the Canal closure upon Blue Funnel's ships has already been illustrated, but one more detailed example may be given. The *Perseus*, an 18 knot 10 000 ton ship of the 'P' class, set out on her twentieth voyage on 23 June just before the Suez crisis erupted. She loaded at Glasgow, Birkenhead and Rotterdam, and passed through the Canal and on to Japan via Singapore and Manila, returning to Liverpool at Glasgow via the Straits and Colombo on 22 October, virtually coinciding with the closure of the Suez Canal. The round trip took 122 days of which over half were spent in port, and the net profits for the voyage were £37 593. The next voyage was an identical itinerary except that the route was necessarily via Dakar and the Cape. This time the journey, starting on 23 October, 1956, took an extra twelve days. But the total net profit, £59 354, was the highest for a single voyage by the *Perseus* since the Korean War. We may add, though, that the Conference system tended to dampen liner freight rates in booms just as they propped up rates in depressions. During the Suez crisis tramp rates rose some 40 per cent whereas liner rates rose by half this figure.

Sandwiched between the peaks of 1952 and 1957 came a period of lower Blue Funnel earnings which were caused not only by the absence of such special factors as Korea and Suez, but by a long and costly struggle with the powerful Mitsui shipping line. The question of the return of Japanese ships to the world's trade routes had been exercising the minds of politicians and shipowners since 1950. In that year Butterfield and Swire became Chairman of the newly-formed Foreign Steamship Association of Japan, a body organised to protect foreign shipping interests in Japan as that country re-established its merchant fleet and as occupation forces withdrew from their control of Japanese ports and port operations. In particular the Association was anxious to see the return of Japanese shipping within the framework of the conference system, a matter of some concern in view of Japan's pre-war activities and the dominant role of the United States in post-war occupied Japan. The Far Eastern Conference, reflecting the views of Holts and Butterfield and Swire, hoped to contain Japanese competition by reaching agreement with the old-established lines NYK (Nippon Yusen Kaisha) and OSK (Osaka Shosen Kaisha). By this policy it was hoped to exclude Mitsui, who late in 1952 applied to join the Conference. Reporting a meeting with Mitsui representatives in December of that year to Eric Price, of Butterfield and Swire, John Nicholson wrote, 'I thought it as well to point out that their combination of trading and carrying interests was one of the principal reasons why both Merchants and Shipowners had learned to dislike them in the past and were still apprehensive of their activities.' Mitsui's bid to enter the Conference was rebuffed, and the result was a bitter three-year struggle which was solved only in 1956 when Mitsui vessels were allowed to operate within conference agreements 'underwing' of the Japanese member NYK. During the 'Mitsui Fight', as it was called, freight rates fell by 20 to 30 per cent, and considerably squeezed profit margins at a time when costs were generally rising.

We may say with the advantage of hindsight that the prosperity of 1957 was a high-water mark not only for Blue Funnel but for British shipping generally. Thereafter came a long and unavailing struggle against rising costs at home and growing competition abroad, so that, notwithstanding considerable growth in overall world trade carried by merchant ships, both the absolute and relative shares carried by British ships shrank. A particular problem for British lines was that while

Table 12.4   Merchant shipping tonnage (non-tanker) 1939–69
(000 gross tons)

Year	UK	World	% UK
1939	13863	49060	27.8
1959	13060	79109	16.5
1964	12624	92474	13.7
1969	12059	115733	10.4

Source: Lloyd's Register of Shipping.

conference developments were making national lines increasingly dependent upon the trades centred on their own home bases, the United Kingdom's economy did not experience very rapid expansion in its own overseas trade in the 1950s and 1960s. And British costs rose more than most. Table 12.4 shows the overall position of Britain's merchant fleet at various dates.

A problem which arose in the post-war years for Britain's ships was the increasing proportion of time which they were obliged to spend in port, a significant factor when one considers the growing costliness of ships and the loss of earning power involved. Delays in loading and unloading were the product of many factors, including port congestion, labour disputes, labour shortages, and general declining levels of efficiency. Port delays were by no means confined to the United Kingdom and in the post-war years were often especially bad in Australia. The result was an inexorable rise in the cost of liner operations. An example of the deteriorating situation is provided by the *Nestor*. On her first voyage in 1913 she made the return voyage to Australia in exactly four months. Her final four voyages, which ended in 1950, averaged 5 months 20 days for exactly the same mileage. Exacerbating the familiar problems of labour difficulties and port congestion was a great decline in the quality of bunker coal which had occurred since *Nestor's* early days. The last voyage of *Nestor*, it may be added, which began from Liverpool on 23 December, 1949, was not only the final bow of a famous Blue Funnel liner, for long boasting the tallest funnel in the world, but the end of the Australian passenger service which had begun in 1910. Launched in 1913, *Nestor* played a distinguished part in both World Wars, including carrying large numbers of Australian and New Zealand troops in the First, and bringing evacuated British children to Sydney in 1940. The *Helenus* class ships which entered the Australian service after 1950, catered only for around 30 passengers, compared with *Nestor's* final capacity of 175.

As the financial results show, Ocean was by no means as badly or as quickly struck by the worsening commercial predicament in which many shipping companies found themselves after 1957. This was partly due to good fortune. The principal sufferers were tramps, where competition was unrestrained and in whose domain the practice of 'flags of convenience' made considerable inroads in the late 1950s and 1960s; and those liners dependent upon passenger traffic which were hard hit by competition from airlines. Between 1957 and 1960 the number of passengers travelling by air increased by 35 per cent, while those travelling by sea fell by 17 per cent. On North Atlantic routes the percentage of passengers crossing by sea fell from 52 per cent in 1957 to only 9 per cent a decade later. Some cargo trades in particular regions, too, suffered more than others. American business was hit not simply by fierce competition in weakly regulated trades, but by the growth of a subsidised US merchant fleet after the war.

Ocean was by no means immune from these developments. The American trades were hampered after the war by the world dollar shortage and later suffered from growing competition from subsidised tonnage and anti-conference legislation. As far as passengers were concerned, Blue Funnel's never very strong commitment to this traffic declined sharply in the post-war period. And the company's once great trade with the Chinese mainland, the basis on which the company was founded, was largely lost in the post-war years. Already before the war, from 1937, especially, the China trade had been upset by political upheavals. After 1949 trade with mainland China, though carried on in a desultory fashion and on occasions promising significant revival, was never more than a marginal addition to Blue Funnel business (though of more significance for the Glen ships). During 1953 Blue Funnel vessels began calling once more at Shanghai, from 1954 handled by the China Ocean Shipping Agency. One of Butterfield and Swire's last acts as Blue Funnel representatives in Shanghai had been to negotiate the sale of Holt's Wharf at Pootung, a note in the Manager's Minutes for April 1952 drawing attention to the 'high rate of losses' there and recording the decision to sell – if 'sell' is the right word for a transaction which yielded only £100 000 and the release of the Wharf Manager, Captain Umpleby, who had been held as virtual hostage by the Chinese authorities.

Post-war political troubles in Asia were by no means confined to China, of course, as we have discussed already. At one time or another Indonesia, Malaya, Singapore, and the less consequential (for Blue Funnel) areas of French Indochina, Thailand and the Philippines faced political upheavals of one sort or another. But as so often seems to have been the case in Ocean's history, the portfolio of interests was sufficiently diverse to ensure that when one trade waned another waxed. At root, South-east Asia and the Far East were, from the mid- or late 1950s, areas of growing world trade which tended to counteract for Ocean the effects both of political upheavals and competitive pressures. The world needed Malaysia's tin, rubber, and palm oil; it wanted South-east Asia's valuable tropical hardwoods, vegetable oils, minerals, sugar, tobacco, fruits and other primary commodities; it wanted Australia's wool, wheat and meat. Japan, industrialising apace by the late 1950s, generated different sorts of trades, being an importer of raw materials and an exporter of industrial products. Hong Kong, followed by Singapore and other countries became important producers of textiles and other manufactures. With an entrenched position in the trade between Europe and the Far East, a closely regulated conference, and with an excellent fleet in terms both of ships and manpower, the Blue Funnel and Glen ships were better placed than most British lines to benefit from developments in world trade during the post-war decades.

Some of John Nicholson's comments to shareholders in 1961, at a time when the outlook for many British lines was bleak, underline the situation: 'Hong Kong's zestful commercial activity is unabated'; 'The tide of Japan's extraordinary prosperity is still rising and we can be pleased with the strong position which our agents have built up for us in that country. . .'; 'In Malaya and Singapore the qualified optimism which we expressed last year was borne out. . .' Commercial results for the year 1960 were especially favourable because the break between the Indonesians and the Dutch left Blue Funnel for some months as the principal carriers of all Indonesian trade with Europe.

Nevertheless, there is no disguising the gradual erosion of prosperity which affected the company perceptibly from the late 1950s and with more insistence as the new decade set in. Some perspective on the circumstances facing Ocean is provided by the amounts retained out of earnings and transferred to depreciation and reserves. This sum was the life-blood of the company, since it was the basic source out of which purchases of new vessels were to come (Table 12.5).

Although the sums set aside were naturally influenced from year to

Table 12.5 Depreciation and amount transferred to reserves, 1957–64 (£ million, consolidated figures)

Year	Transferred
1957	5.9
1958	5.0
1959	4.2
1960	4.5
1961	4.1
1962	3.7
1963	3.9
1964	5.7

Source: Ocean Steamship: Annual Accounts.

year by particular factors, such as taxation, trading conditions, or transfers to the pension fund, Table 12.5 shows a rather dismal picture. In 1958, £5 million 'still adequately covers the cost of our normal rate of ship replacement', shareholders were told. In 1959 £4.2 million 'is barely sufficient to cover the cost of normal ship replacement'. The average for the next four years was below even this level. And all the time the costs of new vessels was rising, so that increasing sums had to be added to the conventional depreciation of book values, the life expectancy of ships was shortening as technical advances occurred, and operating costs were also on the increase.

Let us look at these cost increases and their impact on Ocean's financial situation in a slightly longer perspective. If we take the 30-year period between 1938 and 1968 the situation appears rather starkly. Labour-related costs rose far more rapidly than shipbuilding or fuel costs, especially after 1956, while these in turn grew faster than the retail price index. Freight rates nowhere near kept pace with the increases in costs. Overall, the costs facing Blue Funnel (that is, the annual expenditures on ship purchase and operating) rose some sixfold over the 30 years, while freight rates rose by about four-and-a-half fold. This situation could have been contained if the volume of cargo had risen sufficiently and was carried in the same number of ships, but this was not the case. As a result it became increasingly difficult to provide sufficient out of ships' net earnings to cover replacements, while the overall return on the capital employed was also unsatisfactorily low.

Table 12.6 shows various cost changes in more detail, and displays clearly the dramatic rise in labour-related costs after 1956 when both Birkenhead loading costs and the earnings of seamen more than doubled.

Table 12.7 shows in different perspective the changing proportional costs facing the company in the post-war years. The rising share of cargo-handling costs and crews' wages is all too evident, these two items being 28.4 per cent of total expenses in 1947 and 46.5 per cent in 1966. By this latter date fuel costs were but 7.7 per cent whereas when

Table 12.6   Index of various cost increases, 1938–68

	Fuel oil (per ton)	New ships	Victualling	Birkenhead loading	A/B earnings
1938	100	100	100	100	100
1947	189	163	196	268	307
1956	371	254	380	368	500
1968	376	367	481	785	1050

Source: Ocean Archives.

Table 12.7 Composition of voyage costs, 1947–66
(percentages)

	1947	1950	1966
Agents' commission	5.2	5.2	5.2
Cargo handling	17.7	21.8	27.6
Port and canal dues	6.5	8.6	10.1
Fuel	17.5	14.5	7.7
Crew wages	10.7	10.2	18.9
Provisions and stores	5.8	4.8	6.4
	63.4	65.1	78.1

Note: Figures are proportions of total disbursements, including administration, repairs, etc., but excluding dividends, taxation, and depreciation.
Source: Ocean Steam Ship Co., *Staff Bulletin*, 'Trade Reports'.

the *Gleniffer* made her 44th voyage in 1936 fuel costs (coal) were no less than 22.7 per cent of total costs, while crew costs were only 7.9 per cent. The cheapness of the Chinese crews in these pre-war years was striking: the *Gleniffer* paid £2 038 in wages to the European seamen on this voyage and £710 to the Chinese.

While costs rose, revenues did not keep pace. Table 12.8 shows outward freight rates for the Far Eastern and Australian trades, and also the homeward rates of two key commodities.

Although there were, indeed, significant increases in freight rates these nowhere near kept pace with the growth of costs. For some major commodities, such as motor cars, the combination of competitive pressure and the collective power of the shippers meant that rises were even lower. In these years the organisation of shippers became distinctly stronger, with the formation of first Shippers' Council in the United Kingdom in 1955, to be followed by similar Councils in other European countries and an overall European Shippers' council. The overall result of a variety of factors was growing pressure on profits. In some years, when full allowance was made for replacement of ships at current prices, total net earnings were less than the amount distributed to shareholders. In 1966 such net earnings were £2.3 million, while

Table 12.8 Index of Blue Funnel freight rates, 1938–1968

	Far East out	Australia out	Rubber	Wool
1938	100	100	100	100
1947	217	182	195	194
1956	296	273	293	297
1968	485	463	415	434

Source: Ocean Archives.

Table 12.9   Return on capital employed, 1963–7 (%)

	All companies	Ocean
1963	11.5	6.5
1964	13.5	7.5
1965	13.0	6.4
1966	12.1	4.8
1967	12.5	4.9

Source: Ocean Archives: Memorandum 'Liner Shipping at the Crossroads', 1968.

shareholders' dividends were £2.4 million. The net return on the vast capital employed was also low and declining (see Table 12.9).

Ocean certainly did better than the majority of shipping companies in the 1960s, and Ocean's return on capital was around double that of the average for all shipping companies, but there is no disguising the dilemma presented by rising costs on the one hand, and inadequate freights on the other.

Under these circumstances it is not surprising that increasingly powerful themes during the 1960s became the reduction of operating costs where possible, the raising of conference rates by negotiation, the rationalisation of services, and the diversification of activities into more promising and remunerative avenues. Substantial improvements in conference rates were not obtained until the early 1970s, by which time, as we will see, significant steps had been taken both to diversify operations and to transfer the major trades to container shipping. In the meantime the company went through a phase of strenuous cost-cutting and detailed investigation into all aspects of running its business which recalled the earlier economy drives of the 1930s.

The measures taken, mostly from around 1960, were many and varied. Outside consultants were called in to review the operations of repair work at Odyssey Works in 1962, into the linen department, the accounts system, and mooring arrangements. That same year the Managers decided to remove passenger accommodation from the Far Eastern services, with consequent savings on stewards and other costs. And in 1964 the company set up its own operational research department to investigate a large number of areas of possible economy and improvement. Wide-ranging as these efforts were, they were concentrated above all on improving fleet efficiency and cargo handling at the docks. Among the victims were the Holt warehouses at Blundell Street and Gladstone Dock, both of which had made steady losses in recent years. First to go was Blundell Street, which closed down in 1965, and the remaining warehouse followed in 1967. Cutting down on traditional manning levels in the ships, which had remained largely unchanged

48 Vittoria Dock Berth, 1967 (*photograph courtesy of Liverpool Museum*)

since pre-war days, became urgent as costs mounted. Among many steps taken were the introduction of labour-saving devices (an example was the Arkas Automatic Pilot, introduced after experiments in 1960 and fitted to 43 ships of the Blue Funnel and Glen fleets by mid-1961), the improvement of cargo-handling machinery, and a host of new crewing arrangements adopted after surveys of work practices aboard ships. Many of the findings were incorporated in the new *Priam* class ships which were able to run efficiently with fewer crew than had ever sailed on previous Blue Funnel ships. These moves were certainly overdue, for Blue Funnel ships had long run with traditionally high manning levels born of a period of low seamen's wages and cheap and abundant Chinese crews. In the late 1930s the crews' wages had formed only around 10 per cent of the total voyage costs, whereas by 1963 they had risen to 26.7 per cent.

Mention of the crews should remind us of the very important contribution made to Blue Funnel's services not only by the Masters and European officers but also by the crews. The ordinary seamen were, of course, an integral part of Holts' quality of service and fine reputation. As we have seen already, recruiting was conducted to exacting standards and the fleet was renowned for the excellence of its crews. As far as Asian ratings were concerned, a special department looked after the recruitment, training, posting, and paying of Asian crews, most of whom were Chinese. From the earliest days these Chinese crews formed the majority of seamen on Blue Funnel vessels. Originally many had come from the Chinese mainland but after the Communist victory in 1949 the majority came from Hong Kong. At the end of the 1950s there were around 3000 Asian seamen serving on Holt vessels, roughly 300 being Malays recruited in Singapore (though doubtless of Chinese origin), and a further 300 resident in Liverpool. The Malays were used for ships on the Straits–American services. The Asians provided all engine-room ratings in Blue Funnel ships, stewards in most of the passenger-carying vessels, and full deck rating complements in the Dutch and pilgrim ships. In Liverpool an Asian Boarding House was maintained with close links with India Buildings, while the company also employed some half-dozen Asian clerical workers to deal with the various matters which affected the Asian crews.

To return to the company's cost-cutting endeavours, in 1964 both the operational research department and a cargo-handling superintendent were installed. The latter's function was to investigate costs of cargo handling with different types of handling gear in ports used by Ocean throughout the world, and to make inroads into the £5 million or so spent annually on stevedoring costs. The operational research department carried out a large number of investigations, large and small, into the work and functioning of different departments. As a result of its work many changes were introduced into an increasingly cost-conscious organisation. Some idea of the work of this department can be seen from the numbers of its reports, 8 in 1964, 16 in 1965, 23 in 1966, 21 in 1967 and 18 in 1968. The range was a broad one, but the majority focused on possibilities of rationalising labour and improving productivity, and a few examples will illustrate the scope: May 1965, 'Crew Operations, SS *Jason*, voyage 39'; December 1965, 'Deck Crew Study, 4 Deep Sea Voyages'; December 1966, 'Computer Feasibility Study'; August 1967, '*Perseus* deck ratings work study'; March 1967, 'Victualling Study No. 3, Manning in port'; May 1967, 'Marine Shoregang, pay structure'.

These investigations were followed by a quite dramatic fall in the manning levels of Blue Funnel ships, though such reductions never took precedence over maintaining traditional standards of service. The numbers of ratings especially were cut significantly. In 1960, for example, an 'A' class ship like the *Dolius* carried a total crew of 67 (50 European and 17 Chinese); in 1969 the numbers had been reduced to 46 (37 European and 9 Chinese). The 'P' and 'H' class ships had each carried 62 deck, catering, and engine room ratings in 1960; in 1967 the number was 38. And by this latter date the number of ratings on the new *Priam* ships

Table 12.10   Changes in manning levels, Blue Funnel ships, 1960–7

Crew	Number carried	
	1960	1967
Purser/Radio Officers	2	1
Doctors, male nurses	1	0
Deck ratings:		
'A' and 'M' – 32 ships	20	12
'P' and 'H' – 8 ships	22	15
'Priam' – 4 ships	–	11
Catering ratings:		
'A' – 26 ships	13	8
'M' – 6 ships	15	9
'P' and 'H' – 8 ships	22	11
'Priam' – 4 ships	–	8
Engine room ratings (all Chinese):		
Older 'A' Class – 20 ships	14	12
Newer 'A' Class – 6 ships	12	9
'M' – 6 ships	12	9
'P' and 'H' – 8 ships	18	12
'Priam' – 4 ships	–	10
Electrical engineers' Officers	2	1

Source: Ocean Steam Ship Submission to Committee of Inquiry into Shipping (Rochdale Inquiry) 1967. (Report published 1970).

was only 29. A more comprehensive list of changes in the manning levels of typical Blue Funnel ships between 1960 and 1967 is shown in Table 12.10. The numbers of deck and engine officers, though, remained unchanged; twelve were carried on 'A' vessels in both 1960 and 1967.

In interpreting Table 12.10 it should be noted that the manning reductions came partly from improved efficiency and better use of labour on board ship, but partly also from the ending of the passenger services, so long a feature of Blue Funnel services. In May 1962 the decision was taken to end the passenger services on the 'A' class ships trading to the Far East. This service had begun just after the First World War, and the carriage of 12 first-class passengers had proved a moderate source of profit and a considerable source of goodwill. However, by the 1960s the growth of air services and the decline of the British colonial services and 'planter classes' who had been mainstay of the travellers was making the continued carriage of the passengers an anachronism. Moreover the relative increase in labour costs and the growing need to provide more space on the ships for the crew and crew facilities added to the arguments for 'depassengerisation' as the process was barbarically termed. Accordingly in 1963 and 1964 the 12-passenger accommodation on the 'A' class ships was removed, and this was followed by abandoning the passenger service on the 'M' ships (12 passengers) and 'P' ships

(32 passengers) over the next few years. Thus by 1967 not only was it possible to run these ships with far fewer catering ratings, but with no separate purser or doctor. At this latter date only the *Centaur*, the fleet's sole surviving passenger ship on the West Australian service, carried a medical officer. As Table 12.10 shows, the greatest economies were achieved in the numbers of catering ratings.

Another area of cost-cutting was victualling. During the 1950s costs had risen inexorably, expenditure on victualling for 'cargo and semi-passenger vessels' rising from £467000 in 1950 to £942000 ten years later. As a result of economies, including lower crew numbers, expenditure had fallen to £755000 by 1966.

Fleet changes needed to achieve such manning reductions were not without costs. Between 1961 and 1967 nearly £2 million was spent on fleet modernisation (including removal of passenger accommodation) and on labour-saving equipment for running the ships and for handling cargo. Additional sums were spent on improving amenities on the ships for members of the crew, which included the provision of swimming pools, and the installation of air-conditioning and better methods of fire-detection.

The post-war period as a whole saw considerable improvements in the standards of crew accommodation. Blue Funnel ships had tradition-ally provided superior accommodation by comparison with other com-parable lines, and from the 1950s most of the older ships were refitted to higher standards. Newly built ships had considerably better conditions, and the figures below show the differences in the cabin space allotted to the various ranks of the crews between the *Peleus* class, built 1950, and the *Priam* class, built in 1966 (figures in square feet per man).

	Peleus (1950)	*Priam* (1966)
Junior officers	83	226
Petty officers	78	195
Ratings	38	82

II

A singular, though muted, event of the post-war years was a step taken in 1961 which led eventually to the ending of fleet insurance. Apart from wartime, Holts had run their ships 'self-insured' since 1874, and the tradition had long been seen as the very core of what the company stood for – pride and absolute confidence in the perfection of its vessels. This confidence was sustained by the quite exceptional record of the fleet, which, apart from war casualties, suffered only five losses in the

Table 12.11   Blue Funnel losses since 1871

Ship	Date built	Date lost
Hector	1871	1875
Orestes	1875	1876
Sarpedon	1870	1876
Teucer	1877	1885
Ulysses	1871	1887
Priam	1870	1889
Ulysses	1888	1890
Normanby	1892	1896
Ixion	1892	1911
Ping Suey	1902	1916
Knight of the Thistle	1917	1917
Menetheus	1929	1953
Calchas	1947	1973

Source: Ocean Archives.

twentieth century. The record is given in Table 12.11, and shows that two of the losses occurred during the First World War at a time when the stress and abnormal conditions of war may have taken its toll. The newly acquired *Knight of the Thistle* had to be abandoned in the Atlantic after the loss of a rudder in December 1917, while the previous year the *Ping Suey* (the original China Mutual purchases kept their former names) went ashore near the Cape because of 'the most flagrant disobedience of the Company's rules for navigation', as shareholders were told. No loss of life occurred in these disasters, although the crew of the luckless *Ping Suey* had to survive on penguins' eggs until rescued.

Among the other casualties, it may be mentioned that the *Normanby*, a small 950 ton steamer, had become part of Bogaardt's fleet in the Straits in 1891, operated by him for Holts. In 1892 Holts became owners, and transferred the *Normanby* to NSMO to form part of the new-formed company's fleet trading with Java. The loss occurred in December 1896, when the ship was wrecked near Pulu Bintang.

The loss of the *Menestheus*, the first total loss for 36 years, was caused by fire. Just after midnight on 16 April 1953 a violent explosion occurred as the ship was moving past the coast of southern California on her way to Long Beach. Within minutes the fire was out of hand, the engine-room ablaze, and there was no alternative but to abandon ship. The crew were picked up by the *Navajo Victory* and landed safely at San Diego. On the afternoon of 17 April the Master and Chief Officer returned in a salvage tug to find, as an account a few months later retold:

Menetheus still blazing fore and aft and [they] were not able to board her, but by fixing the towing gear to the rudder they were able to tow her stern first to Magdelana Bay on the Mexican Coast, where they got on board on 20 April. By this time everything that would burn had been consumed, except the ship's cat, which by some miracle was still alive and has since recovered.

An interesting aftermath of this episode was John Nicholson's visit to the stricken ship, which he recommended should be scuttled as a total loss. But W. H. Dickie would not hear of it, and wired at once to stop any such action. The imperative, overriding any cost considerations, was to find out why the fire started and whether lessons could be learned for the future. So at considerable expense the hulk was towed back to Long Beach and a full inquiry held. Dickie's action was in true Holt tradition, and as a result of lessons learned he was able to tighten up considerably fire precautions and drill on board ship, to introduce substantially improved fire-fighting equipment, and to improve subsequent ship design.

The final casualty, the *Calchas* in July 1973 was also a victim of fire, this time as a result of an accident to a fork-lift truck working in a cargo hold. The ship was docked at Port Kelang in Malaysia, and was completely gutted by a five-day blaze.

Short of total losses there have, of course, been a number of other misfortunes of varied gravity, though again these have been few and far between. Fire at sea is naturally the greatest hazard facing merchant ships in their normal operations, and all outbreaks of fire, extensive or otherwise, have always led to the fullest post-mortem investigations and detailed reports. One such fire occurred in 1964, not at sea but in the company's own home territory in Gladstone Dock, Liverpool. The *Pyrrhus* fire, which blazed uncontrolled for 12 hours and was not finally extinguished for a further 12 hours, started on the afternoon of 16 November. Twice it became necessary to abandon the ship as the huge volume of water being pumped caused the ship to list badly. For a time there was danger of an engine-room explosion and of total loss, although eventually the fire was overcome and the damage to ship and cargo, though extensive, contained. This particular fire led to what one member of the Ocean staff called 'a welter of retrospective analysis'. The great impact made by this episode may be explained by three factors. First, the cause of the fire remained a mystery; sabotage was suspected by some. Second, the speed and intensity with which the fire spread was a puzzle in view of the relatively non-flammable nature of the cargo. Third, there were obvious lessons to be learned from the actual methods of fire-fighting, where the very actions of the firemen threatened to undermine the stability of the ship. It was widely recognised that had the same accident befallen a Blue Funnel vessel of a different class, such as the newer Mark 5 or 6 ships of the 'A' class, there would very possibly

49 The *Pyrrhus* fire, 1964

have been a total loss. As a result of the *Pyrrhus* fire a considerable numbers of changes were made to fire-fighting equipment and methods, and steps taken to deal with the problems of stability under similar circumstances.

Modern communications and improved methods of weather forecasting considerably reduced the hazards of unexpected storms which had taken considerable toll of merchant ships in the days before wireless.

50 After the hurricane: *Phemius* with makeshift funnel, at anchor at Kingston,
Jamaica (*photograph courtesy of Liverpool Museum*)

But in 1933 the *Phemius* possibly because of an inaccurate forecast of the
path of the storm, ran headlong into a hurricane of terrifying intensity.
We are fortunate to have an account written by the Master of the ship,
Captain D. L C. Evans, and this vivid tale is reprinted here as Appendix
IV. Not the least interesting part of the story, however is its sequel, the
Master reprimanded by the company for having run his vessel into a
hurricane. Holt standards were never more in evidence. For while
Captain Evans was given a reception by the Lord Mayor of Liverpool
and received numerous awards for his courage in the teeth of extreme
danger, Lawrence Holt decided 'He will be required to forfeit his
insurance deposit of £200 to us on the ground that he failed to exercise
due forethought and prudence thereby bringing his ship into manifest
hazard and causing the company a heavy loss.'

Fires, storms, and collisions may be considered 'normal' risks of the
business of merchant shipping. Unfortunately a variety of other
hazards, political in origin, have made their appearance even in peacetime. One such incident occurred in the closing stages of the civil war in
China in 1949. With Communist victory almost complete the *Anchises*
was making passage up the Whangpo towards Shanghai on 21 June
when she was attacked by a Nationalist Mustang fighter-bomber. No
one was killed, although three crew-members had minor injuries, and
an 18-year old deck-boy was seriously wounded by bomb splinters

(though he subsequently made a good recovery). The Master, the same Captain J. E. Watson who had written his impressions of post-war Shanghai to 'his' school in 1947, was forced to beach the ship while repairs to the damaged engine room were carried out. His report mentioned that on 22 June 'Diver commenced and worked all night on patch port side, great difficulty experienced in keeping Chinese in sampans away from ship's side whilst diver at work. These people are salving oil leaking from damage. Police gave good assistance but had to open fire on several occasions to enforce authority.'

Political troubles of a different kind brought the losses of the *Melampus* and the *Agapenor* in 1967, although since these losses were eventually covered by the Liverpool and London War Risks Association, they have not been included in the earlier table. Both ships were stranded in the Suez Canal when the Arab–Israeli 'Six-day War' broke out on 5 June. The Egyptians promptly blocked the Canal, which remained closed until May 1975. For a long time the ships were kept in good condition with a monthly change of crews, but at the beginning of 1971 both ships were finally abandoned to the underwriters.

The mention of compensation brings us back to the decision to abandon full self-insurance in 1961. The initial step in this direction was only a small one, for Ocean set up an Insurance Fund, registered in Bermuda, which underwrote most of the fleet's insurance. Reinsurance with the outside market (i.e. 'real' outside insurance) was at first only taken against the contingency of having to meet more than two total losses in a single year. But with the arrival of the costly *Priams* more reinsurance was taken, and the step away from self-insurance widened. The decision taken in 1961 was prompted partly by the rising costs of modern ships which made insurance cover prudent, and partly by the considerable tax advantages which would be obtained by establishing a wholly owned insurance subsidiary in Bermuda. In March 1961, Odyssey Insurance, with a capital of £2 million was set up and so broke a tradition lasting nearly 90 years. Yet very little publicity marked this epoch-making event. The annual report to shareholders contained no mention of the move, nor did the house journal *Staff Bulletin*, carry any comment. This coyness was in part due to a certain reluctance to publicise overseas-based tax-saving enterprises (Odyssey Trading, established in Bermuda in 1957 to act as trustee for some small pension funds and to charter vessels received similar inattention). But mainly there was a concern that the move should not be seen as any departure from traditional standards of fleet maintenance, an interpretation which might lower the morale of Masters and crews. Sir John Nicholson's explanatory letter to Masters made light of the change, emphasing 'with all the force at my command that we have taken this step simply as a matter of practical convenience and that it represents no kind of departure from the policy which we have inherited and intend to perpetuate for ever of assuming the fullest possible collective and individual responsibility for the consequences of our own mistakes.'

III

For all the loyalty and pride engendered among the staff, Holts were for long not particularly notable for the various paraphernalia of corporate-identity-seeking measures and institutions. There was, for example, no house magazine until 1947, when the initially rather uninformative and slender *Staff Bulletin* was published twice a year. The company never made special note of the long service records of staff, and, with characteristic belief in the values of individualism, was tardy in intro-ducing a pension scheme. After the Second World War there were signs of change. In 1952 the 'Blue Funnel Reunion Association' for retired Masters, Chief Engineers, Superintendents and Chief Stewards was formed with company support (and with assistance from the P. H. Holt Trust). The Association grew from occasional reunion dinners which had been organised by Lawrence Holt since the 1930s and Lawrence Holt became the Association's first President. In 1961, on the initiative of C. D. Storrs, an association for retired office staff, the Nestorian Associa-tion, was formed, with Storrs the first President. Another association, outside company auspices, was formed in 1927 and has continued an active existence ever since. 'The Blue Funnel Club, Newcastle' was formed as a dining club with membership restricted to those who had served as medical officers on Blue Funnel ships.

As regards pensions, the initial scheme set up through the initiative of John Hobhouse remained the basis of the company's arrangements until the early 1960s. Under the scheme, introduced on 1 February 1936, contributions from staff, sea and shore, were made by individuals towards a pension on retirement or cover in case of death. Before this, non-contributory pensions had been paid to Managers and some Mas-ters and Chief Engineers on an individual basis at the discretion of the Managers. The aim of the pension scheme was eventually to provide an annual sum of half the final salary after 40 years service. In 1941 a Pension Fund Trust was set up into which the company paid an annual lump sum to cover service prior to 1936, while additional sums were added from 1946 to cover current service, in view of rising wage and salary levels. By the early 1960s the pension arrangements had become both costly and cumbersome, there being a great many different arrangements for different categories of staff, anomalies, and 'hard-luck' cases (the latter the special responsibility of George Holt). In 1962 Harry Chrimes arrived as Assistant Manager after a career in industry and with a strong economics background, and he was at once given responsibility for staff matters and pensions (he was 'Chief Steward' in the parlance of the Managers' Minute Books). Chrimes, who was appointed Manager in 1963, initiated a drastic overhaul of pension arrangements, the Pension Trust and Insurance Fund were wound up, and a new fund, the Nestor Pension Scheme inaugurated which was both simpler and less costly to operate.

We should at this stage pause to look more closely at the company's financial position in the 1960s. In the decade or so before the acquisition of Liner Holdings in 1965 there were two outstanding features: the continued ability of the company to maintain its earnings and dividends, in contrast to the overall performance of the shipping industry, and the high ratio of liquidity. Table 12.12 shows the net trading profits and investments income of the Ocean Steam Ship and subsidiary companies over the period 1956–64.

A striking feature of Table 12.12 is the very high ratio of investment earnings both to trading profits and to total capital. In 1964 investment income alone represented 10.3 per cent of total capital, which was a healthy basis for the dividend payment that year of 12 per cent. Ocean's sound position was reflected in the increasing value at which its shares were traded, both when unquoted before March 1965, and subsequently. Table 12.13, adjusted for the 1959 scrip issue of 29 for 1, shows the price range and average index of Ocean stock over the period 1959–66, and also shows by comparison the Investors' Chronicle Index for the shares all shipping companies.

Thus in a decade the market value of Ocean stock had trebled while

Table 12.12   Ocean Steam Ship Co., net trading and investment income, 1956–64 (£)

Year	Capital stock	Net trading profits (before tax)	Investment income	Dividend rate[1] (%)
1956	425 337	2 444 700	767 373	2.5
1957	425 337	6 086 786	864 533	3.2
1958	425 337	2 137 722	962 013	3.7
1959	425 337	1 892 844	833 581	7.8
1960	12 760 110	2 937 444	927 218	8.5
1961	12 760 110	2 014 743	1 100 642	8.5
1962	12 760 110	2 537 598	1 100 654	9.25
1963	12 760 110	2 970 037	1 140 147	11.5
1964	12 760 110	3 829 380	1 316 470	12.0

[1]Adjusted for 1959 scrip issue of 29 for 1.
Source: Report on Ocean Steam Ship Company by Tilney, Sing, Parr and Rae (members of Northern Stock Exchange), 1966.

Table 12.13   Ocean Steam Ship Co., variations in stock prices, 1956–66

Year	Highest	Lowest	Index	All shipping index	All industrials index
1956	17s 5d	15s 5d	100.0	100.0	100.0
1957	21s 5d	17s 5d	118.3	101.9	105.4
1958	24s 5d	16s 5d	124.9	91.6	116.6
1959	42s 0d	24s 7d	202.5	115.9	159.2
1960	42s 3d	39s 3d	248.2	119.0	193.8
1961	43s 3d	38s 3d	248.2	118.7	199.3
1962	38s 3d	30s 3d	209.6	72.0	184.7
1963	30s 5d	35s 3d	260.9	92.1	196.3
1964	51s 0d	48s 7d	303.0	93.1	209.4
1965	57s 3d	44s 0d	308.1	99.1	197.1
1966	53s 9d	50s 0d	315.7	103.3	198.1

Source: As Table 12.12.
Annual Abstracts of Statistics (HMSO).

that of the shipping industry generally had stagnated, and the index for All Industrial Shares no more than doubled. The most dynamic feature of Ocean's business, though, at this aggregated level, was income from investments. On 31 December 1965, subsequent to the takeover of Liner Holdings, the market value of investments, a considerable proportion of which were held in gilt-edged securities, together with cash and short-term loans, stood at no less than £41 059 160. This was equivalent to 48 5d a share, which was almost equivalent to the market price of the shares, leaving but a few shillings to represent the fleet and other assets. Small wonder that Ocean began to feel increasingly vulnerable to take-over bids, especially as by the 1960s shipping companies were by no means immune from the attention of a rising generation of specialist 'asset-stripping' holding companies.

IV

We saw earlier that the programme of accelerated post-war ship replacement was pursued vigorously throughout the 1950s. Table 12.14 summarises the post-war building programme which was finally completed with the delivery of the *Melampus* from Vickers on 10 June, 1960.

The fleet numbers were supplemented and modified by a number of transfers to and from the Glen Line, and by the addition of the three ships originally laid down for the Silver Line which were acquired in 1949–50. When the *Melampus* was delivered, therefore, the Blue Funnel fleet consisted of 60 vessels with a total gross tonnage of 496 345.

The decision to order the last two 'M' vessels in 1956 for delivery in

Table 12.14 New building programme, 1947–60

Class	Number	Period delivered	Average tonnage	Average speed
'A'	27	1947–57	7500	16
'P'	4	1949–50	10000	18
'H'	4	1949–51	10000	18
Neleus	3	1952–55	7800	16
'M'	6	1957–60	8500	16½

Source: Ocean Archives: Fleet Books.

1959 (though in fact they were delivered late, and were handed over the year following) brought to the fore the question of future replacements for the Blue Funnel fleet. Now the management had not only to consider the 'normal' replacements for remaining pre-war vessels and the sturdy but ageing Sams and Liberties, but turn their collective thoughts to the whole future shape of the company's operations into the 1960s, 1970, and beyond.

This was no easy task in the shifting sands of Britain's maritime enterprise. From around the middle of the 1950s, as the end of the post-war programme was in sight, the Managers began to consider plans and designs for a new class of ships which were to spearhead Blue Funnel's fleet development in the new era. 'Priam' was the name given to the class, and the naval architect's department was asked to develop basic designs for a series of vessels which would be larger and faster than anything yet built for the Far Eastern trades. Among the Managers, R. S. MacTier played a leading role in all technical and design matters, and he worked closely with both naval architects and engineers on the project. At this stage, around 1956, some eight or nine ships were envisaged which would have a speed of 17 knots or more, and which would come into service around 1960. By the early 1960s the Managers were already considering the new group of ships to succeed the 'Priams', planned to be as advanced technically and efficient commercially as any conventional ships afloat. But the 'Troilus'-class never left the drawing board, overtaken by events. Certainly in the mid-1950s no one could have dreamt that no new Priam would be in service before 1966, that the programme would not be completed until 1968, or that the ships would have service speeds of 21 knots. Still less could anyone have forecast that these magnificent and technically advanced vessels, built for a working life of around 24 years, would serve only half their term as Blue Funnel or Glen ships, nor, sadly, that they were to be the final conventional liners ever constructed for the company's traditional trades.

With the container revolution still unperceived, the decision to embark on a new class of ships was, of course, a most important one for the company. These were the ships which would take Blue Funnel into a new and uncertain era, and their operating costs and standards of service would determine Holts' future competitiveness. The ambitious plans for a stream of large, fast ships meant inevitably that the costs involved would be enormous. The last 'M's had each cost around £1.5 million and costs were on the increase. There were certain to be, too, a host of crucial technical and commercial decisions to be made on the designs of the ships, motive power, manning levels, and the like.

Initial groundwork was prepared in 1956. Preliminary designs were drawn up and options on shipbuilding berths and machinery taken. The following year the Managers had to decide whether to go ahead with these options, and, if so, to confirm the specifications of the ships and place firm orders with the yards. From the minutes of Managers' Meetings and the various memoranda circulated among the Managers and between departments, it is clear that the new building programme was considered with painstaking and meticulous thoroughness. The eventual decision to postpone the orders may seem in retrospect a pusillanimous one, and one, moreover, which could hardly benefit the company. In many ways, despite their excellence the new ships of the fleet, 'A's, 'P's, 'H's, and 'M's, were all rather conventional vessels, relatively small and relatively slow compared with the newest vessels of some of the company's rivals. Blue Funnel needed advanced new tonnage. The delays in constructing the *Priams* meant that although the fleet did acquire a cluster of brilliantly designed and highly efficient ships, they were extremely expensive, had been subject to long and costly delays in construction, and had only a few years in which to operate before containerisation drove them from the trades for which they had been designed.

We should, therefore, consider the circumstances which led to so much vacillation and eventual lengthy postponement of the new building programme. Two points, in particular, deserve emphasis. First, 1957, the year in which the first major decision to cancel or not the reserved building berths had to be taken, was not an auspicious one for shipowners. It was certainly bad luck for Sir John Nicholson that the start of his tenure as senior Manager should coincide with the first serious downturn in shipping prosperity since the war, and a year which, in retrospect, can be seen as the beginning of a long period of decline for the industry as a whole. There were many circumstances urging caution, some of which we have considered already. They included fears that a new generation of fast tramp ships would be able to attract cargoes traditionally carried by conventional liners; that newly independent Third World countries would build and subsidise their own national lines and for changes in the conventional closed conference system; that American anti-conference legislation would be taken up by other countries; and that the grave political situation in Malaya, Singapore and Indonesia might destroy some of Blue Funnel's main trades. Thus, to those willing to take a gloomy view, the very future of Blue Funnel's services seemed overcast and uncertain.

The second factor was the position of the Glen Line. It will be recalled that Glen, from 1935, was a wholly owned subsidiary of the Ocean Steam Ship Company, trading, like the Blue Funnel ships, primarily to the Far East. Both immediately before and after the war the fleet programmes of the two lines had progressed largely independently of one another, while Glen had continued to operate as an autonomous entity with management from London. Before the war the Glen fleet had been considerably strengthened and modernised, while the Blue Funnel fleet remained almost untouched. After the war the situation was reversed, as Blue Funnel commenced its 'A' class rebuilding programme while Glen built up its operations largely on the basis of rehabilitating the pre-war ships. However, from the late 1950s the issues of further fleet replacement for the two fleets became increasingly intertwined.

Glen's trading circumstances were very different to those of Blue Funnel. The Blue Funnel vessels enjoyed a substantial monopoly of the west-coast sailings to the Far East, carrying around 90 per cent of the total cargoes to and from the west coast, and able to mount a service of 7 or 8 sailings a month. Glen, on the other hand carried only around a quarter of the cargo to and from the east-coast ports, with three sailings a month out of the total of 13 or so operated by UK owners. Glen, therefore, faced more vigorous competition, though to be sure west-coast lines were themselves to some extent in competition with the east coast as well as in providing continental services to and from the Far East. But at any rate in the 1950s the competitive pressure faced by Blue Funnel was not so severe as that confronting Glen. Moreover, by the end of the decade this competition was becoming increasingly severe, particularly as new fast ships operated by the Ben Line entered the trade. The east-coast ships, therefore, needed to be rather faster than those of the west coast to remain competitive. And from these various factors the issue therefore naturally arose as to whether Glen needs should take precedence over Blue Funnel, whether the programmes of the two lines would be developed in harmony, and whether it was appropriate anyway for Blue Funnel to order new ships in view of the uncertain political and other circumstances.

During 1957 the Managers, with an eye on troubled events in South-east Asia, were drawn towards a compromise. The compromise meant cancelling the options on berths and machinery already taken and examining anew the designs of future Blue Funnel ships which might, if events demanded it, be flexible enough to be transferred to the Glen Line. If they were to have this flexibility they would need to have extra speed, more than the 17 knots initially envisaged for the *Priams*, and with additional refrigerated space for the homeward trade.

By mid-1957 the 'Glen Hedge' had become firm policy and a programme of 8 ships of 19 knots decided upon, the orders to be placed in two batches of four each. MacTier was requested to make a particular study of the new powerful Ben ships and the Managers by September were ready to make preliminary costings. These showed that with delivery in late 1960 each vessel would cost around £2.8 million, when an extra £200000 for the 'Glen additions' had been allowed for. The Managers made a specific note that the Glen option was of 'very great importance' in view of the long-term risks for conventional liner operations from subsidised national lines (or 'government trading' as the Managers called it), anti-conference legislation, the increasing strength of the Japanese fleet, and 'competition from the 14 knot tramp'.

The forebodings of the Managers made them understandably cautious, and they instructed the naval architect's department to undertake a major redesign of the *Priams*, while W. Threlfall, nearing the end of his long career with the company, was to look closely at the financial operations of the ships under the new specifications. By the close of 1957 it was anticipated that if the programme were confirmed within three months, one ship could be ready in 1960, and three in 1961. By now, however, a sharp deterioration in trading conditions was making the Managers have second thoughts about the whole programme. While it was recognised that the company had large cash reserves of £4 million for building, as well as £16 million in quoted investments, and high-class tonnage was certainly needed (especially from Rotterdam, where Ben competition was being felt), other factors were pointing in the opposite direction. For one thing, the inflation of shipbuilding costs was less than it had been, so that the impetus to build quickly lest future costs were excessive was distinctly less. Also, the slower growth of Blue Funnel's trade after 1957 now indicated that the volume of trade could be handled throughout the next decade with the addition of only four, rather than eight new ships (with the exception of the new *Centaur* for the West Australian service). Four similar ships could be ordered for Glen.

The anxiety uppermost in the Managers' minds at this time was the series of political disturbances which took place in 1957 in Indonesia, Malaya, Singapore and elsewhere. If trade shrank, fewer ships would be needed. The situation in South-east Asia was indeed of the utmost significance for the entire Ocean Far Eastern service, for it was estimated at the time that the combined trade of Malaya and Singapore was worth 'probably not less than 50 per cent of the total Blue Funnel and Glen Trade, outward and homeward'. The Managers sought the views of John Hobhouse, recently retired, and an expert on the South-east Asian region. Hobhouse was pessimistic, and forecast 'A dark age and an authoritarian future', and the rapid spread of 'communism and communalism'. The Managers swayed by political circumstances decided in January 1958 that 'we should shrink from ordering now', and a few months later it was decided to abandon the berths reserved for two *Priams* in 1960 and to reconsider the whole question the following year. It is interesting to reflect that the Managers thought in terms of 'classes' of ships as they had always done. This was a contrast to the approach of, say, the Ben Line, who ordered their ships one by one and so kept up with the most modern developments, as well as avoiding the huge investment outlays faced by Ocean. But the Holt Managers, of course, had much larger services to consider, and were conscious of the cost of savings brought by standardised orders.

At this stage the affairs of Glen took centre stage. The Glen Chairman, McDavid, who had succeeded Wurtzburg in 1952 (and who had initially made his mark as leader of a team of outward cargo canvassers established by Leonard Cripps in the early 1930s) was convinced that competition from Ben and other lines could not be met with *Priam*-type vessels. It may be added that McDavid's constant pressure on the Ocean Managers to make Glen considerations a major priority had at times annoyed Nicholson and other Managers. McDavid now argued that Glen should have four specially designed ships capable of at least 19½ knots, which might at a later stage be supplemented by *Priams*. The development completely altered the situation as far as the Blue Funnel fleet was concerned. As we have seen, the Managers were coming round to the view that the Far Eastern trades could manage as they were with only four new ships at some future date, which vessels could be dovetailed with a similar order for Glen. In the meantime, it was decided in March 1958 to go ahead with an order for a new class of four notable motorships for Glen, each of 11 500 tons and with a service speed of 20 knots. These were the *Glenlyon* class, ordered in October 1960 and delivered between 1962 and 1963. Two ships were ordered from Fairfield's yard on the Clyde, but the other two orders went to Dutch shipyards. The orders were obtained at fixed prices which, shareholders were told, were 'considerably below the variable quotations received two or three years ago, but it is regrettable that even now only one of several British firms to whom we sent enquiries was able to match tenders from the Dutch yards'. This theme of poor competitiveness by British yards is one which was occurring with increasing frequency in Holts' dealings and which was to recur in striking fashion as the history of the *Priams* unfolded. The *Glenlyon* ships were the last ships designed by Harry Flett. He died, still in the company's service, on his 71st birthday in 1961, sadly not living to see his last ships enter the Holt fleet.

Meanwhile the Managers had also considered the needs of the Singapore–Fremantle service, since the *Gorgon* and *Charon* were approaching the ends of their lives. For historical reasons Blue Funnel's trade with Australia had grown up along two distinct paths, the west coast being served by the joint sailings to Singapore, and the east coast by ships turning round at Adelaide and not calling at Fremantle. By the late 1950s the formerly profitable West Australian service, operated by three small ships, was becoming increasingly unremunerative. Until 1952 Blue Funnel ships had been able to carry the bulk of the monthly cargoes of roughly 2000 tons to the Straits and 3000 tons to Indonesia, but from that time a number of new lines entered the services between Fremantle and Singapore, mostly for onward journeys to India or the Far East. After mooting various possibilities the Managers determined upon the *Centaur*, a splendid vessel specially designed for the idiosyncratic West Australian trade. The idea of a single replacement for the *Gorgon* and *Charon*, a ship which would be both large and fast, emanated from ROC Swayne, and Swayne played a leading role in the subsequent detailed discussions and decisions. The matter was first formally discussed at a Manager's Meeting in February 1961 and in April the Managers decided to go ahead with the plan. There followed a long period in which specifications were drawn up, negotiations conducted with the Australian government on a number of issues (including the number of sheep to be carried), and the type of propulsion and speed of the ship considered. Eventually, in December 1961, it was decided to place the order with John Brown and Co, the *Centaur* to have a speed of 20 knots, to be driven by Burmeister and Wain diesel engines, and to cost £2.5 million. The ship was delivered on the last day of 1963 and

51 *Centaur* III, 1964, built specially for the West Australian trade

sailed on her maiden voyage for Australia on 20 January 1964. The decision to build the *Centaur* was a bold one, and there was considerable scepticism that such a large and expensive vessel could be profitably employed on the Singapore–Fremantle run. In fact the *Centaur* quickly justified the Managers' confidence and soon attracted a large passenger business as well as the traditional cargoes. In 1973 *Centaur* passed to the managership of Straits Steamship, though still sailing under Blue Funnel colours, and made her final departure from Fremantle in 1981.

Having satisfied Glen's immediate needs the Managers returned to the task of deciding when four new *Priams* might be needed for Blue Funnel. In the light of the continually changing competitive environment the Managers decided to proceed with updating and improving designs of the *Priams*, but to delay placing firm orders while the commercial outlook remained uncertain. In 1963 the orders were 'deliberately held over', as shareholders were told, and only in May 1964 were orders eventually placed for eight *Priam*-type ships at a cost of some £18 million. Four of the ships were destined for Blue Funnel, and four, classified as the *Glenalmond* class, for Glen. The ships were 12000 tons, powered by single screw diesel engines, and capable of 21 knots. They were equipped to carry a wide range of bulk liquids and refrigerated cargoes as well as general cargoes, possessed greatly improved crew accommodation, and represented a very great advance on previous Holt ships. All were for delivery in 1966.

The *Priam* and *Glenalmond* orders received considerable national publicity. Newspapers seized on the fact that of the eight orders, two were being placed in Japan. Hitherto, with the exception of the handful of ships built in Hong Kong at the time of the First World War, all the Blue Funnel liners had been built in British yards. Glen it is true, had gone to Denmark and Holland for some of the *Glenearns* before the Second World War, but this was because of the exceptional pressure on British shipyards at the time. Before the Holt order, no British line had ever placed an order in Japan for a complex, fully-automated cargo liner, though Japan had already shown the capacity and skills to build tankers in half the time and at 15 per cent lower costs than British yards could manage. None of the eight engines for the Holt ships, moreover, was to come from Britain. The six British-built ships were to have Danish Burmeister and Wain engines (long used, and indeed pioneered in Britain by Holt ships), while Mitsubishi, builders of the two Japanese ships, were to supply Sulzer diesels under licence. Of the six British ships, no fewer than five of the orders were placed with Vickers, while the other was ordered from the Clydeside firm of John Brown.

National publicity dwelt not only upon the foreign orders but upon the way Holts chose to announce their decision. In making the orders public at the end of May, the company warned that they would turn increasingly to foreign shipyards if British builders could not become more competitive. The Japanese ships would cost 'very substantially less', and Sir Stewart MacTier, commenting on the orders, said that the company had decided to place the orders with British yards at a 'considerable financial sacrifice'. In fact, although this was not divulged by Holts, the Japanese tenders were some 12–20 per cent lower, around £320 000, than those of Britain's yards. The *Financial Times* chose to link what it called 'the fine patrician tone' of Holt's warning to Britain's yards with a comment on the Liverpool company itself. The comment, which extended to the newly acquired *Centaur*, is worth quoting at length.

> When Holts in the same breath take British shipyards under their wing, and warn them to pull their socks up, they are quite unselfconsciously taking up their natural attitude inside Liverpool – not so much a business as a way of trade, a private company which is also a public institution.
>
> At first sight it is an almost Victorian way. In India House [sic], the seat of more concentrated maritime power than any other building in Liverpool, traders still have to queue up to book space for their cargoes. The first of the group's 20-knot ships still had upright funnels, though the new ones will give way to fashion. Holts get their ships built like ships used to be built – it was their private wartime boast that it took two torpedoes to sink a Blue Funnel ship, and awed shipyard men, who get the most detailed orders from India House, say that it still would.
>
> The structure of the management also has a whiff of the best Victorianism. The group has no directors in the ordinary terms – only full-time managers who are sometimes referred to as partners. None is acknowledged as head man; only one, Sir John Nicholson, holds any important outside business seat (in Martins Bank, another bastion of the Liverpool way of business). But all partners are expected to do their term with the Mersey Docks and Harbours Authorities and the Port Employers. They feel responsible for this side of things.
>
> It is all uncommonly efficient though. The new liners will give the group a long lead over much bigger lines in really fast ships; it also has such highly specialised vessels as the Centaur, designed for the Singapore–Australia trade with accommodation for 200 passengers and 4500 sheep. This gives a fine example of the Holt passion for detail. To get all those sheep off the ship at Singapore could be problem. Holt has solved it neatly by using Judas rams on the shore – not Holt-owned, of course, but hired.[1]

Although Holts publicly drew attention to their 'financial sacrifice', there were other less altruistic reasons for the decision to order from

British yards. One was the taxation advantage stemming from government investment allowances, which were expected to reduce the gross costs of the ships by over 20 per cent. Although the allowances were not confined to British-built ships, there was some feeling among the managers that it would be impolitic to place an even more substantial order overseas. Another factor heavily emphasised by MacTier, himself largely instrumental in persuading the company to place its orders with Vickers, was that the many new features involved in the design, developed as they were by Holts' own technical departments, made it advisable to have the ships built under the company's eye and control.

The history of the *Priams'* construction was an immensely sad one as far as British shipbuilding was concerned. While the Mitsubishi ships were built efficiently and more or less on schedule, the British ships from the outset began to suffer from delays and poor quality work. Just what was wrong with British yards defies any simple analysis or summary. Certainly the company's comment to shareholders that there was 'a restless labour force and a shortage of skilled labour' did little more than touch the surface of a problem which lay deep under the skin of British industry. The John Brown ship, *Glenfinlas*, was delivered around six months late at the end of 1966. By this time the first Japanese ship, the *Glenalmond*, had been delivered satisfactorily the previous September although Mitsubishi's second vessel, the *Pembrokeshire*, was not handed over until the following March, three months behind schedule. But Vickers was appalling. Despite constant pressure from Holts, visits from Managers and technical staff to the Vickers yards, and personal pleas by Nicholson to the Vickers Chairman, Sir Charles Dunphie, the situation deteriorated steadily almost as soon as the orders had been placed. Table 12.15 summarises the sorry saga of the five Vickers ships.

The last ship, the *Radnorshire*, sailed on her maiden voyage to the Far East on December 1967. So it was 1968 before the Priam and Glenalmond fleets were in their full operational timetable. By this time, and despite the advance the *Priams* represented on previous Holt ships, rival

Table 12.15 Priam and Glenalmond vessels: promised and actual deliveries, Vickers-Armstrong

Ship	Originally promised	Actual delivery	Months late
Priam	31.3.66	18.11.66	7½
Peisander	31.5.66	28.2.67	9
Prometheus	31.7.66	5.6.67	10½
Protesilaus	30.9.66	31.8.67	11
Radnorshire	30.11.66	16.11.67	11½

Source: Ocean Archives: Directors' Minute Books.

Table 12.16   Time taken between keel-laying and
delivery, Priam class

Ship	Keel-laying/delivery (months)
Vickers	
Priam	26
Peisander	25
Prometheus	16
Protesilaus	18
Radnorshire	18
Mitsubishi	
Glenalmond	12
Pembrokeshire	12
John Brown	
Glenfinlas	21

Source: Ocean   Archives: Directors'   Minute
Books.

concerns like Ben and a number of continental companies had taken
delivery of ships every bit as good, if not better. The deplorable
performance of the British yards is also shown starkly by the comparative figures shown in Table 12.16.

As we have mentioned already, there are no simple answers to the
vexed question of the British yards' dismal performance, and the issues
are both complex and controversial. Some commentators have emphasised weak management practices, others labour relations and work
practices on the part of a highly unionised labour force. One point,
though, is worth stressing because it has received insufficient attention.
This is that, by comparison with rival yards, British shipbuilding
concerns were small, fragmented, and under-capitalised. At the time of
the *Priam* orders Mitsubishi alone had a shipbuilding capacity equal to
almost one half of all Britain's many yards, while Mitsubishi's order
book about equalled total British capacity. Britain's shipping companies,
Holts included, were not blameless in sustaining this fragmentation
over the years. Orders to small subsidiary companies, such as Caledon
and Scotts, had for a generation helped to preserve an inefficient and
overmanned atomistic structure which was growingly uncompetitive
once European countries and Japan rebuilt their own shipyards after the
Second World War.

The Ocean Managers were, of course, most distressed by the course of
events which took place against the background of worsening commercial conditions anyway. As it became clear that the Vickers yards would
be unable to complete the *Priam* programme without serious delays the
Managers, early in 1965, threatened to transfer the *Radnorshire* to a

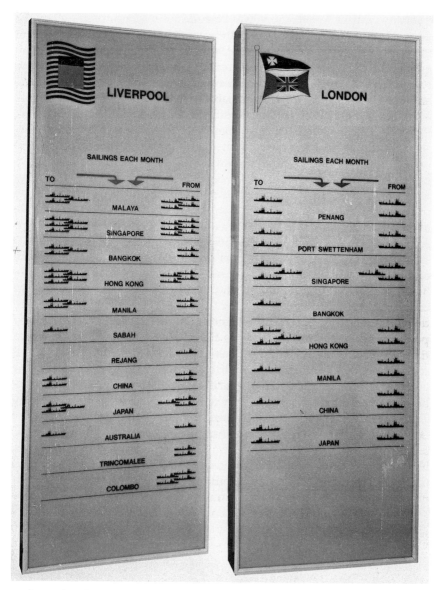

52 List of sailings, Bangkok Exhibition, 1966 (*photograph courtesy of Liverpool Museum*)

Dutch or Japanese yard. They were dissuaded from doing so partly by a strong plea from the Vickers management and also by the successful advocacy of MacTier who at that stage had faith in Vickers' ability to transform its operations. But the situation did not improve. Publicly Holts were moderate in their criticisms of Vickers, fearing that adverse publicity would lower morale at the yard further and so delay the programme more. But privately they were scathing. At a meeting in India Buildings in October 1966, attended by senior managers and technical heads (W. H. Falconer and M. Meek) and senior Vickers representatives, a memorandum resulted which produced a sorry catalogue of Vickers' shortcomings: absenteeism and poor timekeeping among the workforce, which 'no effort is made by Vickers to correct', time wasting, bad organisation, poor discipline, 'weaknesses in leadership', planning 'especially weak throughout', 'no team spirit between management and men and departments', senior Vickers managers never seen in the yards, and so on. The memorandum, which was almost as much a reflection of where Holts thought their own managerial strengths lay as it was an indictment of Vickers, ended curtly, 'we must see an improvement in the organisational structure in the shipyard and in the effectiveness of the management'.

The financial losses of the delays both to Holts and to Vickers were considerable. Holts estimated their loss of earnings from non-delivery of the various ships to total nearly £4 million, although nearly one-half of this was made good by the penalty compensation from Vickers and by saved depreciation charges. The losses were not only financial. The long traditional connection between the Blue Funnel fleet and British yards had come to an end. The Priam and Glenalmond vessels were the last conventional cargo-liners ever built for the Ocean and Glen fleets, but they were also virtually the last British-built ships ever to enter Blue Funnel service. After a long interval, in 1980, the final three, bearing Greek names though in Elder Dempster livery, emerged from Scotts of Greenock, an almost poetic end to a story which had begun in the same place with the *Agamemnon* in 1865. For their tankers, containerships, bulk carriers, and other specialised vessels the group now turned to Germany, Japan, Holland and Scandinavia. The Priam class marked in more than one sense the end of a tradition and of an era.

13

Organising a Group

One begins a shipowning life with the idea that management has everything to do with success. I wonder whether anyone ever ended one without acknowledging that luck had more.

(Alfred Holt, *Diary*, Dec. 1876)

Ocean's centenary year, 1965, was as portentous for the Company as the year of its birth. In one single year a new course was charted, and, whatever the future might hold, the Ocean Steam Ship Company had undergone an irrevocable transformation. Three events, each in their own way fundamental, distinguished the year. The decision to seek a public quotation on the Stock Exchange made untenable the family business organisational structure, and set in train a process of diversification which was to take the company to land and London. The acquisition of Liner Holdings, bringing with it the Elder Dempster fleet and other shipping interests was more in keeping with traditional business; yet the expansion of operations into new areas, new conferences, and also the control of another publicly quoted company meant a far more considerable diversion from historic interests than, say, the acquisition of China Mutual or Glen. And the decision to form Overseas Containers Ltd, OCL, was the most far-reaching of all, spelling, as it did, the end of conventional liner operations for Alfred Holt and Company.

The impact of these decisions was not immediately cataclysmic. The Managers who implemented them, indeed, were only dimly aware of just how dramatic the changes would eventually be. But a chain of events was set in motion which led inexorably to diversification, increasingly into land-based operations and culminating in the acquisition of Wm Cory & Son in 1972, to the collapse of conventional liner trades, to a shift in the company's centre of gravity from Liverpool to London, and to the demise of a unique style of business management.

To explain the many changes of 1965 and the subsequent years is perhaps best achieved by a slight juggling with chronology. In this chapter we will look at the decision to seek a public quotation and three consequent developments, the purchase of Liner Holdings, the early stages of diversification, and the changes in management structure these changes impelled. In the subsequent chapter we will look at the formation of OCL and the acquisition of Cory, at the beginning and end of the chain.

In the metamorphosis from specialised shipping company to multi-enterprise group, Ocean's decision to seek a public stock-market quotation in 1965 must be regarded as one of the great turning points in the company's history. Before such a step, takeovers by Ocean of other large concerns were well-nigh impossible, since it was hardly appropriate to offer stock in an unquoted company for shares in a quoted one. Equally, a takeover of Ocean while an unquoted company was also unthinkable, since the Managers, at least in theory, could simply refuse to part with stock to an unwanted bidder. In any case Ocean's ability as a non-quoted company to hide from public gaze behind sparse annual reports and somewhat summary balance sheets meant that takeover interest from the outside world was improbable.

After the public quotation Ocean's cosy world was transformed. Now exposed to public gaze, Ocean could easily attract attention from possible bidders, especially in view of the company's very high liquidity. Immediately, therefore, safety dictated a programme of building up the company by acquisition and diversification, a process now made easier because quoted Ocean stock could be offered. The purchase of Liner Holdings in 1965 was only a first step on a long road of such expansion which eventually was to shift the centre of gravity of the company from sea to land, and from Liverpool to London.

The reasons which drove the Managers to seek a stock market quotation were by no means bound up with the desire for future expansion, however. In some ways cause and effect were the other way round. The principal factors were two, one long-term and the other immediate. The long-term factor was a striking change taking place in the pattern of the company's stock ownership. It will be recalled that when the company was founded the number of stockholders was only 26. Thereafter the numbers grew, but remained relatively small. At the beginning of 1902 there were 86 stockholders, in 1927 some 365, and by 1955 still only around 630. At this latter date the list still reflected very much the family origins of the concern, with many holders being descendants of the original shareholders. Prominent in the list of names were to be found Holt, Hobhouse, Stapledon, and Swire, as might be expected, and still many representatives from such well-known Liverpool families as Eills, Durning, Rathbone, Melly, and Maxwell. There were no large institutional holdings, and the pattern of stockholding remained in the control of the partners, who put out circulars for tenders when stock came on the market, and who carefully vetted applicants. At this stage roughly half the stock was still owned by some 80 individuals, many of them Holt relatives whose identification with the company was sometimes distant. Some were urging public quotation, and the Managers, feeling that such a step was inevitable before long, decided that stocks should be steered towards institutions. Thus by 1965 much had changed. There were now around 2300 stockholders, including several large institutional investors such as insurance companies and pension funds. The changes stemmed partly from the further consequences of death and subsequent division of inheritances, and also from the inroads of death duties. Undoubtedly, too, an impetus to dispersed holdings came from the decision implemented in 1959 to increase the company's capital from £425 337, where it had remained since 1902, to £12 760 110. This was done by the creation of new ordinary stock units of £1 which were issued to existing holders at the rate of 29 new units for each £1 of stock held. This led some holders to appreciate the size of their holdings and so encouraged a market for the stock. At the same time the Managers encouraged institutional buyers to purchase stock in the company. This inevitably altered the composition of ownership, while pressure for a public quotation began to arise from these new institutional holders. In any case the very expansion of transfer activity made the traditional tender system of sale and purchase increasingly difficult and unwieldy for the Managers to operate.

Supplementing this pressure was a more immediate factor which hastened the decision to seek a public quotation in 1965. This was the new Capital Gains Tax foreshadowed by the Chancellor in December 1964, and introduced in the Budget the following April. The Managers felt duty-bound to demonstrate that the price of Ocean's shares was as high as possible. Interestingly, the initial market quotation in 1965 valued the company at exactly the rate of the last tender, after which the price fell somewhat.

Accordingly, an extraordinary general meeting gave approval in February 1965 for the company to seek a stock-market quotation on the Liverpool and London exchanges, and Ocean shares were quoted the month following. The decision certainly precipitated further changes which took Ocean further away from its traditional style and range of activities. Almost immediately, in March, the Managers constituted themselves as a Board of Directors. John Nicholson was appointed Chairman, and the first Managing Directors were George Holt, R. S. MacTier, R. O. C. Swayne, J. L. Alexander, H. B. Chrimes, R. H. Hobhouse, K. St Johnston, and R. J. F. Taylor. Later in the year, after the acquisition of Elder Dempster, P. J. D. Toosey, Chairman of Liner Holdings, was added to the Board.

More fundamentally the way was now open for Ocean to attract and pursue takeover offers. Ocean was in a rather vulnerable position, having considerable liquid reserves (that is, investments very largely in gilts which were a high proportion of total market valuation). Partly as a defensive measure the Directors, as we now must call them, decided

53 Sir John Lindsay Alexander, Manager and Director, 1955–86; Chairman, 1971–80

within weeks of entering the stock market lists to increase their already large stake, some 25 per cent, in Liner Holdings. This was done by purchases through a nominee as stock became available. The action, though, coincided with some substantial sales of Liner Holdings stock, and when the price of the shares failed to fall, brokers rightly suspected that Ocean was responsible for the firm price. In these circum-

54 George Palmer Holt, Manager and Director,
1949–71

stances, Ocean felt compelled to make a full bid for the outstanding shares.

The acquisition of Liner Holdings, as well as the formation of OCL later the same year, were both stepping-stones towards diversification and expansion. For example, Liner Holdings brought in addition to Elder Dempster's entrenched position in West African trades and the Henderson Line's Burmese trades, interests in car transportation (through Seaway Car Transporters, later Seaway Ferries), container operations (through African Container Express), and various property and agency interests in West Africa. Seaway's first car carrier, *Carway*, came into service early in 1967, ferrying cars between Felixtowe and Scandinavia, and by 1970 the company was operating three ships.

A further move towards diversification, though very much in the traditional mould, came in June 1967 when Ocean bought a half share in Swire's China Navigation Company. A prime motive here was to broaden the base of shipping operations in anticipation of the probable diversion of Blue Funnel's ships from their traditional fields, and there were initially high hopes that China Navigation services might employ Blue Funnel ships and men. However, the main reason for the share-holding was to protect China Navigation's position in Far Eastern trades when containerisation took place, the link with Ocean giving China Navigation so-called 'grandfather rights' in the conference arrange-ments. At the same time Ocean was able to participate in a more

55 Sir Ronald Oliver Carless Swayne, Manager and Director, 1955–82

diversified range of activities through the expanding horizons of subsidiaries and associated companies, for China Navigation had a substantial interest in the Australian–Japan Container Line, while Straits, during 1969, became increasingly involved in offshore oil servicing.

These stepping-stones towards a more diversified structure lay across an aquatic route of course. Dry land, or even air, in the form of non-marine activities had scarcely been touched prior to 1969. From that year onwards, though, the momentum towards diversification showed a marked and conscious acceleration. From early 1970 Lindsay Alexander was in charge of a small group which investigated future diversification plans, and as a result of these deliberations a more coherent view of the future group structure emerged. In the ensuing years a number of additional marine-based activities were undertaken, the *Financial Times* in November 1976 rather inelegantly noting that 'between 1969 and 1972 Ocean vigorously elbowed its way into a variety of new shipping ventures'.[1] These ventures included an ore–oil carrier, tankers, bulkers, chemical carriers, offshore services, and an LNG carrier.

The various moves towards diversification were made in the light of

Ocean's great and historic resources of shipping management, technical expertise, experience in cargo-handling, and arguably the finest cluster of Masters and seamen in the merchant navy (for it was known by 1967 that some 1500 officers and engineers would not be required in their traditional roles). With containerisation due to remove the traditional trades and operations it was natural for the Holt management to seek new paths which could build on the foundations already in place. Thus, new types of ships for new trades was an obvious avenue; a move which would both make use of Ocean's technical and management expertise and employ seamen diverted from traditional spheres. Then there were various marine-related ventures, which again could call on resources, contacts, and experience already developed. Air cargo handling was another natural development, as was the move to rationalise and improve such assets as overseas property.

The ore–oil carrier was the *Tantalus*, ordered from Nippon Kokan in 1970 and delivered in 1972. A joint enterprise with NYK, *Tantalus* had a deadweight tonnage of 215 680 tons and was driven by steam turbines and capable of 15½ knots. The first oil tanker, *Titan*, was built in Sweden for Ocean Titan, and delivered in 1970; the second, *Troilus*, arrived in 1974. Five bulk carriers were also ordered in 1970, each around 26 725 tons deadweight with speeds of 15½ knots. Two were delivered in 1972 and the remaining three in 1973, the bulkers being entered in two consortia, Scanscot trading in timber products and Atlantic Bulk Carriers trading on the open market.

The move to chemical carriers was a joint venture with P&O, Panocean Shipping and Trading being set up in 1969. As early as 1965 a joint feasibility study by the two companies had been set up to invest-igate the possibilities of a joint tanker service for the carriage of liquid chemicals and edible oils. The new company started operating in 1972 with three small tankers engaged in a monthly service carrying palm oil from South-east Asia to Europe. Panocean, after a modest start, ex-panded quickly and by the end of 1973 four new tankers carrying liquid chemicals and vegetable oils were in service, and tank storage terminals had been developed at Antwerp, Rotterdam and Liverpool.

Another joint enterprise at this time was developed with the Inchcape group. Ocean Inchcape was incorporated in January 1971, with a 60 per cent holding by Ocean and 40 per cent by Inchcape. This large-scale enterprise operated as a marine supply and servicing company for enterprises involved with undersea oil and other mineral exploration. Four ships for this company were ordered from Holland and were delivered in 1972.

To the varied marine and marine-based enterprises were added a growing number of non-marine ventures. One was Repcon, a company specialising in the repair of containers and their trailers. Repcon became a wholly-owned subsidiary in late 1970, and early in 1971 Ocean established Repcon International to operate overseas. By the end of 1972 some £1.7 million had been invested in facilities and development costs. Another development came in a new geographic area for Ocean, when a majority interest in Barnard and Sons was taken in 1969, a company specialising in tourist-related activities, with 'a useful mixture of merchandising and transport business in St Lucia with possibilities of extension in that island and elsewhere in the Caribbean', as Nicholson explained to shareholders. Expansion did, indeed, follow. The following year a controlling interest was taken in a similar enterprise in Barbados and in 1971 a stake was taken in a large Jamaican trading house. The Caribbean venture was peripheral both to Holts' traditional spheres of operations and resources of expertise and experience. The promising move was taken largely on the initiative of Kerry St Johnston, but never proved to have the capacity for expansion or profitability which the Board had hoped.

Property was another obvious source of land-based diversification, since Ocean already possessed valuable properties in Hong Kong, Singapore, and elsewhere. In 1968 Ocean became actively involved in planning the redevelopment of their magnificent harbour site in Singapore, and purchased adjacent buildings in preparation for reconstruction. In 1971 Ocean Properties was established in which Straits Steamship held a 75 per cent share, 'to erect one of the largest office blocks in the city on the site of our former building in Singapore', as shareholders were told. This vast project was completed in 1974. But while a new and lucrative property interest was being developed in Singapore, the valuable Kowloon property in Hong Kong was relinquished. The decision was made in 1971, and the lease on Holt's Wharf was sold for £9 million. The sale was certainly a rational one, for trades were about to be transferred to containers which would operate from a newly built container terminal. In view of the soaring land values which subsequently took place in booming Hong Kong Ocean's directors may well have looked back on this decision with regret.

Ocean had had links with air transport which went back to pre-war days through Mansfields and through various joint booking arrangements. A revival of this interest came at the beginning of 1968 when Ocean took a stake in Transglobe Airways, an operator of air charter services with particular emphasis on air freight. The original approach came from Barings, who strongly recommended acquisition. After only a few weeks the Board decided to acquire a stake, and shares were purchased in mid-February. That same year, upon the reorganisation of Bahamas Airways, managed by John Swire and Sons, Ocean took a 16 per cent holding through the China Navigation Company. Another, though very different, joint air venture with Swire came with the formation of McGregor Swire Air Services, also in 1968. Ocean, through McGregor, Gow and Holland, originally took a 70 per cent share in the operation, but by 1971 they had increased their holding to 90 per cent, with the remainder being held by John Swire and Sons, (Swire's holding being bought by Ocean in 1982). By this stage MSAS had overseas investments in Hong Kong, Singapore, Malaysia, Australia, and Japan, and at the end of 1972 was represented in 11 countries by which time Ocean had invested around £2 million in building up the enterprise. A related move was the purchase in 1970 of a 50 per cent holding in Clarkair International, a small-scale but well-established air-broking concern. Being concerned exclusively with broking, and nothing to do with cargo handling, this venture, like the Caribbean interests, did not fit easily into Ocean's self-perceived diversification scheme.

Unfortunately none of the new ventures, marine or non-marine, was without teething troubles, and some had to be extracted rather painfully from the corporate gum. Among the disappointments were the aviation ventures. Transglobe Airways went into liquidation in 1968 after less than a year as part of the Ocean group, with a loss of around £800 000. A similar loss was incurred when Bahamas Airways collapsed in 1971, the shareholders having been informed only a few months earlier that the venture 'should become a thoroughly profitable operation'. Ocean's experiment with roll-on roll-off car ferries was also unrewarding, and after a series of heavy losses the Board decided in 1972 that Seaway's ships and services on the Felixtowe to Scandinavia run were no longer competitive, and this operation was abandoned at the beginning of 1973. Tanker operations, too, were a bitter disappointment in view of the high hopes entertained at the end of the 1960s. Two of the three large tankers, *Troilus* and *Titan*, were sold in the spring of 1975, both at heavy losses, the *Troilus*, the company's biggest tanker, was sold at a net loss of £2 million. The jointly-owned China Navigation Company holding was eventually sold back to John Swire and Sons in 1975, having also failed to live up to expectations. The trading operations of China Navigation during Ocean's period of partnership were rarely profitable and Ocean's investment was marginally profitable only because of China Navigation's stake in Swire's buoyant Cathay Pacific.

Other ventures, even when eventually successful, sometimes experienced unexpected difficulties. For example, McGregor Swire Air Services, ultimately to become a major profit-earner for the group, went through a long period of substantial losses until it was able to turn the corner during 1976. Panocean too got off to an unfortunate start when in 1970 the company was forced to cancel its initial order for four chemical tankers from the troubled Cammel Laird yard. This was a sad episode, for the original orders had been for very fine advanced ships to be built by a yard anxious to diversify into such new ventures as its naval operations ran down. The decision to cancel the orders was not reached easily, and the two principals involved, Nicholson for Ocean and Sir Donald Anderson for P&O, may have been influenced by the prospects of substantial losses should Cammel Lairds become bankrupt, indicated to them by Canadian Pacific, who also had ships on order from the yard. There was some feeling in the Ocean Board that Nicholson and Anderson had been outmanoeuvred by Canadian Pacific (who got their own ships) in this affair. None the less, Panocean, together with MSAS, Ocean Inchcape, and a number of other new ventures begun in the years before the acquisition of Cory were to prove valuable investments for the Ocean group.

Undoubtedly the costliest failure in the new phase of diversification was the LNG (liquified natural gas) carrier *Nestor*. Ocean's interest in the carriage by sea of liquified natural gas, then in its infancy, took shape in the late 1960s and by 1969 Ocean had undertaken a lengthy investigation into such a project. At that stage it appeared that the potential demand for seaborne LNG was considerable, and Ocean's directors considered that the company had the necessary technical skills to launch such a development. A major problem at that stage appeared to be to secure a suitable building berth for constructing such a complex and technically advanced ship, for there were considered few shipbuilders capable of handling such a venture. Discussions took place between John Nicholson and representatives from Shell, who already had several LNG carriers engaged between Brunei and Japan, and who suggested that Ocean 'might be five years ahead of the demand'. But Ocean's concern not to lose one of the few possible building berths to a competitor led to the decision in 1970 to place an order with Chantiers de L'Atlantique in France as part of a joint project with the major Dutch shipping company Nederlandsche Scheepvaart Unie, who placed a similar order for the vessel *Gastor*. The Dutch ship was to be ready in 1976, Ocean's as late as 1977, but, as Lindsay Alexander told shareholders in 1972, 'early entry into this field should enable us to take advantage of major opportunities in due course'.

Already by the end of 1972, however, circumstances had changed radically. The main change was that the trade for which *Nestor* was destined, the shipment of liquified gas from Indonesia to the United States, was now threatened by an agreement for such trade between Algeria and the United States. But this was only one aspect of the problem. The oil price rise of 1973 hit the world energy market while the crucial American market proved to be impenetrable, the gas interests there shrouded in protective layers of regulatory bodies, patriotism, environmentalism, and general protectionist sentiment. If the United States was closed, opportunities in Europe and Japan were similarly inaccessible, for purchasers in both regions had long contracts with gas suppliers and it was impossible for a small independent carrier like Ocean to break into the chain. Even when the Algerians pulled out of their deal with the Americans this did not rebound to Ocean's advantage, for there were now some ten additional LNG ships on the scene competing for contracts. There must remain a question mark as to whether the Ocean Board, by this time deep in the new phase of reconstruction in the aftermath of the Cory acquisition, should, or could, have abandoned the project. Certainly the issue was raised at the board but the order was allowed to remain.

The *Nestor* was delivered on schedule in 1977. However by this time the cost, originally estimated at £25 million, had reached nearly £63 million. Meanwhile the fears which some had envisaged in 1972 proved justified. The ship was unable to be used for the trade envisaged, and despite protracted negotiations with the American charterers and the US Department of Energy, *Nestor* went into immediate lay up at Loch Striven in Scotland upon delivery. There she remained throughout the 1980s.

When reviewing the rather chequered experience of diversification prior to the acquisition of Cory, several points emerge with clarity. First, although any such newly acquired and venturesome portfolio might be expected to carry some prospective failures, avoidable mistakes were made. The air ventures were embarked upon rather impulsively, in part, as Nicholson said in a subsequent memorandum, due to 'our the anxiety to gain experience of air operations'. It should be added that some initiatives came very largely from Nicholson himself, who was too readily influenced by Swire's success with Cathay Pacific, and who assumed that Swire management of Bahamas Airways was sufficient guarantee of sucfcess. In the case of Transglobe a rather chagrined Nicholson told stockholders in 1969, 'This Company proved to be less well-based than we had been led to suppose, and it became evident that its recently acquired Canadian aircraft would, before long, be no match for even medium-sized jet machines in carrying affinity groups and inclusive tours and in the slower-growing freight business.' The problem as far as Ocean was concerned was partly a lack of experience and financial expertise in appriasing such projects, and partly the lack of a coherent policy to investigate appropriate avenues of expansion and diversification. A Board post mortem on the air operation failures led to moves to strengthen the company in both directions, and in 1969 Lindsay Alexander, together with Ken Wright and others, helped to set up Ocean's diversification programme. Among other troubled Ocean investments, McGregor Swire Air Services was a source of heavy losses. Accumulated losses were in the order of £3 million by the end of 1973, a substantial part of which was due to the heavy costs of establishing an ambitious and sophisticated world-wide enterprise. However mistakes were made in the organisation of the company in the early years, and a loss of £500 000 was put down to 'administrative losses' in 1973. The following year the management was reconstructed, but not until the end of 1975 did MSAS become a source of profit to Ocean. More arguably, as we have discussed already, the *Nestor* catastrophe could also be regarded as an avoidable mistake.

In retrospect, however, the Ocean Board can shoulder only a small proportion of the blame from the disappointments encountered during the early phase of diversification. The villain of the piece was quite simply a series of unforeseeable events, political and economic, which would have brought to nought the best-laid plans. The early 1970s were difficult times for British business. Rampant inflation coincided with serious labour unrest, while labour costs rose by leaps and bounds and there were periods of world economic recession. Labour troubles flared at different times, and were a constant source of anxiety, imposing, as they did, both delays and greater costs throughout Ocean's UK-based operations. In October 1973 came war between Israel and the Arab countries, a world oil crisis, and consequent soaring oil prices and world depression. Meanwhile the march of the container revolution had released a flood of redundant tonnage which inevitably depressed freight rates at a time when total trade was not expanding fast enough to take up the slack.

These various events could not but affect every corner of Ocean's enterprise, and some of the fledgling enterprises succumbed as a direct result of the disturbances. The tanker trades were particularly badly hit, first by generally low rates from 1970 (a result of over-building in the era of previous growth) at a time of rising costs, and then by the world oil crisis of 1973–4. The quadrupling of oil prices had two unfavourable consequences for Ocean. Total demand for oil was reduced, leading to an excess of tanker tonnage, and so to the loss-making sales of *Titan* and *Troilus* in 1975. Also, the economics of marine fuel costs reversed a trend towards very large bulk ships which had been a feature of the 1960s. The relative competitiveness of smaller, less thirsty, ships was improved, so damaging the prospects not only for the giant tankers in Ocean's fleet, but also those of OCL's large container ships which from 1974 faced fierce competition from non-conference operators of smaller vessels. At the same time significant improvements in marine diesel engines also altered the economics of shipping operations. Steamers, like *Titan* and *Troilus*, became uncompetitive, while the early generation OCL ships had eventually, in 1982, to be re-engined with diesel engines.

Despite the many and varied moves towards diversification it is appropriate to recall for how long the group's activities remained focused on traditional operations. The mainstay of activities and of profits remained rooted in conventional liner shipping and related activities into the early 1970s. Marine-related business (if we include OCL) remained the dominant feature of Ocean's profits. The first significant shift to land-based operations did not occur until the acquisition of Wm. Cory & Son in the spring of 1972. This was the year that containerisation of the Far Eastern trades was largely completed, the month of August seeing the containerisation of both the Singapore and Hong Kong trades. In terms of Ocean's history, here was a turning-point with a vengeance, a landmark and a sea-change at the same time. On the one hand the Darwinian-like move to land was a major step in a process to become increasingly dominant in the following years. If Ocean was still predominately a sea creature it was at any rate by now amphibious. And on the other hand the core of the old Blue Funnel trades, the China and Straits services had now passed beyond the realms of conventional cargo liners and beyond the exclusive realms of India Buildings too.

On the eve of the Cory acquisition shipping activity still dominated Ocean business. Some idea of the extent of this dominance can be gleaned from the figures of group turnover shown in Table 13.1, although turnover figures of themselves say nothing about profitability. As the table shows, in 1970 rather more than four-fifths of total turnover was provided by the main shipping lines.

Shipping operations also continued to produce most of the operating profit. Table 13.2 shows a breakdown of operating profit in the years prior to the Cory takeover. The table helps to bring into focus the dilemma facing Ocean in these difficult years, with both 1970 and 1971

Table 13.1 Ocean Group turnover, 1970 (£000)

Company	Turnover	%
Ocean/Mutual	30715	40.11
Glen	10802	14.10
NSMO	4007	5.23
Elders/Guinea Gulf	18260	23.85
Seaway	784	1.02
Ocean Port Services	5780	7.55
Rea Towing	672	0.88
India Buildings	339	0.44
McGregor, Gow	1007	1.31
MSAS	170	0.22
Repcon	160	0.21
Mansfield and subs.	1143	1.49
W. African Companies	1463	1.91
Meyer	476	0.62
Caribbean	817	1.06
Total	76595	100.00

Source: Ocean Archives: Directors' Minute Books.

56 Ocean Buildings, Singapore, c.1970

showing greatly reduced profits compared with 1969, and several component parts of the group recording substantial losses (though the
profits for 1969 were a record). With freight rates determined internationally, subject to intense competition and resistance from shippers
and some national governments, it is not surprising that the impetus to
diversify became irresistible. Taking the year 1970 again, the total
turnover for the main fleets was rather higher than in 1969, yet trading
profits were only one-third of their previous levels. Table 13.2 brings out
the poor result in 1970 and the significant recovery the year following
during a period of more settled industrial conditions.

The operating profits shown in the table were, of course, supplemented by the group's investment income which formed an uncomfort-
ably high proportion of total profitability. Over the three years 1969–71
investment income (including income from Odyssey Insurance) formed

Table 13.2   Group operating profits £(000), 1969–71

	1969	1970	1971
Ocean/Mutual	2958	832	1862
NSMO	237	14	144
Glen	772	243	−05
Elders/Guinea Gulf	1661	851	819
Titan	–	−10	640
Seaway	−65	−226	83
Odyssey Insurance	292	318	340
Ocean Port Services	76	148	148
Rea Towing	56	31	70
India Building	174	131	274
McGregor, Gow	66	177	−9
MSAS	−103	−453	−370
Repcon	–	−40	−22
Mansfield	50	74	29
West African Companies	199	209	205
Meyer	39	−30	−43
Caribbean	–	–	18
Sundry	5	−33	35
Adjustments	49	164	–
	6466	2417	4128*

*This was a preliminary figure; the final total was £4 306 000, but the broken-down totals were not available.
Source: Ocean Archives: Accounts.

no less than 39 per cent, 69 per cent, and 52 per cent respectively of total pre-tax profits. These high figures were partly a reflection of Ocean's large liquid reserves. It was, of course, traditional Ocean policy to finance new shipbuilding from internal accumulation, while the high levels of liquid reserves were also influenced by the effects of government policy, which restricted dividend payments, permitted subsidised shipbuilding loans, and produced large sums due to reduced taxation from the accumulated investment allowances for the *Priam* ships.

Ocean's high liquidity was also, and at root, a manifestation of the commercial environment confronting the company. One element was the faster than expected rundown of conventional cargo services with the onset of containerisation. Another was the failure of new marine endeavours to provide an adequate substitute for these conventional services. And a third was the disappointingly slow progress of non-marine ventures.

The situation naturally left Ocean unstable and vulnerable. The Board itself was not satisfied with the *status quo,* and was on the lookout for new areas of expansion, while Ocean's cash reserves and investments were always likely to invite takeover bids from outside. From the late

1960s, and especially, during the years 1970 and 1971, there were a number of market rumours of possible bidders for Ocean, with Rank, Inchcape, and European Ferries among those mentioned in the national press. Two events may be taken to exemplify the uncertain future which hung over the Ocean Board in this period, a proposed alignment with P&O in the late 1960s and a near takeover bid from Ranks in 1970.

The P&O episode was a curious one. For one thing the proposal appears to have come from the prospective bride, John Nicholson coming to an arrangement with Sir Donald Anderson, Chairman of P&O, before consulting the Ocean Board. The scheme was partly a by-product of a close friendship between the two chairmen which had developed over the years. Nicholson and Anderson had, of course, worked closely together in a number of arenas, such as conference affairs, containerisation, and various national shipping organisations. Sir Donald Anderson, one of the great figures in the modern British shipping industry, was Chairman of the Far Eastern Freight Conference from 1938 until 1955, an unmatched tenure, and Chairman and Managing Director of P&O from 1960 until 1971. During 1966, with the future direction of Ocean very much on his mind, Nicholson appears to have concluded that a large grouping was in his company's best interests. Possibly the move could be interpreted as a merger which, especially once Sir Donald departed as P&O chairman, would lead to the aggrandisement of Ocean and its board. Moreover Nicholson, as he neared retirement age (he was 55 in 1966) was naturally concerned about the succession at Ocean. Like many dominant personalities, he had less than full confidence in his most obvious successors, and thought that the management resources of P&O would be to Ocean's ultimate advantage.

Ocean's Board, when confronted with Nicholson's proposal in January 1967, were not persuaded. To some, the amalgamation was not a merger between equals, nor a takeover of P&O's management by Ocean's management, which was how Nicholson presented the case, but a simple takeover by P&O; a takeover, moreover, by a loosely organised concern with a number of interests, like passenger trades, which Ocean was well out of. The Board was virtually evenly split on the issue, but the strength of opposition thus revealed, especially since the opponents included most of the senior Board members, was sufficient to kill the scheme.

The future was once more called into question at the end of 1970 when Nicholson became aware of a possible approach from the Rank organisation. Rank was apparently anxious to utilise Ocean's large liquid resources but apparently prepared to leave the Ocean management intact to run its shipping business. Nicholson promptly contacted Ocean's merchant bankers, Barings, and arranged to meet the Rank chairman, John Davis. In the meantime intensive discussions in India Buildings took place, and

Ocean's directors made it clear that they would not countenance any such takeover, and would resign *en bloc* if a Rank bid were successful.

The meeting between the two sides took place in December, Nicholson being accompanied by John Greenwood, the Finance Director. Nicholson's objective was to head off a possible bid, and he suspected that Rank was lured by knowledge of Ocean's large liquid reserves disclosed to the Rochdale committee two years earlier. He therefore made it clear to John Davis and his colleagues that the Rochdale evidence was now hopelessly out of date, with most of Ocean's once-large cash reserves now spent or pledged. Moreover none of Nicholson's colleagues was prepared to work for Rank, and there was no common ground between the Rank and Ocean business. This last point was a telling one, since Davis had recently stated to his own shareholders that Rank's interests were planned to lie within a narrow and clearly defined range of activities (which did not include shipping). The implication of Nicholson's arguments was in effect to say to Rank: you do not want our ships; we have no cash for you; then what sort of case can you make for a takeover? The tactics were wholly successful, and within a few weeks of the meeting Rank informed Ocean that they intended to pursue the matter no further.

I

By the time of the Rank episode there had been some significant changes in Ocean's managerial structure. The events of 1965 had made change both desirable and urgent. One impulse had come from the acquisition of Elder Dempster's fleet. If Elder Dempster, with its forty or so ships, were to be fully integrated with Blue Funnel, how could an already overstretched management cope? Another spur to reorganisation was the prospect of further acquisitions. If these were to be made, especially outside the traditional realms of cargo liner business, how could they be grafted on to the Ocean organisation? The Managers' workloads, involving as they did a great deal of day-to-day business and undevolved responsibility, were increasing all the time as the business became more complex. Becoming a public company had increased the workload. So had the formation of OCL, which in 1965 absorbed much of the time of R. O. C. Swayne and R. S. MacTier, and so threw additional burdens on the others.

Longer-term considerations also suggested that changes were overdue. John Nicholson, as chairman, was becoming aware how outdated in a shifting world was the traditional paternalist mentality exemplified by Richard Holt between the wars. The Holt tradition relied very much upon ideas flowing downwards from the experienced elder statesmen.

Now it was time for ideas to flow upwards from the younger Managers and other senior employees. But this was difficult when the day-to-day workloads were detailed and pressing. By the early 1960s the Managers were also aware that the range of expertise necessary for shipping management was increasing all the time. The recent launch of Ocean as a public company had exposed it to public gaze, and had also exposed the company's lack of accounting and financial expertise at top levels.

These varied considerations, then, in 1965, impelled Ocean to examine its own managerial structure, and in the next four years the company underwent a number of significant changes. Perhaps nothing more typifies the sea-change which came over the enterprise between 1965 and 1972 than that in the former year, responding to the acquisition of Liner Holdings, the Managers themselves felt able to arrive at an appropriate system in their own inimitable and quarrelsome manner. As Nicholson said in 1965, 'we must find a way through this quite perplexing maze, and I hope we can do so without seeking outside advice'. But by 1971, as the problems posed by expansion and diversification grew more involved, an outside consulting group was brought in to chart the main directions the managerial structure should take.

In July 1965 Stewart MacTier presented a carefully prepared paper to his colleagues in which he confronted the problem put clearly by Swayne in his comment on MacTier's paper: 'Do we want to try to perpetuate our traditional form of direct management in all matters (family business style) or must we adopt a system of delegated management (big business style)?' MacTier's proposal (developed in conjunction with Geoffrey Ellerton, a retired senior East African administrator who had recently joined Elder Dempster, appointed with a view to his succeeding F. L. Lane as chairman) was to combine a main board, the Ocean Steam Ship Company, with overall responsibility, and to which two separate companies, Ocean Management Ltd, and Ocean Maintenence Ltd, each with directorial representation, would be responsible. George Holt was lukewarm to the whole concept of expansion and diversification, and argued against the integration of the Elder Dempster and Blue Funnel fleets. He proposed a management structure more in keeping with the family business traditions, with a series of Divisions each responsible to the main Board.

The end result of a lengthy period of further study and suggestions and counter-suggestions was that by the middle of 1966 the Directors, as they were termed after 1965, were ready to agree to the integration of the Ocean and Elder Dempster fleets. At the same time they set up a new management structure to take effect from July 1967. This was based very much on the original MacTier proposals. At the core of the new arrangement was a central board of the Ocean Steam Ship Company Ltd, with a Chairman, Sir John Nicholson, and ten managing directors who included the original Ocean Managers and the Chairmen of both Glen (W. H. McNeill) and Elder Dempster (F. L. Lane). Reporting to this Board were four fleet operators, Blue Funnel Line Ltd, Elder Dempster Lines Ltd, Glen Line Ltd, and NSMO, each with its own board of management. The last two companies, of course, were already in existence and were largely unaffected by the new arrangements. Blue Funnel and Elder Dempster were responsible for all the commercial operations of their respective fleets, for the financial results, and for the activities of the various agencies. Lindsay Alexander became Managing Director of Blue Funnel alongside Richard Hobhouse (also called Managing Director) who was his deputy. There were additionally six Blue Funnel directors, John Nicholson, George Holt, H. R. Disley, H. W. Garton, C. Lenox-Conyngham, and K. St Johnston, each charged with particular spheres of responsibility. Thus Nicholson and Alexander remained responsible for the Conference and Far Eastern departments, Holt for the Indonesian department, and Lenox-Conyngham for Australian affairs, and in none of these arrangements was there any change from the previous situation. Disley was in charge of Outward Freight, Garton of Inward Freight, while St Johnston took charge of the American department (though for a brief period only since he, like R. O. C. Swayne, was by now in London, fully engaged in the establishment of OCL), and Hobhouse and Lenox-Conyngham were responsible for Voyage Operations and Agency Accounts. Elder Dempster Lines was rather differently constituted. The Chairman was F. L. Lane with four Directors (Nicholson, P. Toosey, M. B. Glasier, and A. M. Bennett) having no executive responsibilities, these falling largely on A. E. Muirhead and G. J. Ellerton supported by an array of Assistant Directors.

Supplementing the fleet operating companies were two service companies, Ocean Management Services and Ocean Fleets. These new creations, drawing on personnel both from Blue Funnel and Elder Dempster, were to provide common services for the fleets. Thus Ocean Fleets (with George Holt and Glasier as Managing Directors) had a Marine Division, an Engineering Division, and a Naval Architects and Ship Repair Division, each headed by one or several directors. The Marine division looked after such matters as victualling (under H. B. Chrimes) and Personnel (under Julian Holt) and the Nautical Adviser's Department (Captain D. R. Jones); while Engineering (W. H. Falconer) and the Naval Architects' department (M. Meek) remained much as constituted before. There were in addition two non-executive directors, MacTier and Nicholson. The fourth body, Ocean Management Services was largely the secretarial, financial, legal, and accounting arm of the group, headed by Nicholson and Chrimes as Managing Directors and with four executive directors, J. Greenwood (secretarial and financial),

T. Kennan (accounts and salaries), H. N. Smythe (legal and shore staff), and Lindsay Alexander (cargo handling and operational research).

The opportunity was also taken to reorganise the Glasgow agency. Blue Funnel's agency in Glasgow had been handled since the 1880s by Roxburgh, Colin Scott and Co., a company which in 1947 had been acquired by Ocean. Another old-established Glasgow firm, P. Henderson and Co., were managers for the Henderson Line (part of Elder Dempster), as well as acting as agents in the New Zealand, West Indies, and Calcutta trades. Henderson Line had been brought into Ocean when Liner Holdings was acquired in 1965. Ocean therefore merged the two Glasgow agencies in October 1967, as Roxburgh, Henderson and Co. The new firm was the largest shipping agency business in Scotland, acting as agents for 21 lines using the port of Glasgow as well as acting for Blue Funnel services and carrying out the commercial management of the Henderson line to Burma. The new arrangements did not affect Blue Funnel's Australian services from Glasgow, which continued as before in the hands of Ocean's other Glasgow agency, Aitken, Lilburn and Co.

In announcing the impending changes to the Annual Meeting of shareholders early in 1967 Nicholson expressed the hope that the new arrangements would 'help to raise the efficiency and lower the cost of all our operations without devitalising any parts of the whole body'. To the staff, in more dramatic terms, and using phraseology reminiscent of Lawrence Holt, Nicholson commended the proposals 'in the conviction that they offer today the best opportunity for each of us to play his or her part in our endless adventure of service'.

A sad by-product of the reorganisation was the decision, announced concurrently, to end the name of Alfred Holt and Co., although the familiar A. H. symbol was to remain the emblem of the house flag of Blue Funnel ships. The thinking behind this decision, announced 'with infinite regret' was that even the rather tenuous logic that Alfred Holt and Co. had existed as 'managers' of the Ocean and China Mutual fleets, was now largely irrelevant. It was now more true to say that the Ocean Steam Ship Company were managers of the Blue Funnel fleet, Elder Dempster, Glen, and so on. It was expected that in time the company would simply be known as 'Ocean' rather than 'Holts'. Thus 'the company that never was' disappeared as quietly as it had arrived.

The management reorganisation of 1967 brought a much-needed element of rationality into the system. The provision of common services for the fleets, amalgamation of the Glasgow agencies, and the division of functions among the various executive directors was clearly likely to improve overall efficiency and produce a more coherent pattern of managerial workloads. At the same time the creation of new departmental titles, with an expanded array of managing directors, directors, assistant directors, departmental managers, and the like, gave to the group a more modern-looking image, a matter of growing significance in the commercial world of the 1960s, with the company's affairs increasingly in the public arena.

But the reorganisation did little to cope with Ocean's fundamental managerial dilemma of the 1960s, and the new structure, dramatic as it seemed at the time, was destined to last only a short time. The urgent problems of the 1960s, as we have seen, were to cope adequately with future diversification in the face of an inevitable run-down of traditional trades, lighten the workload on the Managers so that they could play a full part in ensuring a sensible programme of diversification, and, crucially, to provide a structure which would bring greater financial and accounting expertise at the highest levels of management. This latter consideration was of the highest significance when it came to appraising possible targets for takeover and integrating new acquisitions. Although the new structure did bring about some strengthening of Ocean's financial and accounting departments, the want of expertise in this area continued to be felt. In other respects, too, the new structure proved inadequate. Above all, the structure remained geared to the main-tenance of fleet operations, while the logic of Ocean's predicament in the age of containerisation suggested a run-down of these activities. Nowhere was there a suitable niche for new non-marine acquisitions, nor for the generation of new ideas as to possible fields for expansion. Unfortunately, too, all the managing directors carried considerable executive responsibilities, and hence a heavy workload which tended to increase from the consequence of technical changes, from the demands of OCL (absorbing the energies of Swayne and St Johnston), and from the calls of wider national shipping interests which at various times made considerable demands on Nicholson, MacTier and Alexander.

Small wonder, therefore, that by the middle of 1969, only two years after implementing the initial reorganisation, the directors began de-bating once more what form of restructuring was necessary. In a paper circulated at the end of December that year Alexander suggested that even 'to stay where we are', major changes were necessary. Agreeing with both Nicholson and St Johnston, Alexander argued that 'the degree of overstrain which is imposed on key people both at Board level and below is insupportable and must be relieved. Far too much of our present and future lies on the shoulders of a tiny handful of presently irreplaceable lives.' As a minimum, Alexander wanted two additional managing directors, one to be in charge of finance, and the other to take control of diversification plans. He wanted also the recruitment of three more accountants 'to add badly needed capacity in the financial and company analysis field', – an obvious reference to the recent misfortunes Ocean had experienced in some of their essays in diversification. Since Ocean could not remain where it was, and would certainly have to expand and diversify, Alexander considered that even more radical changes were necessary. He thought that 'we have reached the limits of our development under our present system in which all the Managing Directors have specific and onerous managerial as well as directorial duties. The body would become unmanageable if we were simply to add new Managing Directors as we acquired new activities.' One outcome was the appointment to the Ocean Board of David Elder, formerly head of Supply Operations, Economics and Programming with Shell. Elder joined the Board in 1971, and his experience was to prove invaluable in the restructuring necessary to absorb Cory within Ocean.

Well prior to the Cory takeover, however, a new company structure was taking place. The directors now felt that they would have to turn to outside help, and after making some improvements in budget control and strategic planning in 1971 they called in a well-known firm of management consultants, the Boston Consulting Group. By coincidence the work of this consulting firm was still in progress when the Cory takeover was completed in the spring of 1972, and the final report was able to recommend a structure appropriate to the very much enlarged organisation.

II

Sir John Nicholson retired in September 1971, after nearly forty years with Holts, fourteen of them as senior manager and Chairman. The succession had lain between R. O. C. Swayne and Lindsay Alexander, and for a long time the natural choice had appeared to be Swayne. Certainly Swayne's claims were in many ways persuasive. He was senior to Alexander, just, both in age and service with the company, although both had become Managers together in 1955. In fact Alexander had been appointed by Holts while at Oxford in 1947 because the company then needed a 'pair' for Swayne, the original appointment having left unexpectedly after only a few months. Swayne had shown the ability to combine charm of manner with natural intelligence and a willingness to assert his views where necessary in a sometimes awe-inspiring environment. His responsibilities within the company were wide-ranging, but he became particularly involved with the Australian side of Holts' affairs and this, in the era of containerisation, made him a key figure in both the preliminary stages and in the formation and running of OCL. Swayne's weakness was, perhaps, a certain disorganisation, exemplified by the story that the stationmaster at Liverpool station always kept a £10 note handy when 'Mr Swayne' would rush breathlessly and penniless for the London train. Lindsay Alexander was, as we saw earlier, another academically gifted 'crown prince' who had worked closely with Nicholson on Far Eastern trade generally. Alexander had been largely responsible for the more coherent approach to diversification, set in train in 1969, and he had also done much to modernise Holts' business networks in the wake of 'going public' in 1965. He was briefly Deputy Chairman in 1971, succeeding George Holt who retired in April of that year. Alexander was widely respected in shipping circles, and in 1976, five years after becoming chairman of Ocean, he was described by the *Financial Times* as 'closer to being a public figure than anyone else in the shipping industry'.[2]

The looming choice between Swayne and Alexander, not an easy one, was in fact overtaken by events at OCL. The container consortium by the beginning of 1969 needed someone at managing director level who had experience of the Australian situation. Swayne had been involved with OCL from the outset and was the obvious choice. In January of that year he became OCL's Deputy Chairman, and later Chairman in 1973, and Lindsay Alexander succeeded Sir John Nicholson in September 1971. H. B. Chrimes became Deputy Chairman in succession to Lindasy Alexander. Swayne and K. St Johnston, who had both been working full time with OCL since 1967, resigned from their remaining executive functions with Ocean.

14

Blue Funnel Contained: New Ships and New Enterprises

> The container system, whereby a large container is filled with smaller items of cargo before loading and emptied after discharge, holds promise.
>
> (*Shipbuilding and Shipping Record*, 1964)

'A hurricane from America has finally hit British shipping', recorded *The Economist* in January, 1966.[1] The hurricane was containerisation, a development destined to sweep aside Blue Funnel's conventional liner services in a timespan shorter than anyone could have foreseen. It is to Ocean's credit, recalling the pioneering traditions of the firm's founders, that when the hurricane headed for Britain's shores, Blue Funnel was well to the fore in Britain's enterprising response.

In retrospect containerisation can be viewed as very much a natural development in the evolution of merchant shipping, though why it came exactly when it did, and in the form it did, is another matter. The movement of cargoes in standardised containers had massive potential advantages over traditional break bulk methods. For one thing, great economies could be achieved in the handling of cargo, especially if the cargo could be loaded into the container as early as possible in the transport chain and unpacked again as near as possible to the final destination. This meant that at the dock terminal, apart from the handling of the actual container itself and loading it on to a lorry or railway wagon for its journey to an inland depot, the work of stevedores, tally clerks, warehousing, and so on could be dispensed with. Huge savings in time and labour costs were in prospect, while the problem of pilfering at the dockside, a constant source of anxiety, was also reduced. Shipping costs, too, could be lessened considerably. With rapid unloading of containers at the ports ships could spend a far larger proportion of their voyage times at sea, with commensurate increases in earnings. Moreover the quick discharge of cargo now broke at last the enduring constraint on the size of the cargo liner, so that economies of scale could be reaped. Thus the way was open for large container ships, with no more crew-members than traditional liners, yet carrying vastly more cargo per voyage.

356

Containerisation was not the only factor compelling changes in the pattern of Blue Funnel's traditional services in the 1960s. Competition, political circumstances and various other factors also led to a number of variations. Among several relatively minor changes to routes and services we may note the ending in 1969 of Blue Funnel calls at China's ports, after two decades of unequal struggle to make the trade pay. So ended, after more than one hundred years, the original link which had started it all. Not all was decline, for in the same year a new service was begun between Rotterdam and Thailand. But most significant were changes taking place in Blue Funnel's Java trades.

As a consequence of Indonesia's disputes with the Dutch, Dutch shipping was banned in 1960 from Indonesian ports. This affected Holts' operations directly in two ways. First, the NSMO ships had to be diverted to other trades, while British-flagged ships replaced those on the Europe–Java route. Second, Holts' long standing joint Java–New York service, run in conjunction with the two Dutch Mails lines and the Holland–America Line since 1919, had to be recast. New partners for the line were found in the Swedish East Asia Co. of Gothenburg, already partners with Holts in the De La Rama service. A new service, called the Malaya–Indonesia Line, was established, covering the former trade from the American Gulf and Atlantic ports to Malaya, Singapore, and Indonesia.

At the same time the opportunity was taken to rationalise services further. Two unprofitable westbound services were abandoned, the De La Rama outward service from New York and Kingston to Japan, Hong Kong, and the Philippines, and also the homeward service from Indonesia and Malaya via Suez to New York. The remaining eastbound section of the De La Rama service, between the Philippines and New York was strengthened, in effect linking in New York with the Malaya–Indonesia Line to give a twice-monthly eastwards round-the-world service. Later, in 1963, confrontation between Malaysia and Indonesia saw the British now banned from Indonesian ports while Dutch-flagged ships, including NSMO, were allowed to return. The following year the Malaya–Indonesia Line service was renamed the Blue Sea Line, and in 1965 this name was applied also the De La Rama service. As part of the arrangements in 1960 Ocean had to make a number of changes to their American service agencies, among them being the severing of the long-standing connection with Booth American when the general agency in New York passed to Funch, Edye.

The Java–Europe trade was particularly vulnerable to the claims of the Indonesian government for a greater Indonesian share of conference allocations, and in 1970 the number of Blue Funnel ships on the route was reduced from 7 to 5. Lindsay Alexander, in his first annual report to shareholders in 1972, drew attention to the political nature of some of the attacks on the conference system made on behalf of the developing countries, and he wrote 'it remains to be seen whether the extremists will be prepared to run the risk of permanently destroying the delicate structure of world liner services'. Of course, by this time the delicate structure was in the process of being transformed by the liner companies themselves.

After containerisation of the Far Eastern trades in 1972, Blue Funnel's *Priam* ships were introduced into the Blue Sea Line, while the rapidly changing circumstances brought further changes in 1974. In that year Barber Blue Sea Line was formed by a merger between the Blue Sea Line Service and Barber Lines in Oslo. The management of the new service was established in Oslo, and Ocean held a 30 per cent stake. All eight *Priams* were allocated to Barber Blue Sea, and for a few years plied familiar routes through Suez and across the Pacific.

In the meantime containerisation of the Far East trades made it necessary to reduce the size of the Glen–McGregor organisation in London. While the Glen ships continued to be operated from London, management was centralised with Blue Funnel in Liverpool. The speedy run-down of conventional services soon precipitated further changes, and in 1974, the year which saw the formation of Barber Blue Sea, all remaining conventional services to the Far East operated by Blue Funnel, NSMO, and Glen were combined with those of the Ben Line to form the Benocean joint service. Each company continued to man its own vessels, but the commercial management of the new service was centralised in Edinburgh, with a small marketing team based in India Buildings.

Before looking further at Ocean's role in the development of containerisation, and the impact of containerisation upon Ocean, it may be useful to put a few figures on the savings brought by containerisation so that the extent of the revolution can be glimpsed. In the Australian trades prior to containerisation Blue Funnel vessels typically spent 50 to 60 per cent of their time in port. Incidentally only 15 per cent or so of this time was spent in actual cargo working; the rest was weekends, night-time, waiting for clearances, and other periods of inactivity. The new container ships, when working to their maximum efficiency, would spend less than 20 per cent of their time in port. With this sort of productivity increase, coupled with the larger size and speed of the container vessels, one container ship could do the work of at least eight conventional ships. Only nine container ships were needed to carry the cargo of 75 of the 80 or 90 conventional ships which had hitherto carried the trade between Australia and Europe (the remainder consisting of cargoes which were unsuitable for containers). The reductions in port handling charges could be even more dramatic. At a conventional berth around 1970 it took about 100 men seven or eight hours to handle

1200 tons of cargo. At a container terminal twenty men could handle 5000 tons of cargo in the same period. In practice the productivity increase brought by containers often fell short of the full potential, but the impact was none the less dramatic. The changed mix of operating costs was also revolutionary. Stevedoring and crew costs in conventional liner operations in the mid-1960s had formed around two-thirds of voyage costs. Containerisation brought the relative share of these items to less than 10 per cent, with significant consequences for labour relations and negotiations with the unions involved.

We have said that containerisation was a natural development. In fact attempts at 'unitising' cargoes so as to speed handling and ease storage have a long history. In the period immediately after the Second World War, with handling costs rising, various methods of unitising cargoes were tried. One method was palletisation, with the cargo fixed to standardised pallets which could be loaded and unloaded by fork-lift trucks. Another system was the so-called roll-on roll-off, or 'roro' vessel. Here the containers were loaded on chassis which could be driven or wheeled on and off the ship (exactly the same principle as a car ferry), thereby greatly speeding up loading and unloading in much the same way as the lift-on lift-off container ships.

There was no reason, of course, why containers could not be carried on a conventional cargo-liner, and they were. Ocean's *Priams* and some other Blue Funnel ships were, indeed, adapted to take cargo loaded in this way. But nothing like the full benefit of containerisation could be reaped. Too few containers could be carried while small numbers could in turn justify neither special unloading machinery on the ships or at the docks. The true container revolution lay in the construction of specially designed vessels and port facilities to handle the containers.

The question naturally arises as to why the container revolution in ocean cargo shipping came in the 1960s and not before. To answer this question it is useful to understand just how formidable a project the containerisation of deep-sea trades was, and, as a corollary, why for a long time it was considered that the new method was best suited to short-distance coastal trades between well-developed ports.

Ideally, for maximum economies to be obtained, containers should be full in both directions, the port terminals should be equipped with gantry cranes and other equipment for loading and off-loading, and the terminals themselves should have sufficient facilities for efficient onward movement of the containers so that bottlenecks did not occur. On the ships, the containers themselves and their accompanying structures used up around 20 per cent of potential cargo space (for roll-on roll-off containers the figure was around 40 per cent), a loss of capacity which became relatively more significant the longer the journey. In short distance trades, too, terminal charges made up a higher proportion of total charges, so that apparent gains were relatively greatest here. Moreover the commercial pressures to containerise initially, somewhat deceptively, made themselves felt in coastal trades. This was because of the very rapid development of road, and in some cases rail, transportation after the Second World War which, combined with rapidly rising costs of handling cargoes at the ports, led to a decline of coastal traffic in some developed countries. One way of countering increased costs at ports was to introduce various forms of unitisation to reduce handling charges.

Around the year 1960 Ocean, in common with a number of other shipping companies, began to investigate the possibilities of containerisation. A note in the minutes of a Manager's Meeting held in March 1960 recorded tersely that containers were 'probably not required substantially for 10 years, although experiments are needed'. A subsequent study instigated by John Nicholson concluded that long-distance trades could not be containerised. Indeed, R. T. Crake, head of the Cargo Handling Department, published a misleading paper in 1963 in which he argued that 3000 nautical miles was the effective limit for commercially viable containerisation. Nicholson himself admitted that his technical experts 'drew wrong conclusions', basing their judgments solely on the carriage of containers in existing Blue Funnel vessels. What no one at that stage realised was that the containerisation of deep-sea trades was certainly commercially viable if a totally new type of organisation was established, involving massive capital investment and an end to the control of the trade by separate shipping companies working within a closed conference system.

Another error made by Ocean's early studies was a considerable under-estimate of the proportion of trade which could be containerised. Some doubted whether wool could be containerised without serious deterioration, a major factor in the Australian trade where wool amounted to about one-third by volume and one-half by value of Australian exports in the early 1960s. By 1966, when OCL's plans were underway, it was estimated by Ocean that some 75 per cent of Australia's trade could be containerised, and 50 per cent of the Far Eastern trade. In fact, within a few years of containerisation, over 90 per cent of both trades were containerised. As a result the Dolphin Line, formed by the OCL members to handle the Australian residual cargoes in conventional ships, proved an unprofitable and short-lived venture. The under-estimates were significant not simply because they dimmed realisation of the full commercial possibilities of containerisation, but because they also led to undue optimism on the part of Ocean and other shipping companies as to the extent to which conventional services could survive in an age of containers.

The fundamental point was that, unlike most marine technological innovations, containerisation of a particular trade could not be adopted in a slow, gradual fashion. It needed to be instituted in a single operation. This was partly because container ships could carry so much more cargo than conventional liners that one container vessel could replace seven or eight conventional ones. But to operate at all the container operation needed huge investments in terminal and depot facilities. In order to make such investments worthwhile, and to provide a service of sufficient frequency to attract shippers, the container operation needed to absorb a large proportion of the existing trade from the outset. Some idea of the colossal capital involved can be seen from the cost of OCL's ships. The first six, built between 1967 and 1969 for the Australian trades, each capable of carrying some 1300 containers, cost a total of £30 million (almost double the cost of the eight Priams). The five larger and faster vessels (carrying 2300 containers), delivered in 1972, cost around £60 million. Each container cost around £1000 and each ship required around two and a half sets for simultaneous use on sea, on land, and in port (and their life-span was only one-third that of the ship). Gantry cranes at port terminals cost around £250 000 each. In all, one authoritative estimate concluded that in total throughout the world over £700 million was invested in the maritime sector of containerisation between 1966 and 1970.

These considerations raised a number of fundamental problems for shipping companies like Ocean. For containerisation meant nothing less than a radical reorganisation of the whole structure of the shipping industry. The huge capital investment and high proportion of existing trades involved made inevitable the organisation of giant consortia to finance and organise the enterprise. These consortia necessarily had interests far beyond those of traditional shipping companies, being involved not simply with shipping operations but with terminal and depot facilities, inland transport enterprises, and so on. Moreover, in order to make fully effective use of terminal and other container facilities considerable standardisation and interlocking arrangements *between* rival users was also necessary. This was not simply because of the huge sums involved, but because containerisation meant the interference with so many established interests. It was inconceivable that other companies would stand idly by and watch their trade taken by the containers; hence consortia were necessary if only to head off crippling competition. Although we cannot enter into the subsequent history of containerisation, it should be added that some successful independent container lines operating outside the main consortia, such as Evergreen and Maersk, emerged as powerful forces in the 1970s.

It is now a matter of history that the container revolution swept the world in the decade after 1965 – a veritable hurricane as *The Economist* put it. In 1966 there were just five container lines operating, all of them based in the United States; by the end of 1967 there were 38 in various parts of the world, and in 1969 no fewer than 88 lines serving over 200 ports. These services, though, were nearly all operated with converted ships; only a handful of United States lines were running specially constructed cellular container vessels. The pattern was transformed in 1969 with the start of OCL's Australian service, running its six Bay class ships between Europe and Australia, and by 1973 a number of giant consortia were organising fully containerised services to ports throughout the world.

The origins of the container revolution have been the subject of some rather unrewarding speculation. The Greeks, after all, entered Troy in a container, and innumerable other historical precursors can be found. Landmarks in the pre-history of deep-sea containerisation certainly include the Korean War in the early 1950s, when America used 'Dravo boxes' to supply her large army. Subsequent experiments led in 1956 to the use of converted partly cellular ships in the trade between New York and Puerto Rico. This service was pioneered by the Sea-Land company, the creation of an enterprising individual, Malcolm McLean. In 1963 Sea-Land extended the service to the West Coast, serving California and Los Angeles via the Panama Canal. By 1966 Sea-Land had 19 container ships in operation, as did the Matson Navigation Company. The latter company had pioneered the world's first oceanic container service between the US West Coast and Hawaii in 1958, the *Hawaiian Merchant* carrying 20 containers from San Francisco to Honolulu. Early in 1966 Sea-Land started a new trans-Atlantic service. On 23 April *Fairland*, with capacity for some 220 35ft containers, left New Jersey for Rotterdam, Bremen, and Grangemouth. Another crucially significant event took place in Australia in 1964, when the newly-formed Associated Steamships Company put the *Kooringa* into a weekly service between Fremantle and Melbourne, the world's first purpose-built fully cellular containership.

The Australian connection was important because Australia, for whom international maritime communications were vital, had long suffered from rising cargo-handling costs, port congestion, and disastrous industrial relations. The Australian government therefore took a close interest in the American and *Kooringa* experiments, and by the beginning of 1965 was considering seriously the prospects of containerising Australia's main trades. Since Australia had no major national line of her own, the possibility was that the Australian government would turn to Sea-Land to inaugurate the service. Since Sea-Land were by this time transporting vast quantities of war supplies westwards across the Pacific to Vietnam, American entry into a major Blue Funnel region seemed an imminent danger.

R. O. C. ('Ronnie', as he was always called) Swayne, the Ocean Manager in charge of the Australian trade, was in close contact with these events. Moreover he had been in charge of measures to achieve possible crew reductions, while Lindsay Alexander had overseen studies to improve the speed and reduce the costs of loading and unloading. In the spring of 1965 Swayne and Alexander drafted a memorandum to the other partners suggesting that the key factor in raising productivity lay in reducing the time ships spent in port, especially in the Australian trade. This in turn needed the mechanisation of cargo operations which could be achieved by containerisation. Almost simultaneously Ocean learned that Sea-Land, due to start their trans-Atlantic service early in 1966 with converted ships, were designing some very large and fast container ships with a capacity capable of covering the entire Atlantic trade. Obviously the containerisation movement was reaching a new stage. Moreover the trans-Atlantic developments could not but affect Blue Funnel's major trades, since there was a likelihood of over-tonnaging in the Atlantic with redundant vessels seeking employment elsewhere. In the words of a Blue Funnel department manager:

> we concluded that Sea-Land would be extremely likely to push westwards across the Pacific with whatever tonnage they found impossible to trade profitably across the North Atlantic. They would rely on the same inland network of trucks and depots in USA, already had the necessary Pacific coast terminals, and were experienced in pivot transhipment. If they did do this, they would at once be able to offer to shippers in Europe and Japan through carriage in containers by their own services across the Atlantic and Pacific Oceans. If, as seemed possible, this proved attractive to shippers because of the low carrying costs implied by the very large and fast vessels they were designing, we might find ourselves by 1970 under severe attack in one of our own main trades.

There was no ready-made solution to the wholly novel dilemma in which Ocean found itself placed. Swayne and Alexander immediately undertook a joint investigation of the implications of the Australian situation, and Alexander presented a paper to the Board in which he advocated the creation of a consortium to containerise the trade. The mechanised loading of containers would, of course, make possible large economies of scale, a situation which carried the danger of cut-throat competition among individual shipowners. A solution was the formation of a consortium. Members of the consortium would have shares in the group, which would necessarily from the outset conduct its operations on a large and comprehensive scale.

Even at this very early stage in Ocean's thinking two elements were at once apparent. First, the suggested consortium would mean a radical break with the company's traditional methods of operation and organisation. Blue Funnel's conventional liners would depart from the Australian trades, to be replaced by huge container ships in which Ocean would simply have a holding. Relations with the conference, with agents, with shippers and with a host of other bodies would need to be re-thought and recast. Second, containerisation itself could not be limited, and would spread inevitably to Far Eastern services. In other words, in the spring of 1965 the Ocean Steam Ship Company was looking down the barrel of a gun which was threatening to end a type of operation which the company had been pursuing for a century.

Understandably, John Nicholson, with his deep sense of the history of the company and its responsibility to its seafarers, was reluctant to accept the proposed move. But, like other hesitant Managers, he quickly became persuaded that there was no alternative to containerisation and that the consortium method was the best approach. Characteristically, Nicholson then threw himself wholeheartedly into the project and became one of the leading figures in the initiation and formation of OCL.

Affairs moved with some speed. On 4 May 1965, Nicholson sent a confidential letter to the P&O Chairman Sir Donald Anderson. In many ways P&O was a natural partner for Ocean in any joint container venture. The two companies had worked closely over many years in the Far Eastern and other conferences and relations had always been cordial. Moreover P&O was the major participant in the Australian trade, commanded large financial resources, and was the only other British line with a substantial stake in both the Australian and Far Eastern trades. On a personal level cooperation was facilitated by the long-standing friendship which existed between Nicholson and Anderson. The letter to the P&O Chairman outlined Ocean's thinking: the operation must '(a) be conducted on a sufficiently large scale to sustain what may be quite a large investment; (b) avoid competition in this field between individual Conference members'. Nicholson added, 'it therefore looks to us as though it might have to be undertaken by a combination of at least four if not all six of the largest Groups, probably with the addition of Ben and Blue Star who, if left outside, might be driven to competing in their own trades by undercutting the established through rates'.

Anderson replied immediately, making it clear that P&O also were actively considering some sort of container consortium, and adding, 'If we are to work with anyone else, it is Alfred Holt & Company whom we certainly would choose, and the sooner we have a talk about possible collaboration with you, the better'.

In a discussion which followed Nicholson pressed successfully for the inclusion in the consortium of both the British and Commonwealth Line and Furness Withy, and a joint approach to these companies was made. The four companies, which between them controlled two-thirds of the deep-sea general cargo-liner tonnage of the UK, agreed to form Overseas Containers Ltd. The four concerns had much in common, for all were public companies with similar responsibilities in terms of dividend performance and accountability. Ben was excluded, against Nicholson's wishes who recalled in a subsequent memorandum that 'I was fearful – unnecessarily as it turned out – of the harm they might do if excluded.' As well as Ben, the other British participants who were in the Australian trade and excluded from the consortium were Blue Star, Cunard, Ellermans, and Harrisons. In excluding these other conference members from OCL, Sir Frederic Harmer, Deputy Chairman of P&O, had argued persuasively that a larger grouping would run foul of anti-monopoly legislation. Predictably, these five agreed to form a consortium of their own, and in January 1966, four months after the formation of OCL, Associated Container Transport was incorporated. In practice, such was the nature of investment in terminal and other facilities that the two consortia worked closely together in the establishment of full containerisation.

The relationship between Ocean and OCL was necessarily a close one. Each of the four Principals took shares in the consortium in proportion to their shares in the trades containerised. Thus Ocean's share of the original capital of OCL was only 19 per cent, but this rose to 49 per cent in 1971 when the Far Eastern trade was containerised. As Ocean's leading investment, the fortunes of the consortium were naturally of great significance to the company, but Ocean itself was also to play a direct role in those fortunes. Two of Ocean's directors, Swayne and K. St Johnston, were seconded to OCL's board. Swayne later became Deputy Chairman of OCL in 1969 and Chairman from 1973 until 1982 when he was succeeded by St Johnston who had been an executive director since 1966.

The inaugural meeting of the OCL board was held in London on the first day of August, 1965, under the Chairmanship of P&O's Sir Andrew Crichton. A new epoch in shipping history had begun, but the progress of the new enterprise was destined to be far from smooth. Not surprisingly, in view of the novelty and complexity of operations, some mistakes were made and unexpected and uncontrollable difficulties encountered. The result was an anxious period of two or three years when significant losses were made until the foresight and confidence of Ocean and the other Principals began to pay dividends.

Fundamental to the eventual success of the new enterprise was of course the provision of appropriate ships, and here, at the frontiers of marine technology (since no container ships had yet been designed for deep-sea trade) Ocean played a crucial role. In April 1966 the OCL Board resolved 'that Alfred Holt and Co. should be asked to undertake the design study, put out the tenders, and supervise the building of any new tonnage which OCL might be authorised to order'. As a result of this work, in which MacTier, Taylor and Meek were closely involved in Liverpool, OCL agreed in June 1966 to order a number of large 32 000 SHP (shaft horse power) vessels with a speed of 21½ knots, each capable of carrying 1300 containers. The orders were placed the following year, five ships from German yards, and one, as a result of political pressure and against the wishes of the OCL board, from the Clydeside yard Fairfields. The German-built ships were all delivered on schedule, and the *Encounter Bay* inaugurated OCL's Australia service when the ship sailed from Rotterdam to Australia on 6 March 1969. A leading article in *The Times*, the day *Encounter Bay* was launched, paid tribute to 'three men of vision and courage' who had set Britain's container revolution in motion six years before – Sir Donald Anderson, Sir Nicholas Cayzer and Sir John Nicholson. However *Jervis Bay*, the Fairfield ship, was plagued by a series of delays which were exacerbated by the virtual bankruptcy and reconstruction of the yard in the middle of 1969. The situation at Fairfields naturally caused much concern, and in the spring of 1967 a Principals' Meeting decided to try to persuade Fairfields to withdraw from the contract should Sir Stewart MacTier, visiting Fairfields on 4 May, conclude that the target date could not be met satisfactorily. It was felt that OCL could not unilaterally withdraw from the contract 'in view of the repercussions on HMG.' But the contract remained and eventually *Jervis Bay* was delivered over one year late, in May 1970, and the shipbuilders accepted liability of £224 000 for late delivery.

The German built ships were not without technical problems which were mostly ironed out during the tortoise-paced construction of the *Jervis Bay*. Bay class vessels were found to suffer from severe vibration, and there were also problems with gearing, turbines, and boilers. These faults caused costly delays and repairs, and could only partly be attributed to the builders: the technical specifications, necessarily produced quickly and with no past experience of such vessels were also at fault. For well over a year Taylor, MacTier, Meek and others from Ocean were involved in trying to eradicate the faults. Taylor was asked to attend the OCL board on a number of occasions to discuss the 'recurrent troubles with the Bay Class vessels', as a minute in April 1970 recorded.

Technical troubles with the ships were compounded by expensive mistakes in the manufacture of the containers, many of which were found to taint the cargoes. The first frozen consignment landed at Fremantle in March 1969 was severely tainted with styrene used in the insulation. As a result all refrigerated containers had to be withdrawn from service, and OCL was obliged to carry large quantities of refrigerated cargo in conventional ships chartered at peak rates.

OCL's commercial performance was also damaged by a series of industrial disputes in the United Kingdom which forced drastic changes for the first eighteen months of operations. Worst of all was the refusal of workers at Tilbury to operate the costly new terminal (as part of a national dispute). Unable to use Tilbury, OCL was forced to use Rotterdam and Antwerp until 1970, with consequent disruption of schedules and greater costs. As a result of what Nicholson called 'the tragedy of Tilbury', and even more the need to charter refrigerated space (which accounted for around half of the losses and reduced earnings), OCL sustained losses of some £656,000 in the financial year 1968–9 and some £2 million in the year following.

Meanwhile, events were hastening the containerisation of the Far Eastern trade at a pace far quicker than originally envisaged. This trade was naturally of far greater direct significance to Ocean than the Australian trade, and even before the formation of OCL Ocean and P&O had begun a joint investigation of the prospects for containerising this trade. As early as January 1966, R. J. F. Taylor, together with Swayne and St Johnston, visited Sea-Land's Newark Terminal to investigate the potential of containerising the Far Eastern trade. Noting that 'Sea-Land even pack 40-gallon drums as well as light bulbs into containers', Taylor, in a subsequent memorandum, inferred 'that a very high percentage of our Far Eastern cargo could go via containers'. He also made cost comparisons between the conventional *Priam* ships and the expected costs of containers, and argued strongly in favour of containerising the Far Eastern trade. With the mushrooming of container schemes in various parts of the world, the issue became pressing. An OCL board minute of 20 December 1968, recorded that:

> Ocean Fleets and the P&O had agreed on the feasibility of containerising this trade and, bearing in mind the threat from Sea-Land, had recognised the urgency to develop in the first instance a container system in the Japan–Hong-Kong service. Existing Conference Lines were to be invited to join and an approach had been made to Ben Line which had resulted in their acceptance.

Once again Ocean played a leading role in the development of OCL's container fleet. Five second-generation Bay Class ships were ordered from German yards in 1969 for delivery in 1971–2, the new vessels being considerably larger and faster than the Australian ships. The vessels were each 36 000 deadweight tonnage, with a speed of 27 knots and a capacity to carry more than 2000 containers. The expected costs of these vessels were initially to be about DM 100 million (over £10 million, before sterling's big fall against the German mark in 1971), but the cost was subsequently increased by some 10 per cent on account of modifications to the original designs. As in the case of the earlier ships, Ocean was called upon to be the consultant for design and development, and Ocean Fleets was appointed by OCL to undertake the management, manning, and maintenance of the ships. In the design of these huge second-generations ships (each about the size of the QE2, and the largest vessels which could use the Panama Canal) OCL were considerably indebted to the outstanding skill and pioneering work of Marshall Meek, Ocean's Chief Naval Architect. Like the Australian ships, they were powered by steam turbines, but the Far Eastern vessels were fitted with twin screws, giving a total of 80 000 SHP from the two propellors. Four of the five ships were owned by Ocean, one by P&O, and the vessels were chartered permanently to OCL. Parallel with the decision to containerise the Far Eastern trade, OCL also took steps with Ocean and Swires through their joint ownership of China Navigation to containerise the trade between Australia and Japan. Australia Japan Container Line was formed at the end of 1968 with OCL as the principal shareholder and with R. O. C. Swayne a member of the original Board. St Johnston, who later joined the Board, also took a major part in the development of the new company. Two large Japanese-built vessels were in service from the autumn of 1970, and at the beginning of the following year an integrated service was started in tandem with the Japanese lines.

A feature of container shipping organisation was the development of international cooperation and joint services between hitherto rival groups. As early as 1966 Nicholson urged the other principals that such cooperation could both safeguard the conferences and encourage the achievement of economies of scale. Accordingly Nicholson and Swayne met leading German shipowners in Bavaria, and subsequent meetings were held between OCL representatives and the Japanese, French, and Dutch. These moves led to the establishment of a British–German–French consortium in the Australian trade (joined eventually by the Dutch and Italians) under Swayne's chairmanship, and operations began in 1970.

The Far Eastern Freight Conference included a number of national lines, and here early measures to achieve cooperation were also undertaken. Negotiations among the various conference members began in 1969, and in 1971 the giant three-nation Trio consortium was formed to operate an integrated container service between Europe and the Far East. The Trio consortium consisted of OCL, Ben Line Containers, Hapag-Lloyd (a merger in 1970 of Hamburg–Amerika and Norddeutscher Lloyd) and Mitsui OSK and Nippon Yusen Kaisha of Japan. Trio's service began in December 1971, OCL's *Tokyo Bay* making its inaugural journey to Japan the following March. The United Kingdom terminal was constructed by the British Transport Docks Board at Southampton. In 1972 another large international consortium, the Copenhagen-based Scan Dutch, was established to run an integrated Far Eastern container service.

OCL had envisaged that Europe's trade with Japan, Korea, Taiwan and Hong Kong would be containerised first, with the South-east Asian ports developed as a subsequent phase. In fact, though, it had already become apparent by the end of 1969 that the South-east Asian trades would have to be containerised more quickly if OCL were not to be forestalled by competitors. Accordingly all the main Far Eastern trades were containerised in the astonishingly short space of time between the beginning of 1972 and the end of 1973. Already by the end of 1973 over 50 per cent of this vast trade was containerised in both directions, and by the end of 1978 the proportion had risen to nearly 90 per cent. Starting with Japan, the service was extented to Singapore and Hong Kong in the autumn of 1972 and the following year extended further to Korea, peninsular Malaysia, and Taiwan. Later, feeder services were developed with the Philippines in 1975 and with Thailand in 1978.

Containerisation of the Far Eastern trade was indeed a massive operation, involving a total expenditure of well over £75 million in the ships, port facilities, and the multitude of associated investments. The venture also called for intricate and delicate negotiations between the various shipping companies, consortia, shippers, governments, port authorities, and a multiplicity of other interests involved. Much of the credit for the smooth transition from conventional to containerised services without over-tonnaging or violently fluctuating freight rates (as had happened in other containerised trades) was due in no small degree to the firm leadership of Sir John Nicholson, who was Chairman of the Far Eastern Freight Conference from October 1969 to June 1972, and who therefore guided the transition of containerisation through the crucial initial stages.

Containerisation of the Far Eastern trades opened a new chapter not simply for Ocean but for OCL as well. Hitherto the container enterprise had been little short of disastrous. The early woes of OCL were legion; the Tilbury shut-down, technical faults with the early ships, heavy losses on the Australia Japan service, losses on the conventional Dolphin Line residual services in Australia, and losses arising from Australian labour disputes. The floating of the British pound against the German mark in May 1971 caused a huge foreign exchange loss on the contracts with German shipbuilders of some £5 million, obliterating a trading profit for OCL of £480 000 for the year 1971–2. Such a loss naturally raises a question-mark against OCL's financial policies in incurring such large mark obligations without adequate provision for exchange movements, especially since German marks were not earned by OCL trading. The OCL Finance Director, Roger Cornwell, reported to his Board that he found dealings with the various finance directors of the Principals 'very unsatisfactory', and it was either difficult to obtain decisions, or such decisions were 'a compromise of the different requirements of the Principals'. Even this list of misfortunes is far from complete. In all, OCL's losses mounted to more than £15 million before the end of 1971, and already at the beginning of the year OCL had reduced its capital by £8 million, Ocean's share of the loss being around £1.6 million.

At the close of 1971 Ocean's shareholders might have been forgiven for thinking that the OCL light at the end of the tunnel was probably the light of the oncoming train. Yet for a number of reasons OCL's performance improved rapidly from around the beginning of 1972. The full operation of the Tilbury terminal and the excellent technical performances of the new Far East ships certainly helped, as did the predictable ironing-out of many early inefficiencies and bottlenecks which had naturally afflicted such a vast and novel enterprise. Another important factor was a concerted effort to raise freight rates which became urgent as costs continued to rise. Successive increases in 1971 and 1972 raised general rates by around 15 per cent, and for some particular freights, such as cars, by far more. A further favourable feature was a general improvement in world trade, 1973 being a particularly buoyant year. The result was a rapid move of OCL towards profitability. Lindsay Alexander was able to talk of a 'remarkable turnaround' in OCL's fortunes during 1972; in the following year OCL made a significant contribution to Ocean's profits, and in 1974 accounted for no less than 42 per cent of Ocean's profits during what Alexander called an 'outstanding year'. Pre-tax profits were well over £20 million in 1974 and 1975, and substantially greater in the succeeding two years. Although in proportional terms OCL's contribution to Ocean profits never again reached the 1974 levels, OCL proved a most valuable holding for Ocean. Certainly it justified to the full the faith of the Ocean Board which never wavered during the early difficult years.

I

In May, 1972, Wm Cory & Son Ltd became part of the Ocean group. In many respects the merger was an ideal one for both parties. While it would be untrue to say that before the takeover Ocean had money but no business prospects, Cory business prospects but no money, it was nonetheless the case that Ocean, deprived of their traditional business and with considerable liquid resources, was urgently looking for appropriate alternatives. In the circumstances of the early 1970s it was inconceivable that even the most promising offshoots of Ocean's early moves towards diversification held out much promise of reaching quickly significant size through internal growth. Even apart from profit considerations and the need to make effective use of Ocean's many

assets, to remain in the existing situation was to invite possible un-
welcome takeover bids as the recent Rank episode had shown. This
indeed, was the aspect seized upon by *The Economist*, calling Ocean's bid
for Cory 'a marriage of convenience' which would raise the new group's
market value hopefully beyond the reach of unwanted predators,
among whom *The Economist* noted Inchcape, European Ferries, and
Rank.[2]

Cory was likewise an old-established company recently undergoing
considerable changes. Rather like Ocean, Cory had found itself in the
post-war world with its business focused largely upon a great
nineteenth-century enterprise. For Ocean, the enterprise was cargo
liners. For Cory it was the coal trade. And coal distribution, like
conventional break bulk cargo carriers, had no prospects of long term
growth. Cory's pedigree was even longer than Ocean's, for the Cory
family had been London coal merchants in the eighteenth century, and
possibly even before. By the early 1830s the partnership of Cory and
Scott owned 23 barges and handled perhaps 5 per cent of the huge
volume of North East coal which came to London each year. In 1838 the
firm of Wm Cory and Son was established and by the mid-1870s was
handling more than half the seaborne coal arriving in London. Operations included a fleet of colliers, tugs, lighters, and barges. In 1896 Cory
became a public company, with a capital of £2 million and a turnover of
£5 million, and by this time owned 25 tugs and 1200 barges. A series of
major amalgamations in the 1890s extended Cory's wharves and coal-
handling installations in the Thames and Medway areas, and at this time
Cory's distinctive black diamond emblem was adopted.

It is unnecessary to trace Cory's growth in detail, of course, but we
should note that in the inter-war period the company expanded from its
essentially London-based coal distribution operations to embrace both
more regions, more countries, and more diversified business operations. As a result of a merger with Mercantile Lighterage in 1920 Wm
Cory became the largest single lighterage employer on the Thames,
while in subsequent years the firm acquired companies based in
Ipswich, Hull, and Glasgow. A new London headquarters was acquired
in 1922, Cory Buildings, built on the site of the old Ironmongers
Company which had been destroyed by zeppelins during the war. This
office was destined to remain the Cory headquarters for exactly half a
century, until sold by Ocean immediately following the merger.
Between the wars Cory developed a number of overseas interests,
including ventures in France, the United States, South America, and
South Africa. Cory now began to look beyond coal distribution, becoming increasingly involved in oil-distribution. The trend towards oil
gathered pace after the Second World War, when a programme of
vigorous expansion and modernisation was undertaken, first under

Lord Leathers (wartime Minister of War Transport and Chairman of Cory from 1945 to 1951) and then, especially, under his son the Hon. F. A. Leathers, later Lord Leathers. In 1953 the group became distributors for Shell-Mex and BP and oil distribution became of growing significance as central heating markets expanded. By the end of the 1950s the company had become a leader in the new sea-dredged aggregates industry, and its interests had widened considerably in numerous directions, in warehousing, transportation, forwarding agency business, ship towage, and shipbuilding. In 1958 Cory ordered three ocean-going tankers and four ore-carriers, and by this time the group consisted of nearly 100 companies, double the number a decade before.

In the late 1960s Cory strengthened its role in general merchandise distribution with the formation of Cory Distribution Services, a company developed to handle general distribution, especially food. By the time of the Ocean takeover Cory Distribution was operating a £10 million enterprise of depots and handling facilities and dealing in a wide variety of merchandise in addition to foodstuffs. In summary, therefore, Cory had moved successively from coal, to oil, and to general distribution. While no major activity had been shed, and many others acquired, coal-based operations became of considerably less significance, though fuel distribution remained the largest component of turnover and profits, as Table 14.1 shows.

Cory's interests and profts were also widened through a large number of associated companies. These included a 50 per cent share in Smit and Cory International Port Towage (a company formed in 1970 with a Rotterdam firm to undertake harbour towage throughout the world); African Bitumen Emulsions (46 per cent), based in South Africa and owned jointly with Shell; Rea Ltd, (50 per cent), a large concern with a range of activities similar to Cory and which in turn had a 49 per cent holding in Rea Towing Ltd, jointly owned with Ocean; and Freight Services Holdings Ltd (29 per cent) another South-Africa-based company

Table 14.1   Wm. Cory & Son operations 1970–1 (year ending 31 March)

	% of turnover	Pre-tax Profits (£00)	%
Fuel distribution	81	861	37
Shipowning	5	648	28
Wharfage and lighterage	7	230	10
Storage and distribution	6	250	11
Finance and central service	1	345	15

Note: not including associated companies.
Source: Tilney and Co., *Report on Ocean Steam Ship Company* (1972).

formed in 1969 to incorporate Cory's varied South African interests in warehousing, forwarding, air cargo, and the like. These associated companies produced a significant share of Cory's total pre-tax profits as the Ocean deal approached. In 1971 the figure was 40 per cent, and in 1972 as much as 72 per cent, a figure inflated by large losses from the nascent Cory Distribution Services.

The range of businesses developed by Cory was in many respects complementary to Ocean's operations and assets. Not only was there a useful overlap in areas such as shipping and towing but, most important, Cory was based essentially on non-marine transport and distribution and hence strong in just those areas where Ocean was anxious to diversify. Moreover Cory was a suitably large company, large enough to bring the expanded concern protection from predators and with enough promising activities to absorb profitably Ocean's liquid resources. Interestingly, both Cory and Ocean had by the early 1970s found themselves vulnerable to unwanted takeover bids, and a merger between the two was an obvious avenue of protection. David Elder, who joined the Board from Shell in 1971, and Nicholas Barber, who had joined Holts in 1964 as one of the last crown princes, were made responsible for overseeing the integration of Cory with Ocean. That such integration was achieved with remarkable smoothness and good-will on both sides, against the expectations of many in the financial world, owed much to the work of both Barber and Elder.

The Cory takeover was accomplished only after a brief and intensive takeover battle. During March 1972 the Cory group was the subject of rival bids from Jessel Securities on the one hand and Court Line on the other. Barings suggested to Ocean that a bid from them for Cory might be welcome. Ocean, in the throes of its diversification planning had already considered a possible link with Cory, and a hurried Board meeting agreed that Ocean should enter the lists. When the Jessel and Court bids were announced Ocean was ready with a rival offer which was made on 29 March. By this time there had been contact with the Cory board, who viewed the initial bids with dismay. Jessel, who held some 15 per cent of Cory stock, made no secret of their determination to sell off Cory's marine operations if successful. The Cory board, reluctant to see a break-up of their business, was much happier to see a link with Ocean and immediately recommended the Ocean offer to shareholders. This first Ocean offer had valued Cory at around £46 million, which compared with the Court Line bid of around £40 million and Jessel's at slightly less. Court Line (who, under an arrangement with Jessel, would acquire Cory's marine business if Jessel were successful) now dropped out. New bids and counter-bids now took place and Ocean was compelled to raise its offer on 13 April and again on 8 May when finally, at a cost of nearly £57 million, Cory became part of the Ocean group.

Around half of the purchase price was paid in cash, the remainder through the issue of shares. We may compare this valuation of Cory with the market value of Ocean, which stood at about £130.3 million prior to the takeover. Cory thus added very substantially, more than one-third, to the size of the new group, and this single move played a major role in the metamorphosis of Ocean from sea to land.

Epilogue

The sale of OCL was the most difficult of all the decisions I had to take as Chairman – it was like selling the Company's traditions.

(W. N. Menzies-Wilson)

In 1973 Ocean Transport & Trading was still massively a marine-based enterprise. The fleet of 88 deep-sea ships still included 30 conventional liners belonging to the Blue Funnel fleets, as well as Glen and Elder Dempster ships, container ships, oil-tankers, bulk carriers, ore-carriers, chemical carriers, an ore–oil carrier, and offshore supply ships. Shipping then accounted for nearly half of total turnover, and 85 per cent of total profits. Still in 1976 some two-thirds of trading profits were derived from shipping, and the Board's plans envisaged that the share would not be much lower into the 1980s. Yet by 1986 shipping and marine services accounted for less than one-quarter of group turnover.

The march of containerisation, technology and a devastating depression which hit world shipping in the early 1980s took a heavy toll. Already in 1977 the last two ships owned by the Glen Line were sold (the last vessels with Glen names went the following year, one, ironically, to Ben). This was the year which saw the *Priams* leave the traditional Blue Funnel Service. Four were sold, one was chartered, another laid up (later to be placed on the Benocean service), and two sold to the China Navigation Company. By April 1978 the Blue Funnel fleet consisted of only ten ships and when the Bangkok trade was transferred that year to OCL, only the Indonesia trade remained of the Benocean joint service. At this stage NSMO had only three vessels, and it was reluctantly decided to close down the Dutch company, so ending a memorable 87 year history. Declining fleet numbers led in 1981 to the closing of Odyssey works and the leasing to Liverpool City Council of the cadet training centre *Aulis*.

In 1982, a year which saw six Ocean ships serve in the Falklands War, the Benocean service was disbanded and the last conventional liners were sold, the two remaining Priam-class, *Phrontis* and *Patroclus*. In 1983 the large shareholding in the Straits Steamship Company was sold to Keppel Shipyard in Singapore. In 1986 followed the sale of Ocean's stake in OCL so closing the chapter on Ocean's involvement in the historic 'Far East Trade'. Now only Barber Blue Sea and the West African service remained as deep-sea shipping activities, while Barber Blue Sea and the embalmed *Nestor*, laid up in Scotland since her delivery in 1977, were the final preserves of Blue Funnel colours. These, too, hardly survived the decade.

Bit by bit, the recognisable shell of Alfred Holt & Co. fell away. The Company had epitomised British shipping and the port of Liverpool. By the 1980s it was losing touch with both. In 1987, with only six cargo vessels remaining in the group's fleet, and with shipping now accounting for less than 10 per cent of trading profits and 7 per cent of turnover, the decision was taken to 'flag-out' the ships to Isle of Man registry. Also in 1987 *Nestor* was put under option for sale. At the close of 1988 Ocean announced its withdrawal from the Barber Blue Sea service and the two Ocean remaining roll-on, roll-off ships were sold. This move meant the end of Blue Funnel shipping services and it brought with it the severing of Ocean's last commercial links with Swires. In 1989 came the sale of Elder Dempster and hence the end of Ocean's involvement with deep-sea shipping.

By the close of the decade the decline of shipping was such that those attending the annual dinners for retired seafarers must have felt as if they had gate-crashed upon an alien host; 'it is obvious that Ocean is no longer a shipping company', they were told at one dinner by the head of the Shipping and Marine Services Division.

As sea gave way to land, Liverpool gave way to London. Already in 1980, the year Sir Lindsay Alexander was succeeded by W. N. Menzies-Wilson as Ocean's Chairman, the London office became the group's headquarters. In 1988 India Buildings itself, founded by George Holt in 1833 and rebuilt twice, was sold to a London-based property company, though Ocean retained a lease for the occupation of the eighth floor.

The transformation of Ocean under Lindsay Alexander and his successors was not achieved without pain to the seafarers and home staff, a great many of whom became redundant, or indeed to the shareholders. In 1982 survival seemed threatened when a deep commercial depression coincided with the writing-off of over £41 million on the unhappy *Nestor* project. On several occasions the Board was concerned at the threat of hostile takeover bids. In 1986 a bid for Ocean arrived from overseas, and battle then was joined. Following the difficult times of the previous years, many observers expected the bid to succeed, but it failed because the majority of the company's shareholders saw that the reshaping of Ocean's activities and management promised well for the future under the existing Board. Ocean had survived one of the worst crises in its history and would after all be able to celebrate its 125th anniversary in 1990.

The Ocean of today is no longer the single activity of the Blue Funnel Line it is based instead on several of the new businesses developed during the seventies: OIL and MSAS, Cory Towage and Cory Waste, Panocean Storage and McGregor Cory. Yet the links with the past are strong. Ocean is still active at sea in its towage and offshore services markets, it is still handling goods and cargoes to 'Holt's Class' and it is still serving the freight markets especially to and from the Far East and Australia. And it is once more growing strongly and investing well, with good returns for its shareholders and with rewarding careers for its staff, who continue to operate within the traditions of excellence and quality service established by their liner shipping forbears.

As Ocean enters the 1990s, led since 1986 by a Chief Executive who came to Liverpool in 1964 as one of the last 'crown princes', the company's name stands as a reminder of its nineteenth-century maritime origins. The Blue Funnellers have departed. But from first to last they maintained a tradition worthy of the finest standards of British mercantile enterprise. Like the Homeric heroes of ancient Greece whose names the Line proudly bore, the ships may belong to the past but the legend will certainly live on.

Notes

1 Introduction: Ocean to Ocean

1. C. A. Middleton Smith, *The British in China and Far Eastern Trade* (New York: E. P. Dutton, 1919) p. 146.

2 A Unique Style of Management

1. Graham Turner, 'The Ship with Eight Captains', *The Observer*, 9 June 1968.
2. John Wyles, 'How Student Princes led Ocean away from the Sea', *The Financial Times*, 9 November 1976.
3. Richard Durning Holt, Speech at Liverpool Town Hall, 14 April 1939; reprinted in *Fifty Years: A Commemoration of the Association of Sir Richard D. Holt with the Blue Funnel Line, 1889–1939* (1939).
4. P. J. Waller, *Democracy and Sectarianism: A Political and Social History of Liverpool, 1868–1939* (Liverpool University Press, 1981) p. 277.
5. Speech by Major Roland Thornton at the Nestorians' Annual Dinner on 29 April 1965.
6. Sir John Nicholson, 'Some Problems and Personalities of A. H., 1930–1960', unpublished manuscript, 1976.
7. Beatrice Webb's Diary. These and subsequent references are from the manuscript diaries held in the British Library of Political and Economic Science.

3 Sailings and Services

1. W. S. Lindsay, *History of Merchant Shipping and Ancient Commerce, IV* (London, 1874) p. 436.
2. Captain E. W. C. Beggs, 'Voyage in the SS *Titan*'. I am grateful to Mr Julian Holt for drawing my attention to this document and making a copy available to me.
3. A. Jackson and C. E. Wurtzburg, *The History of Mansfield and Company, I, 1868–1924* (Singapore, 1952) p. 5.

4 Representatives Abroad

1. Middleton Smith, *The British in China*, p. 194.
2. S. Marriner and F. E. Hyde, *The Senior: John Samuel Swire, 1825–98* (Liverpool University Press, 1967) p. 61.
3. Ibid, p. 205.
4. K. G. Tregonning, *Home Port Singapore: A History of Straits Steamship Company Limited, 1890–1965* (Singapore: Oxford University Press for Straits Steamship Company Limited, 1967) p. 59.
5. Ibid, p. 55.

5 Eastward Ho! The Beginnings of Ocean Steam Ship

1. J. K. Fairbank *et al.*, *East Asia: The Modern Transformation* (London: Allen & Unwin, 1965) p. 340.

6 Combined Efforts: Holts and the Conference System

1. Marriner and Hyde, *The Senior*, pp. 135–6.
2. Ibid, p. 115.
3. Quoted in Eric Jennings, *Cargoes: A Centenary Story of the Far Eastern Freight Conference* (Singapore: Meridian Communications, 1980) p. 25.
4. *China Mail*, Hong Kong, 22 November, 1879, quoted in Jennings, *Cargoes*, p. 24.
5. D. J. Amos, *The Story of the Commonwealth Fleet of Steamers* (Adelaide: E. J. McAlister) p. 12.

7 The New Century: Profits and Perils

1. *Diary of a Liverpool Dock Worker: Memoirs of Mr Pridgeon, 1883–1940*, unpublished manuscript in the possession of his granddaughter, Miss Cathy Pridgeon, to whom I am indebted for permission to quote from the memoirs.
2. *The House of Dodwell: A Century of Achievement, 1859–1958* (London: Dodwell, 1958) p. 29.
3. Ibid, p. 30.
4. F. E. Hyde, *Blue Funnel: A History of Alfred Holt and Company of Liverpool, 1865–1914* (Liverpool University Press, 1957) pp. 180–1.
5. F. G. Hanham, *Report of Enquiry into Casual Labour in the Merseyside Area* (Liverpool, 1930).
6. Based on information communicated to the author by G. P. Holt.
7. See Brian Heathcote, 'Lawrence Holt', in *Strive* (Outward Bound magazine) winter 1971.
8. Based on information communicated to the author by Sir Lindsay Alexander.
9. *Diary of a Liverpool Dock Worker*.
10. Hyde, *Blue Funnel*, p. 175.
11. *The Economist*, 24 November 1917, p. 833.
12. Ibid.

8 After the War: An Uncertain World

1. A. G. Arnold, *Development of the Marine Diesel Engine in Messrs Alfred Holt and Company's Fleet, 1921–1956* (paper read before the Institute of Marine Engineers) p. 135.
2. Department of Overseas Trade, *Report on the Commercial, Industrial and Economic Situation in China for 1927* (London: HMSO, 1928) p. 27.
3. W. J. Moore, *Shanghai Century* (Ilfracombe: A. H. Stockwell, 1966).
4. Ibid, p. 70.
5. Ibid, pp. 70–1.

9 The 1930s: Collapse and Revival

1. Compton Mackenzie, *Realms of Silver: One Hundred Years of Banking in the East* (London: Routledge & Kegan Paul, 1954) p. 254.
2. Tregonning, *Home Port Singapore*, p. 161.
3. There is interesting information on competition between Glen and Blue Funnel ships in M. Cooper, 'McGregor Gow and the Glen Line: The Rise and Fall of a British Shipping Firm in the Far East Trade, 1870–1911', *Journal*

of Transport History, September 1989. I am grateful to Dr Cooper for letting
me see a copy of his paper prior to publication.

4. *The Times*, 12 March 1932.
   5. *The Economist*, 19 March 1932.
6. Quoted in E. Green and M. Moss, *A Business of National Importance: The Royal Mail Shipping Group, 1902–1937* (London: Methuen, 1982) pp. 185–6.

10   The Company at War

1. S. W. Roskill, *A Merchant Fleet in War: Alfred Holt and Co., 1939–1945* (London: Collins, 1962) p. 20.
2. Ibid.
3. I am grateful to Mr Julian Holt for the loan of this document.
4. R. H. Thornton, *British Shipping* (Cambridge University Press, 1939). A second edition was published in 1959.
5. Roskill, *A Merchant Fleet in War*, p. 30.

11   Picking up the Pieces

1. United Nations, *Economic Survey of Asia and the Far East, 1951* (New York: United Nations, 1952) p. xv.
2. *The Economist*, 20 February 1965.
3. A. McLellan, *The History of Mansfield and Company, Part II, 1920–1953* (Singapore, 1953) p. 9.
4. Ibid.

12   Into the 1960s: Calm before the Storm

1. *The Financial times*, 22 May 1964.

13   Organising a Group

1. *The Financial Times*, 9 November 1976.
2. Ibid.

14   Blue Funnel Contained: New Ships and New Enterprises

1. *The Economist*, 15 January 1966.
2. *The Economist*, 8 April 1972.

Sources and Bibliography

The major primary source used in the preparation of this volume has been the Ocean archive housed in the Record Office of the Liverpool Maritime Museum. This important archive contains minute books, ledgers, Managers' correspondence, Chairmen's reports to shareholders, voyage records, fleet books, and a great deal of additional material. Sadly, though, the bulk of Company material relating to the years before 1941 was lost as a result of the destruction of India Buildings in the war. Another important archival source is contained in the records of John Swire & Sons, deposited in the Library of the London School of Oriental and African Studies. Due to the courtesy of the companies concerned I have also been able to use additional material still retained in India Buildings in Liverpool and in Swire's head office in London. Many documents relating to the Holt family are deposited in the Record Office of Liverpool City Libraries (including Sir Richard Holt's unpublished diary). George Palmer Holt kindly made available to me the unpublished diaries of Alfred Holt and George Mansfield.

Anyone wishing to explore the history of Blue Funnel from published works must start with the major history of the company, F. E. Hyde, *Blue Funnel: A History of Alfred Holt and Company of Liverpool, 1865–1914* (Liverpool University Press, 1957). The history of the Blue Funnel and Glen fleets during the Second World War is told in S. W. Roskill, *A Merchant Fleet in War, Alfred Holt & Co., 1939–1945* (London: Collins, 1962). Also useful are two fleet lists, H. M. Le Fleming, *Ships of the Blue Funnel Line* (Southampton: Adlard Cotes, 1961) and Duncan Haws, *Merchant Fleets: Blue Funnel Line* (Torquay: TCL Publications, 1984). From time to time the Company has published brief historical booklets and other material, of which the best is *The Blue Funnel Line* (Liverpool: Alfred Holt and Company, 1938). Much information can also be obtained in *The Manchester Guardian Commercial*, 'The Blue Funnel Line', 24 June 1924. For particular aspects of Blue Funnel history the various issues of the house magazine *Staff Bulletin*, first published in 1947 (and later renamed *Ocean* and *Ocean Mail*) are invaluable. Technical details of the diesel engines can be found in A. G. Arnold, *Development of the Marine Diesel Engine in Messrs. Alfred Holt and Company's Fleet, 1921–1956* (Liverpool: A. Holt and Co., 1957). W. J. Moore, *Shanghai Century* (Ilfracombe: A. H. Stockwell, 1966) gives a vivid picture of conditions in the 1930s and 1940s at Holt's wharf in Shanghai.

The biography of John Swire by S. Marriner and F. E. Hyde, *The Senior, John Samuel Swire, 1825–98* (Liverpool University Press, 1967) contains much of interest about Alfred and Philip Holt, the Blue Funnel Line, and early conference operations. Also concerned with Swire history are Charles Drage, *Taikoo* (London, Constable, 1970); and *Fifty years of Shipbuilding and Repairing in the Far East* (Hong Kong: The Taikoo Dockyard and Engineering Co., 1953). Mansfields is less well served, although there is much of interest in Eric Jennings, *Mansfields: Transport and Distribution in South-East Asia* (Singapore: Meridian Communications, 1973) and in the two brief pamphlets A. Jackson and C. E. Wurtzburg, *The History of Mansfield and Company, Part I, 1868–1924* (Singapore: Mansfield and Co., 1952) and A. McLellan, *The History of Mansfield and Company, Part II, 1920–1953* (Singapore: Mansfield and Co., 1953).

Histories of other Holt agencies which have relevant material include D. R.

MacGregor, *The China Bird: The History of Captain Killick and One Hundred Years of Sail and Steam* (London: Chatto & Windus, 1961), and Dodwell and Co., *The House of Dodwell – A Century of Achievement* (London: Dodwell and Co., 1958).

For the component parts of Holts' shipping interests, by far the most detailed treatment is that given to Elder Dempster in P. N. Davies's excellent book *The Trade Makers: Elder Dempster in West Africa, 1852–1972* (London: Allen & Unwin, 1973). Another fine study is K. G. Tregonning, *Home Port Singapore: A History of Straits Steamship Company Limited, 1890–1965* (Singapore, Oxford University Press for Straits Steamship Company Limited, 1967). For the Glen and Shire Lines there is useful background information in two small booklets: E. P. Harnack, *Glen Line to the Orient* (London: Glen Line, 1970), which contains revised material originally published in *Sea Breezes*, and W. A. Laxon, *The Shire Line* (London: Glen Line, 1972). There is also a discussion of Holts' acquisition of Glen and Elder Dempster in E. Green and M. Moss, *A Business of National Importance: The Royal Mail Shipping Group, 1902–1937* (London: Methuen, 1982). For the earlier period a recent paper by M. Cooper 'McGregor Gow and the Glen Line: The Rise and Fall of a British Shipping Firm in the Far East Trade, 1870–1911', *Journal of Transport History* September 1989 contains much relevant material.

Surprisingly there is no biography of Alfred or Richard Durning Holt. F. E. Hyde's book, already mentioned, which contains much information on Alfred's life and background, and Alfred Holt's unpublished diary and *Fragmentary Autobiography* (1911) are the most helpful. Information on Richard Durning Holt, especially on his political career, can be found in P. J. Waller, *Democracy and Sectarianism: A Political and Social History of Liverpool, 1868–1939* (Liverpool University Press, 1981).

Appendix I
Ocean Managers and Directors, 1865–1973

Name	Joined	Appointed	Resigned
Alfred Holt	1865	1865	1904
Philip Holt	1865	1865	1897
Albert Crompton	1872	1882	1901
Maurice Llewelyn Davies	1889	1895	1913
Sir Richard Durning Holt	1889	1895	1941
George Holt	1889	1895	1912
William Clibbert Stapledon	1901	1901	1930
Henry Bell Wortley	1893	1908	1919
Lawrence Durning Holt	1904	1908	1953
Sir Charles Sydney Jones	1912	1912	1947 (died in office)
Sir John Richard Hobhouse	1912	1920	1957
Hon. Leonard Harrison Cripps	1919	1920	1944
Roland Hobhouse Thornton	1919	1929	1953
Sir John Norris Nicholson, Bart	1933	1944	1976 (Chairman until 1971)
Charles Douglas Storrs	1921	1944	1960
William Hugh Dickie	1920	1944	1954 (died in office)
George Palmer Holt	1933	1949	1971
Sir Reginald Stewart MacTier	1927	1955	1967
Eric Guard Price	1955	1955	1963
Sir Ronald Oliver Carless Swayne	1947	1955	1982
Sir John Lindsay Alexander	1948	1955	1986 (Chairman until 1980)
Henry Bertram Chrimes	1960	1963	1985
Sir Herbert Gladstone McDavid	1915	1963	1964
Richard Henry Hobhouse	1951	1963	1976
Kerry J St Johnston	1955	1963	1976
Robert Julian Faussitt Taylor	1957	1964	1979
Philip John Denton Toosey	From E.D.	1965	1970
William Hogg McNeill	1929	1966	1969
Malcolm Bruce Glasier	1933	1966	1970
Frank Laurence Lane	1962	1966	1972
Henrik Otto Karsten	From Glen	1969	1973
James Douglas Spooner	Non-exec.	1970	1972
John Greenwood	1935	1970	1978
David Renwick Elder	1971	1971	1980
Geoffrey James Ellerton	1965	1972	1980
Charles Denis Lenox-Conyngham	1960	1972	1985
William Napier Menzies-Wilson	1973	1973	1988
Colin David St Johnston	1970	1973	1988

Notes: E. D.: Elder Dempster

Capt. Malcolm Bruce Glasier resigned in 1951 on appointment as Director of Elder Dempster

W. H. McNeill joined Mansfields in 1932 and became chairman in 1961, and later chairman and managing director of Glen in 1964.

Appendix II
List of Co-owners, 1886

Dedication: To Alfred Holt with a Diamond Necklace for his wife in hearty and grateful recognition of the unselfish and able management of the Ocean Steam Ship Company from his Co-owners, 1886

I. S. Ainsworth
W. M. Ainsworth
L. R. Baily
William Cliff
R. N. Dale
Joshua Dixon
E. B. Drenning
William Eills
Mrs M. L. Foot
Mrs Chas Forget
Miss A. Forget
Miss F. Forget
Ferdinand Forget
H. Gardner
H. B. Gilmour
T. B. Gunston
Bernard Hall

Charles Harding
George Holt
R. D. Holt
W. D. Holt
Mrs W. D. Holt
William Johnson
Charles Langton
Miss G. Langton
Edwin Lawrence
James Maxwell
Hyslop Maxwell
William Maxwell
George Melly
Miss L. C. Melly
John Moore
Mrs I. B. Neilson
F. W. Percival

Richard Peyton
John Phillips
Benson Rathbone
Oswald Rathbone
P. H. Rathbone
E. W. Rayner
Miss I. Rodick
Turner Russell
I. A. Sellar
Miss Jemima Smith
R. T. Steele
C. M. Swire
J. S. Swire
William Swire
James Thornely
William Thornely

Appendix III
List of 'Correspondents', February 1902

UNITED KINGDOM

London	John Swire & Sons (8 Billiter Square)
Manchester *Birmingham*	George Simpson & Co.
Cardiff	Heard & Co.
Portland *Dartmouth* *Plymouth*	Fox, Sons & Co.
Dover	Hammond & Co.
Newcastle-on-Tyne	H. E.P. Adamson
Hull	Thos Wilson, Sons & Co. Ltd
Falmouth	G. C. Fox & Co.
Portsmouth	Garratt & Co.
Queenstown	Sterling & Co.
Scilly	T. Johns Buxton
Southampton	Keller, Wallis & Co.
Milford	H. Kelway
Glasgow	Colin Scott & Co. (94 Hope Street). Aitken, Lilburn & Co. (Direct Australian Steamers).
Newport	G. W. Jones, Heard & Co.
Swansea	Burgess & Co. Ltd
Maryport	Hine Brothers
Barrow	Jas Fisher & Sons

CONTINENT AND MEDITERRANEAN

Hamburg	Knöhr & Burchard Nfl. (Direct Australian Steamers) Hugo & van Emmerik (Direct General Steamers) Sühr & Classen (Transhipment Cargo)
Antwerp	Aug. Schmitz & Co. Ruys & Co. Gondrand Frères (Direct Australian Steamers)
Dunkirk	Do.
Amsterdam *Rotterdam*	Meyer & Co.
Bremen *Bremerhaven*	H. Dauelsberg
Havre *Paris*	J. M. Currie & Co.

Nantes	A. Maujot
St. Nazaire	Ev. Quirouard
Lisbon *Oporto*	} Garland, Laidley & Co.
Marseilles	Savon Frères
Genoa	Carlo Figoli
Gibraltar	Smith, Imossi & Co.
Algiers	Burke & Delacroix
Cadiz	Lecave & Co.
Tunis	F. Ravasini
Malta	Thomas C. Smith & Co.
Alexandria	R. J. Moss & Co.
Port Said *Suez*	} W. Stapledon & Sons
Tangier	Eugene Chappory
Constantinople	Joly & Segar

ASIA

Jeddah	C. R. Robinson & Co.
Perim	Perim Coal Co. Limited
Aden	Luke, Thomas & Co.
Shahah	Abdul Rahman bin Abdullah Bagarsh
Bombay	W. & A. Graham & Co.
Calicut *Cochin* *Tellicherry*	} Peirce, Leslie & Co.
Beypore	Jas. Dalziel
Colombo	Delmege, Forsyth & Co.
Point de Galle	E. Coates & Co.
Karical	Abdul Kadul & Co.
Madras	Wilson & Co.
Mangalore	A. J. Saldanha & Co.
Calcutta	{ Hoare, Miller & Co. Mackinnon, Mackenzie & Co.
Penang *Singapore*	} W. Mansfield & Co.
Bangkok	Windsor & Co.
Kudat	Albert W. Nieuveld
Deli	F. Kehding
Langkat	P. Sandel
Saigon	W. G. Hale & Co.
Rangoon *Moulmein*	} Bulloch Bros. & Co. Ltd
Manila *Cebu* *Iloilo*	} Smith, Bell & Co.
Sandakan	B. Sorentzen & Co.
Batavia	Maclaine, Watson & Co.
Indramayoe	Rupe & Colenbrander
Samarang	McNeill & Co.
Sourabaya	Fraser, Eaton & Co.

Macassar	Michael Stephens & Co.
Boeleling	Zorab, Mesrope & Co.
Padang	Haacke & Co.
Palembang	Richard Lange
Hong Kong	
Shanghai	
Foochow	
Swatow	
Chinkang	
Canton	
Kiukiang	
Wuhu	
Hankow	
Ichang	Butterfield & Swire
Nianking	
Tentsin	
Ningpo	
Newchwang	
Yokohama	
Kobe	
Amoy	
Chefoo	
Hoihow	Schomberg & Co.
Whampoa	Siemssen & Co.
Haiphong	A. R. Marty
Nagasaki	Holme, Ringer & Co.
Shimonoseki	Wuriu Shokwai
Moji	
Hakodate	Howell & Co.
Vladivostock	Crompton & Schwabe
Port Arthur	Clarkson & Co.
Wei-hai-wei	Lavers & Clark

AUSTRALIA

Onslow	James Clarke & Co.
Broome	Streeter & Co.
Derby	Adcock Bros
Shark's Bay	H. E. Bates
Fremantle	
Geraldton	
Carnarvon	
Cossack	Dalgety & Co. Ltd
Port Hedland	
Albany	

Direct Australian Steamers

QUEENSLAND	*Townsville*	W. Bartlam & Co.
	Brisbane Do.	Geo. Wills & Co.
		Thos. Brown & Sons Ltd
		(Outward consignments only)

N. S. WALES	Newcastle	R. B. Wallace
	Sydney	Gilchrist, Watt & Sanderson Ltd
VICTORIA	Melbourne	John Sanderson & Co.
	Geelong	Strachan, Murray & Shannon Proprietory Ltd
	Portland	W. P. Anderson & Co.
SOUTH AUSTRALIA	Adelaide	Geo. Wills & Co.
	Pt. Pirie	Do.
	Pt. Augusta	Young & Gordon
	Pt. Victor	G. H. Landseer
	Kingston	Geo. Wills & Co.
WESTERN AUSTRALIA	Albany	Do.
	Freemantle	Do.
TASMANIA	Hobart	C. Piesse & Co.
	Launceston	Ronald Gunn & Co.
NEW ZEALAND	Dunedin	Murray Roberts & Co.
	Napier	Do.
	Wellington	Do.
	Bluff, Lyttleton, Christchurch, and Auckland	Arrangements left in hands of Murray, Roberts & Co.

AMERICA, &c.

New York	Booth & Co.
San Francisco	Catton, Bell & Co.
New Orleans	Lucas E. Moore & Co.
Halifax	Pickford & Black
Galveston	William Parr & Co.
Norfolk	Castner, Curran & Bullitt
St John's, Newfoundland	Bowring Brothers & Co.
St Thomas	Lamb & Co.
Monte Video	C. R. Horne & Co.
Rio de Janeiro	Norton, Megaw & Co.
Madeira Las Palmas	Blandy Brothers & Co.
St Vincent	Wilson, Sons & Co. Ltd
Teneriffe	Hamilton & Co.
Dakar	Messageries Maritimes Cie
Capetown	James Searight & Co.
Mauritius	Richardson & Co.

Source: Ocean Steam Ship Co., *Notices to Masters*, February 1902.
Note: Spellings as in original.

Appendix IV
The *Phemius* Epic, 1932

Phemius: An Epic of the Sea
by Captain D. L. C. Evans, Master of SS Phemius

A true report to the best of my knowledge and memory of the hurricane experienced by SS '*Phemius*' between the 5th and 10th November and from 10th till 12th November on which latter date *Phemius* arrived Kingston Jamaica 1932.

4th November 1932

Noon. Wind fresh sea moderate. Weather reports gave a hurricane position to the Eastward travelling WNW of small intensity and area of 60 miles in width.

Some time same afternoon (see mate's log) after due consideration I altered course to S. 60° W. true, so as to give storm a wide berth. During the night barometer rose steadily.

5th November

6 *a.m.* Wind freshening barometer started to fall, continuing to fall. At about 7 a.m. I altered course to *South* true, as I wished to give all reefs to the westward a wide berth. Weather reports from Colon gave gentle breeze and fine weather. At about 8 a.m. I discussed the position with the chief officer and mentioned to him that if barometer continued to fall I would heave to when it reached 29.40. I received weather reports which justified my action, one being from SS *H. A. Rogers* stating that at midnight he had a barometer of 29.06 and had been hove to all night. His position from me was NNE about 120 miles. My barometer continuing to fall puzzled me, and at 9 a.m. barometer 29.55 I hove to ship's head NE & N, wind NE strong. The ship hove to without taking a drop of water on board.

Noon. Barometer falling continuously, wind increasing, sea rising. (All loose gear around decks had been secured before then.) I began to realise storm was not in position given or it had recurved.

2 *p.m.* By this time wind was of hurricane force with terrific squalls. Ship up to noon behaved splendidly but now fell off unmanageable. After 2 p.m. wind reached indescribable violence, and owing to flying spray visibility was nil. The first damage was the blowing away of bridge apron, loosening dodgers, and smashing of wheelhouse windows. All endeavours to heave to of no avail, and direction of sea could not be ascertained owing to 'no visibility'. Sea was very high judging from motion of ship.

3 *p.m.* Wind further increased to such a terrific force as to be beyond human conception. I had reduced speed at noon and now put engines on "slow" – as no power could cope with this terrific windforce. At this time during a few moments lull saw some wreckage close alongside and saw some dark objects flying in the air. These proved to be our No.2 hatch covers. Ordered pumps to be put on No.2 hold at once.

3.30 *p.m.* No. 6 hatch stripped, some hatch covered secured, remainder blown

389

overboard. Nothing could be done to No. 2, so ordered chief officer to try and save No. 6 hatch as much as possible. Some of the hatches were saved and what could be done was done. The wind-force was so great that they were unable to return for some hours, which gave me great concern.

No heavy water taken on board up to this time. The force of the wind carried an almost solid sheet of spray over the weather (port) bulwark. This striking on the tarpaulins, the edges over the hatch coamings, caused the tarpaulins to be carried away as if cut with a knife; the pieces remaining in the cleats and wedges afterwards proving this. The ship yawed in the sea to from 4 to 6 points bringing the wind from right aft to about 2 points abaft the port beam. The ship was rolling very heavily, quick jerky rolls which satisfied me as to stability. At about 6.30 p.m. the chief and 3rd officers returned, also midshipmen, having been unable to return before this time. About 6.55 p.m. I left the chief officer in charge on the bridge, as I wished to go down and see for myself how things were going on in the engine room. After considerable difficulty I got down, and found the engineers doing all they possibly could. I have good reason to believe that I inspired them to greater confidence, but pointed out that at all costs the pumps had to be kept going on No. 2 and 6 holds. Whilst I was going down the engine room a steam pipe burst in the stokehold – this was taken to be a super-heat pipe. Main steam started to go back, fumes and smoke came back into the engine room.

Undoubtedly these men were doing their utmost under most difficult circumstances. I left the engine room with a feeling of confidence that everything possible would be done, and everything was done. Whilst down there I ordered them to fill No. 2 port ballast tank, and before I left engine room they started running up the tank.

On reaching the deck I was caught in choking smoke, which I found came out through the fiddley. Then on looking up and around to see the cause of this, I discovered that the funnel had gone. I managed to reach the bridge and saw the chief officer, who had by this time noticed that funnel was gone. As it was then approaching 8 p.m., the funnel must have disappeared about 7.30 p.m. whilst I was down the engine room. The escaping steam proved to be from the whistle steam pipe. I now again left the chief officer in charge and went down to the saloon to see for myself how things were down there. I found the saloon awash and all the rooms on the starboard side flooded; undoubtedly a pitiful state of affairs. I noticed that words of encouragement were needed here, and that was done with good effect.

The chief engineer did everything in his power down below but conditions were so bad that every effort produced no results. I then returned to the bridge and tried to estimate the force of the wind. By putting my arm out into the wind the force of the spray striking the hand was agony, and for some minutes after remained numbed as if after a severe electric shock. My estimate was that the minimum force of the wind was 200 miles an hour. In all this and right through to the end I never gave up hope or allowed myself to be disheartened – although realising to the full our very serious position.

9 p.m. Steam failed, fires blew out with the back-draught and oil ran out through the furnace doors. Ship in total darkness. After this notes were taken at stated times as per mate's log book. These notes were taken with the aid of electric torches, the notes dictated mostly by myself and the chief officer scribbled on wireless forms which were then stowed away in one of the electric switch boxes in wheel house. These notes were written by the 3rd officer, and are the exact statements of the circumstances obtaining at the times they were taken. The constant almost solid sheet of spray, but no heavy water continued to blow over the open hatches throughout the night, and I estimated that ten tons an hour went down the open No. 2 hatch and five tons an hour down the Nos. 1 and 6 hatches. We were now without fresh water.

6th November

2 *a.m.* During temporary lull (centre) secured as well as possible Nos. 1, 2 and 6 hatches. Unable to obtain stores.

4 *a.m.* Hurricane centre passed over wind force 12. Very high dangerous sea continues throughout.

8 *a.m.* Wind force and very high sea continues. Emergency set working. (Original aerial went over with funnel.)

9.30 *a.m.* After due consideration sent out SOS requesting assistance.

10 *a.m.* Reply received from SS *Ariguani*. This helped me considerably to keep everyone hopeful. It was only in reply to his message 'Are you in a sinking condition?' that I gave out to him our true position with hatches gone, for by now, in my own mind I realised that I was being carried on by the hurricane, but still hoping that it would pass over quickly. I want to say this that every message I sent out from first to last was well considered, calm and deliberate.

Engineers continued throughout the day to try to raise steam, including donkey boiler – also emergency engine. All efforts fail. Now feeling the need of food and water. Burning oil running out of furnace doors, and put out by fire extinguishers. Soundings taken of holds where possible (records of these in mate's log). Throughout the night ship rolling and lurching very heavily at times taking a starboard roll of 38°; ship all this time carrying a heavy starboard list. I fully realised that our position was now very serious, as pumps could not be used.

Noon Conditions unchanged.

4 *p.m.* Temporary lull (centre) all hands again securing No. 2 hatch with awnings.

5.30 *p.m.* Gale again increasing and so through the night, barometer showing signs of rising only to fall again, and by midnight hurricane again increased to a fury, and continues unabated.

7th November

7 *a.m.* Nos. 2 and 5 hatches again stripped.

8 *a.m.* Short lull again, secured Nos. 2 and 6 hatches, but before 9 *a.m.* wind again blowing with violence. Rolling heavier and shipping heavy lee water, but so far no large volume entering hatches. All attempts at sending out further wireless messages fail, wireless room wet and set earthing from excessive water. This gave me further concern as I fully realised the anxiety of all concerned. Barometer again falling, ship rolling and lurching. Starboard gangway carried away, dragging rails with it, by lee water. Very dangerous rolls these. Braked quadrant. I now realised that I had an excellent chief officier, ready and anxious to carry out any order. I was, of course, closely observing the rest of the officers and crew, and at varying times made it my business to be amongst them to keep up their spirits.

Noon. Conditions unchanged – tremendous sea.

1 *p.m.* Observed ship to be in soundings and I realised we were passing over a reef. The sea was dangerously high. The wind was now SSE and blowing with terrific violence. Ship's head ENE making our heavy listed side (starboard) the

weather side. Oil in latrines had been got ready for this, the sea was terrific and threatened to engulf the ship. I gave the order to keep the oil going at all costs. One sea struck us and I estimated that one hundred tons went down No. 2 hatch and although that was the case I made everyone feel easier by stating to them that hardly any water went down. I gave the order to put on lifebelts. The time had now come when I had to decide how far I could go on keeping from my men this added seriousness of our position. I personally felt confident that, wind and sea as they then were, we would clear the rocks for I assumed that this shallow water was Serrana Bank. It proved later to be Serranilla Bank. I had ordered cast of the lead, which gave 60 fathoms, but as the force of the wind carried the lead straight out, this sounding I knew was unreliable. I must repeat and report on the sterling qualities of the chief officer. He was cool, calm and fearless, and any orders I gave he carried out without hesitation or question and both then and now I want to say how I appreciate and will always remember his confidence in myself.

This was the true test, for I realised that if anything was going to happen it would happen quickly. It will be realised that the desire for water and food was very much felt and showed itself but in a quiet manner amongst the crew. All possible were now stationed pouring oil and I call attention to the splendid conduct of the two midshipmen and the cadet. The difficulties of carrying along five-gallon drums of crude oil to the forward latrines were terrible, and although the crew, led and assisted by the chief officer to get things going, worked fairly well, I must with pleasure state that the three boys mentioned and the 3rd mate in this case and subsequently, are deserving of the highest praise I can give them. As for myself I frequently went along to give all encouragement possible. My orders were that four drums had all the time to be kept as a reserve, and these orders were carried out. All through the night the pouring of oil was carried out and although suffering from bruises, exhaustion, hunger and thirst the work was carried on without ceasing. The engineers carried on amidships and all worked hard. The 2nd officer superintended the after latrines oil supply. The effect of the oil was almost beyond belief. Towering seas tearing along direct for our exposed (listed) side crumpled up within ten feet of the ship's side, and although we could not entirely escape them, they landed on board in heavy volumes of dead water.

Without question, and in this report I make no random statements, the ship would have foundered had the pouring of oil not been carried on continuously.

Continued pouring oil throughout the day, wind and sea unabating till later in the afternoon. About 5 p.m. wind lulled (centre again) and again all hands endeavoured to secure No. 2 Hatch with awnings. Later took off No. 7 port tank door to try and obtain fresh water. Some water was procured but only a little. Tank door kept off to allow tank to fill from engine room and stokehold, as by this time there were six feet or more water in the engine room. Tank filled as vessel rolled and door replaced.

8th November

Midnight. Pouring oil continues with weather conditions unchanged. In addition to others mentioned I wish to mention the surgeon. He was always ready and willing to tackle and assist at anything. He assisted in getting fresh water, and was particularly helpful later in assisting the wireless operator and throughout he attended to constant wounds and bruises. His helpful willingness is certainly worth reporting. His loss in belongings was the greatest among the officers, for almost everything he possessed was destroyed.

8th November About 6.30 a.m. observed ship to be in deep water by colour of sea which gave me much relief. Barometer risen only to fall again throughout the day.

Noon. Strong gate very high sea.

About 2 p.m. wind lulled (again centre). All hands again securing No. 2 hatch with boards and awnings. As can be seen by this report the ship was repeatedly in the centre of the hurricane, and I have to mention that I had to overcome any depressing effect this gave me. The ship was overwhelmed with all kind of birds, also insects, both large and small, each time we reached the centre. Large birds of the Heron specie landed on the ship in such large numbers that I considered it possible that they added to our danger – as we had a dangerous enough list without adding to it. I do not wish to dwell on this but must mention things as they then were and be done with it. The decks, bridge, etc., were crowded with small birds and it was almost impossible to walk with crunching them under our feet. They landed fearlessly on our bodies wherever they could. Added to this was the oppressive heat, and the dead silence of the centre of a storm, with a depressing barometer. These were moments to test the nerves of any man!

I patiently waited for the blast that I knew had to come from an opposite quadrant and was glad when it came, carrying with it to destruction all those birds and insects. During these times I had seen sharks but kept that to myself – later when weather moderated they were of course seen by all.

About 3 p.m. wind shifted from SSE to SW again increasing to hurricane force. Later in the evening opened No. 6 port tank door to try and obtain fresh water – very little water was obtained. This tank also was allowed to fill from engine room. I now estimated that there was over a thousand tons of water in the holds and engine room. During all this time I would not allow the refrigerator to be opened, as once opened everything would be destroyed and I hoped to save this. What water and biscuits could be saved from damaged lifeboats was used. I think it was the morning of this day that some fruit and vegetables and cheese were taken from cooling chamber. The chief steward did all he possibly could and I must say that what he did later under difficulties is very much to his credit, to try and make every one comforatable. The wind blew very fiercely all the night, but barometer showed real signs of rising.

Midnight Barometer 28.72.

9th November

2.30 a.m. Barometer 29.10 and rising, also wind moderating.

8 a.m. Emergency aerial rigged, trying our utmost to gain communication. I realised the anxious time to all concerned during our enforced silence. I must now report the good work of the wireless operator. The difficulties he had to contend with were surmounted. The wireless room had to be dried up with blow lamps. The operator worked well and the surgeon was here of great assistance. Immediately on the first sign of a spark I ordered him (W/O) to keep on sending out our position, etc. We could not receive with our emergency set but by the aid of a private receiving set belonging to the 2nd Mate – he reported with some excitement that ships had heard us and at 9.30 *a.m.* we were in communication with SS *Killerig*. We were then bombarded with offers of assistance from all ships in the vicinity – to such an extent and wishing to conserve my emergency set, that I had to sent out to all ships 'Do not require immediate assistance.' 10 *a.m.* managed to get observation Longitude.

Noon. Obsn. position 18–12 N. 80.07 W having been carried N.8° W.209 miles from our estimated position by the hurricane. This day 3rd officer went down after store-room, with life belt and line attached. Store-rooms flooded but he was successful in clutching some tins as they washed to and fro. Wind had now moderated to fresh breeze but the high sea and heavy swell continued, the ship rolling very heavily, with very heavy pounding on starboard quarter. This rolling and pounding continued through the night. I was now more concerned over the motion of the ship than when she was in the storm – the lee lurches were dangerous owing to cargo on starboard side absorbing more water. I again gave orders to chief engineer to get emergency engine going or burst it in the attempt. But no results. During the storm the Chinese crew worked fairly well, but some of the time they were terror-stricken. However, when lulls came along they worked well and with a will, but the only ones I could mention were the lamptrimmer and 2nd and 3rd stewards. The 2nd steward particularly so.

10th November

4 a.m. Bar. 29.70 moderate N.W. wind and swell occasional heavy pounding starboard quarter. Clear and fine.

5 a.m . Firing rockets for benefit of *Killerig* making for us and due alongside daylight. Took star observation daybreak.

10 a.m. Sighted smoke bearing N.$_4$°W. which proved to be *Killerig*. I gave him his bearing and he bore down and arrived close to us about 10.45 a.m.

Killerig then circled round the ship from the portside crossing ahead and then rounding starboard quarter and stopped on the port quarter. The chief officer was on the bridge beside me and we watched every move. After some few minutes *Killerig*'s boat was launched and I counted fifteen men leave the *Killerig*'s side in the boat. These were tense moments for me and I at once ordered the chief officer with his men to 'stand by' the ladder and to allow no one on board except Capt. Tooker. I realised the gravity of these orders but was determined to see them carried out, as I resented such a large body of men boarding my ship. As the *Killerig*'s approached the ladder I hailed Capt. Tooker and ordered him to keep all his men in the boat as I would only allow he himself only on board.

11.15 a.m. Capt. Tooker Salvage Officer boarded. I received him on the lower bridge. His first words were – 'Captain I congratulate you'! I then invited him into my room having previously instructed the chief officer to be present. I asked him what the good old ship looked like. His reply was – 'You look pitiful'! then to business – He asked me what I required. I replied 'Lloyds Contract', to which he agreed. This was then duly signed by the two of us – destination Kingston. He now wanted to know the reason why I would not allow his men on board. I replied that no man boarded my ship without my permission. He then wanted to know whether there was any illness on board to justify me in my action. The reply was no. He then said that he insisted on having his men on board. I still stood my ground. He then threatened to tear up the contract. I replied that he could do as he pleased but I pointed out that my copy would not be torn up. I then asked him what bearing these men boarding would have on the towage question. He replied 'none whatever, that it was the usual procedure'. I was still adamant and pointed out that my men could still do all the work necessary. He then gave his solemn assurance that his men boarding would have no bearing on the matter at all. I was then satisfied and now instructed the chief officer to allow all the *Killerig*'s men to come on board. Preparations for towing were then immediately started. We had already made a bridle on each bow leading ten feet forward of the bow with our towing wire. This was altered. Finally a long manila tow rope was attached to the bridle and 0.33 *p.m.* this was completed.

0.55 *p.m.* Towing commenced. Ship was towed at a slow speed, which I desired and although occasionally ship sheered two or three points, I was satisfied that ship was towing nicely and with but little strain on the bits, the bridle being made fast to two sets of bits on each side. The 2nd and 3rd mates were put on four hour watches and our usual method of careful navigation carried out. Throughout this time the engineers were constantly working hard to try and get donkey boiler under steam and also emergency engine to work, but failing all the time to get any results. Eventually we received a tank of fresh water from *Killerig* and this was rationed out by a man stationed at the tank. Great care had to be taken as *Killerig* was also running short of water. We also received some hot food for all hands from the *Killerig* that evening. We now proposed to rig up a temporary funnel, using port stokehold ventilator – and after some time this was rigged, cement being obtained from *Killerig*.

11th November

Endeavouring to raise steam on donkey boiler with oil continued, but even with temporary funnel rigged failed. Owing to heavy list burning oil ran out of furnace and it was at this time that Salvage Officer objected to such risks being taken. Engineers working hard indeed and like everyone else on board they were in an exhausted condition. I used every effort to persuade them to carry on for at 8 *p.m.* all efforts at buring oil fuel was hopeless. We now proceeded to get furnaces ready for wood fuel. Grates were made from tubing. Again the engineers worked with energy through the night.

10.30 *a.m.* Steam on donkey boiler with wood fuel. Steering gear in order and vessel steering nicely.

Noon. Gentle breeze, smooth, fine and clear, and so through the night.

12th November

Record of times and bearings of lights, etc., as per mate's log.

2.30 *p.m.* Received Pratique

4 *p.m.* Anchored Kingston Harbour in 8 fathoms.

Killerig kept alongside till new donkey funnel erected. *Killerig* passed steam hose on board connecting same to our pumps and pumping engine room holds then began.

D. L. C. EVANS Master.
Source: The Java Gazette, vol. 1, no 10, April 1933

Appendix V
Letter from Alfred Holt and Co., to Midshipmen on Appointment, 21 December 1953

Dear Sir

On your appointment as a Midshipman in the Company's service it is well that you should understand what is expected of you. The Managers have accordingly set down the following outline of your duties and recommendations as to general conduct which they expect you to remember. You should, therefore, keep this letter carefully and re-read it at the beginning of each voyage.

The Master of your ship will give you such duties to perform as in his opinion will provide you with the best training for the profession of an officer in the Mercantile Marine. Every kind of duty will be given for the purpose of instilling knowledge and experience of a ship's work, and no such duty must be regarded by you as beneath your dignity. It is imperative that you should have thorough practical knowledge, or you will never have full confidence in yourself, or be able to inspire others with confidence in you. An officer should always be able to show the man how the job should be done; also it is necessary to know how much work may be rightly expected of a man. It is a good rule never to tell a man to do anything which you could not, or would not, do yourself. You are, therefore, expected to obey all orders given to you with alacrity. You have no right to spoil the life of your shipmates by grumbling, and you are lacking in manhood if you do not try honestly to enter into every duty with zeal and cheerfulness.

Take a pride in your uniform and do not allow yourself to acquire slovenly habits in your dress or person. Let your bearing on duty be always smart and attentive. Always address the officers of your ship with respect, asking questions at reasonable moments and without impertinence. Remember that your best chance of learning is from what others will tell you of the work and of their experience and that, if you show yourself troublesome or ungrateful for their help, they are hardly likely to take much interest in you. Above all, be truthful and straightforward.

In your dealings with the sailors and all other non-commissioned ratings be always civil and friendly but without unseemly familiarity. They can often teach you much, and it is part of your duty to learn by sympathetic understanding of their lives how to obtain from those under you willing and efficient service. Moreover the happiness of life aboard ship depends upon a healthy and natural comradeship all round and the Managers will expect your conduct among yourselves and toward the ship's company to show that this is realised by you. Remember that natives do not understand skylarking and resent deeply being struck or otherwise ill-treated. Be careful, therefore, that your conduct towards them, whether on board ship or ashore, is never familiar and always dignified.

Do not imagine that it is enough for you to learn the routine of a deck officer's work. If you desire a successful career, you must have a thorough knowledge of all things relating to the care of a ship and her cargo. The carriage of goods between the different parts of the world is the first and foremost function of the Mercantile Marine, and the Managers attach the highest importance to the skill and care shown by their officers in the safe and efficient performance of this duty. Omit no opportunity of learning all you can about naval architecture, not only from books but from a study of your own ship, the running of the engine-room and stokehold, and the operation of the wireless installation. If you really mean to reach the top of your profession, you should take care not to waste your spare time in reading trash but form a taste for scientific knowledge of the sea, of its living animals and plants, of meteorology and astronomy, and of good literature generally. All these pursuits will add to your prospects of success in your career and will, at the same time, render more interesting your life both to yourself and to those with whom you come in contact. Form the habit of keeping a notebook of your own.

Be careful to report promptly any symptoms of ill-health to the doctor or Master. Do not try to doctor yourself and do not conceal any ailment. In hot climates it is unwise to take ice-cold drinks or to purchase fruit, etc., from native hawkers. When wet, or after heavy perspiration, in the tropics it is most important to change completely your clothing. Make a rule of washing your whole body every day and be careful to have always at hand a change of clean clothing. You can easily make sure of this by learning to do a little washing yourself when on long passages. Abhundant physical exercise of all kinds will not only help to give you a manly bearing, but by keeping you free from illness will help you in your studies and in your enjoyment of a sailor's life.

Perfect eyesight is essential to a sailor. You will be unwise if you endanger it by excessive smoking. Avoid forming the habit of taking alcohol. No officer has a right to risk the safety of his ship and the lives of her crew and passengers by going on duty with faculties in the least imparied by drink. You cannot take too firm a resolve to avoid this risk. Your own self-respect and regard for your future health and happiness will prevent you likewise from becoming stained with other degrading vices which will cross your path in the low resorts of every great seaport. The Managers expect you to have a high ideal of pure and upright manhood.

Certain studies will be set for you to do during each voyage, but this does not mean that you are not expected to do more if you can. A log book has been prepared for the use of the Midshipmen, and a copy will be given to you at the beginning of each voyage. In this book you are expected to note down your daily observations through the voyage, whether at sea or in port, of the state of the sky, of the force and direction (relative and actual) of the wind, of the barometer, of the temperature (sea and air), and of many other natural phenomena which your own observation or the instruction of the Master may cause you to study. The purpose of this record is to train you in the habit of observation. It must be sent in to the office at the end of each voyage, together with your report, and the amount of care and intelligence shown therein will be duly recognized by the Managers. They desire you to give free play to your originality in the manner of arranging your record, and they encourage you to include in it any considered deductions which your observations may suggest to you. Do not collaborate with your comrades, but make and keep your own independent record.

At the conclusion of each voyage the Mangers wish you to report to them in writing what duties you have performed on board ship and how your time has been occupied in port. You should select each voyage for study one or more of the scientific books in your library (for the care of which the midshipmen will be held responsible) and mention the choice in your report to the Managers. You will be expected to answer questions on your work and studies if desired. The Managers expect this report to be truthful in every particular and will take a grave view of any discovery to the contrary. At some time between the time of arrival in the United Kingdom and of sailing on the next voyage the Managers will require you to report in person at this office. If the Managers do not consider that your conduct has been satisfactory or that you possess aptitude for the profession, they will not hesitate to ask your father (or guardian) to withdraw you from their service.

You must always wear uniform when ashore, unless on home leave. Holidays will be given to you, as far as possible, at the conclusion of each voyage, and arrangements will be notified to you on arrival at one of the ports of discharge in the United Kingdom.

Finally, never forget that the secret of happiness is a free and enthusiastic interest in your life's work. Have a pride, therefore, in your work and in the company, especially if it is your intention to throw in your life's lot with us. By so doing you will help best to make the Company's service acceptable to yourself and to all others who seek their livelihood in it. The Managers desire most earnestly that every loyal worker should find lasting satisfaction and happiness in his connection with the Company, and they expect you to contribute loyally to the attainment of this end.

Wishing you a very successful career, we are,

Yours very truly,

ALFRED HOLT & CO.

Index

GENERAL INDEX

'A' Class (Anchises Class) 259–60
Accounts Dept 249
Agamemnon Class 192, 204, 224
Agents and Merchants Ch. 4 *passim*
  Adamson Gilfillan 80
  Aitken Lilburn 56, 299, 352
  Assiut Shipping Agency 301
  Birley and Co. 61, 62
Booth American 55, 357
Borneo Co. 41, 68, 101
  Boustead and Co. 46, 55, 233
  Braund, F. W. 47
  Butterfield and Swire 4, 17, 43, 54,
    58–70, 114, 121–2, 184, 195, 233,
    240, 243, 246, 280, 281, 284, 294,
    302, 304
  China Ocean Shipping Agency 304
  'Correspondents' 55
  Dalgety and Co. 46
  Dare, G. J. and Co. 72
  Dent and Co. 74, 88
  Dodwell and Co. 55, 139
  Funch, Edye 295, 357
  Gellatly, Hankey, Sewell 123
  Gilchrist Watt 47
  Gow, Alan and Co. 227
  Gracie Beazley and Co. 217
  Gribble, Henry and Co. 31
  Guthrie and Co. 55
  Harper, Gilfillan and Co. 55
  Harrison and Crosfield 55
  Heard, Augustine and Co. 60
  Henderson and Co. 352
  Jardine Matheson 64, 86, 88, 233
  Killick Martin 54, 55, 70, 233
  Maclaine Watson 55, 90, 296
  Mansfield, Bogaardt and Co. 40
  Mansfield, George and Co. 72
  Mansfield, W. and Co. 4, 16, 36,
    38, 40, 43, 46, 50, 54–5, 58–9, 62,
    66, 71–2, 74–81, 143, 185, 225, 240,
    282, 284, 296, 298; in Second
    W. W. 276; after 1945 277
  McGregor, Gow and Co. 106, 227–8

  McGregor, Gow and Holland 70,
    228, 232–4, 277, 341
  Meyer, J. B. and Co. 42, 55, 56–8,
    233, 282
  Norris and Joyner 124, 228
  Orr, Dare and Co. 72
  Preston, Bruell and Co. 60, 61
  Roxburgh Scott 55
  Roxburgh, Colin Scott and Co. 56,
    299, 352
  Roxburgh, Henderson and Co. 352
  Roxburgh, John 56
  Sanderson, John 47
  Scott, C. W. 29
  Scott, Colin and Co. 55
  Smith, Bell 55
  Stapledon, William and Sons 54,
    58, 103, 233, 301
  Swire, John and Sons 55, 58–62,
    67, 69–70, 101, 121, 187, 280,
    340–1
  Syme and Co. 61, 75
  The Syndicate 47
  Trinder, Andersons, and Bethell
    44
  Wills, George and Co. 47
  Windsor and Co. 55, 101
American Shipping Act, 1916 134
'Aulis' Training School 271
Australian Oversea Transport
  Association 132

Bahamas Airways 340–1, 343
Bangkok Wharf Syndicate 68
Banks and Financial Institutions
  Bank of England 232
  Bank of Liverpool 112
  Baring Brothers 231, 340, 373
  Liverpool Union Bank 111–12
  Lloyds Bank 111–12, 187, 195, 205,
    232
  Martins Bank 112, 290
  Overend Gurney 74
Bellerophon Class 181

Birkenhead Docks 4, 26, 29, 35, 137, 148–9, 203, 295, 306
Blue Funnel Club, Newcastle 318
Blue Funnel Line (*see also* Trades and Services, Ocean S.S. Co., Cargoes)
  cruises 222
  fleet changes 27–8, 103–4, 111, 113 table, 116, 139, 141, 153, 158, 161, 181, 183, 191, 192, 201, 223, 235, 258–66, 264 table, 312, 320
  origins 6, 83, 91, 94
Blue Funnel Reunion Ass. 318
Boston Consulting Group 354
Boxer Rebellion 138
British Chamber of Shipping 290
British Perlit Iron Co. 163
Burmeister and Wain 192, 194, 326–7
Calchas Fire 314

Cargoes 31–7, 43, 137, 139, 149, 220, 295, 305
  beancake 64
  bulk liquids 34–5
  chilled beef 34–5, 134, 217, 282
  coffee 39
  coffins 36
  copra 215
  latex 34–5
  palm oil 34–5, 304
  rice 40–1
  rubber 31, 34–5, 46, 77, 176, 215, 219, 222, 304
  silk 30, 34
  sugar 34, 39, 111, 305
  tapioca 35
  tea 31, 35, 39, 50, 62, 85–6, 89–90, 97, 100–1, 111, 125
  textiles 31–4, 107, 110–11, 123, 125, 128–9, 136–8, 295, 305
  timber 35, 39, 49, 111
  tin 34, 39, 78, 80, 215, 219, 304
  tobacco 27, 31, 39–46, 71, 88, 131, 305
  wool 46–7, 111, 218, 305
Cargo Handling 35, 148–9, 302, 308–9, 358
Casual Labour 19, 142, 147, 149, 151
Cathay Pacific Airways 280, 341, 343
Cathcart Street Dock, Birkenhead 29
China National Airways 69
Chinese Compradores 114
Chinese Crews 5, 56, 75, 114, 169, 195, 224, 271, 296, 307, 309–10,

  rates of pay 114, 190, 307, in Second W. W. 240, recruitment 309
Chinese Emigrants 36, 44, 49, 65, 79, 86, 89, 280
Clarkair 341
Competition 101, 105–6, 107, 109, 119, 135, 183, 224, 266, 322–3, 325
Conference System, Conferences
  Ch. 6 *passim*, 16, 27, 59, 107, 123, 220
  Australia Conference 131–2, 217
  Batavia Freight Conference 131–2
  China–New York Conference 52, 227
  Criticisms 134–5
  Deli Freight Conference 131, 133
  Far Eastern Conference (China Conference) 32, 33 table, 65, 105, 107, 121–5, 129 table, 139, 142, 176, 184, 216, 220, 348, 368–9
  Foundation 60
  Indo-China 133
  Lancashire and Yorkshire Agreement 128–30, 135
  North Continental Pool 120, 129–30
  Rebate System 119, 121, 123, 134
  Royal Commission (1909)
  Siberian Conference 120, 142
  Straits Conference 120, 124–5, 127, 134
  Straits–US 132
  UK–Calcutta Conference 121
Containerisation 7, 134, 336, 339, Ch. 14 *passim.*
Correspondents (*see also* Agents) 55, App.III
Cory, Wm. and Son 7, 256, 333, 342, 344, 370–3
Cost of Ships 104, 174, 182, 183, 204, 261, 262 table, 272 table, 306, 322, 324
Crews (*see also* Chinese Crews, Masters, Training and Apprenticeship)
  manning levels 311 table
  pay and conditions 113, 233, 241, 257, 269, 307, 309–12, 337
  war casualties 167–70, 238–41, 245–6
'Crown Princes' 10, 18, 59, 71, 283, 289, 354, 373

Deck Boy Training School 271
Diesel Engines 148, 173, 189–194
Diomed Class 181
Dock Labourers Union 150–1
Dock Workers and Stevedores 45,
    149–51
Dunlop Co. 34
Durning Family 144

East India Company 85, 88, 89
Egerton Dock, Birkenhead 29
Elder Dempster Lines Holdings 231
Female employees 18
Financial returns *see* Ocean Steam
    Ship
First World War 27, 30, 51, 57, 80,
    136, 140, 141, 148, 155–70, 304
  losses and casualties 158, 165,
    167–70, 168 table
  requisitioning 166, 183
Freight Rates 28, 35, 36 table, 107,
    109, 113, 123 table, 125, 127, 129,
    131, 139, 155–7, 159, 162, 174, 203,
    217 table, 220, 225, 301–2, 305,
    307–8, 369

General Council of British Shipping
    283
Gladstone Docks, Liverpool 34, 200,
    314
Glenearn Class 234–5, 327
Glenlyon Class 224, 325
'Goal-Poster' Ships 137, 153
Great Depression Ch.9 *passim*, 53,
    155, 173, 209
Greek names of ships 5, 13

'H' Class (Helenus) 262, 269, 303,
    321–2
Holt Family 144
'Holt's-Class' 5, 14, 108, 168
Holt's Warehouse 200, 308

India Buildings 7, 10, 14–17, 24, 71,
    93, 146, 155, 180, 200, 203, 205,
    208, 225, 234, 236, 251, 274, 284,
    289, 299, 301, 310, 332, 348, 358
  first 18
  founded 89
  reconstruction after 1918 163, 195
  second 200
  third 273–4
  war damage 241, 247–8

Insurance 4, 14, 108, 156, 236, 243,
    251, 266, 312, 317
Inward Freight Dept. 18, 203

Jenkins Bros. 163, 164

Korean War 254–5, 280, 295, 300, 362

Lancashire and Yorkshire Agreement,
    1911 128–30, 135

Lester and Perkins 163–4, 194
Liberty ships 250, 261, 262, 321
Lifeboat Training School 271
Liner Holdings 256–7, 290, 319, 333,
    335–6
'Little White Fleet' 80
Liverpool Institute 15, 146
Liverpool Port Employers Ass. 290
Liverpool Steamship Owners' Ass. 24
LNG carrier 342–3
Local Services 26–7, 40–5

'M' Class (Menelaus) 311, 320–1
Malayan Airways 277
Manchester Chamber of Commerce
    139
Manchester Ship Canal 29
Marine Engineers 183, 366
Marine Superintendents 146, 181, 241
Masters 69, 97, 99, 108, 114, 145, 238,
    246, 257, 258, 269, 309, 316–19, 337
McGregor Swire Air Services (MSAS)
    70, 341, 343, 345, 347
Menestheus Fire 313–4
Mersey Docks and Harbour Board 21,
    109
Midshipmen's Dept. 269
Mitsui 'Fight' 284, 302
Morpeth Branch Dock, Birkenhead
    18, 29, 149

Nautical Adviser 181, 351
Naval Architects 24, 143, 148, 153,
    183, 203, 268, 273, 321, 365, 368
Neleus Class 321
Nestorian Ass. 318
Nonconformity 14, 18, 19, 92, 146, 155
North Eastern Marine Co. 163, 164,
    192

Ocean Building, Singapore 80, 165
Ocean Fleets 351, 367–8

Ocean Inchcape 339, 342
Ocean Management Services 351–2
Ocean Steam Ship Co.
  acquisitions in W. W. 1 163–5
  centenary 256, 333
  company reorganisation 352–4
  cost-cutting in 1960s 308–11
  cost-cutting in depression 209–10
  crisis of 1880s 84, 105, 109
  diversification 336–44
  financial policy 104, 108, 140, 203,
    205, 206, 210, 212, 236, 291
  financial results 109, 110 table, 113,
    140, 140 table, 141, 159, 160 table,
    161, 212–13, 221, 243, 299
  management structure and policy
    9–18, 47, 177, 179, 258, 284–7,
    335, 349–52, 354
  managers' meetings 24, 251, 322,
    324, 326, 360
  'new policy' 105, 112–14
  origins and growth Ch. 5 *passim*, 4,
    96, 99
  public quotation 256, 291, 333–5
  redundancies 209–10
  shareholders
Odyssey Insurance 317, 345–6
Odyssey Trading 299, 317
Odyssey Works 269, 271, 272, 308
Oil-Burning Engines 186
Opium 85, 88–9
Opium War 85–6
Outward Bound Schools 14, 147,
  271
Outward Freight Dept. 18
Overseas Containers Ltd (OCL) 71,
  128, 256, 284, 291, 333, 344, 349,
  355, 361, 366–7, 368–70
  formation 364–5

'P' Class (Peleus) 262, 269, 312,
  321–2
Panocean Shipping and Terminals
  71, 339, 341
Passenger Dept. 18, 248
Passenger Services 27, 37, 47–9, 101,
  135, 141, 209, 222, 303, 311
  'depassengerisation' 311
  fares 48, 222 table, 223
  far east 49, 53, 186, 308, 311
Pension Schemes 283, 284, 318–9
Phemius Disaster 316–7, App. IV
Philip Holt Trust 159, 318

Pilgrim Trade 30, 37–8, 49, 70, 79,
  103–4, 157, 195, 199, 207, 215,
  295–8
  fares 38–9
  numbers carried 38–9
  regulation 38–9
Pool Dept. 126
Priam and Glenalmond Classes
  309–12, 317, 321, 322–332, 347,
  358–9, 367
Pyrrhus Fire 314–15

Quarry Bank School 15
'Quarterdeck' 10, 18, 148, 289
Queen's Dock, Liverpool 29, 200

Rank Organisation 348, 370
Rea Towing Co. 163, 164, 345, 347
Repcon 339, 345, 347
Rochdale Committee 349
Royal Commission on Shipping Rings
  (1909) 32, 122, 127, 134, 139
Russo-Japanese War 138, 228
Seaway Ferries 336, 341, 345, 347
Second World War 25, 27, 57, 146,
    Ch. 10 *passim*
  losses and casualties 237–9, 329
    tabl3
Shareholders 7, 10, 11, 12, 13, 59, 69,
  334–5
Shareholders' Meetings 22, 24, 42, 46,
  97, 107, 110–11, 124, 126, 141, 142,
  143, 147, 154, 163, 167, 168, 182,
  184, 196, 203, 221, 258, 273, 305,
  327, 329, 343, 352
Shell Co. 156
Shipbuilders
  Brown, John 326–7, 329–30
  Caledon 163–4, 181, 195, 218, 235,
    262, 272, 330
  Cammel Laird 186, 341
  Chantiers de L'Atlantique 342
  Denny, Alexander 93
  Fairfields 325, 366
  Harland and Wolff 164, 228
  Hawthorn, Leslie 181, 267
  Leslie, Andrew and Co. 101, 104
  Mitsubishi 327, 329–30
  Nippon Kokan 339
  Scott and Co., Greenock 95–6,
    103–4, 163–4, 181, 183, 186, 262,
    272, 330
  Vickers 272, 320, 327, 329–32

Workman Clark 46, 181
Ship Losses and Damage (*see also* First *and* Second World War) 108, 158, 167–70, 237–9, 313–16, 313 table
Calchas 314
Menestheus 269
Phemius App. IV, 316–17
Pyrrhus 314–15
Shippers and Merchants 127, 132, 229, 307
Shippers' Councils 307
Shipping Controller 157, 165–7, 183
Shipping Lines and Companies (*see also* Ocean S.S.Co.)
Aberdeen 187
African S.S.Co. 228
Anchor 150
Ben Line 32, 106, 123, 129, 130, 135, 240, 323, 325, 358, 365, 367
Blue Star 130, 217, 224, 365
Booth S.Co. 105, 108, 142, 147
British and African S.N.Co. 228
British and Commonwealth Line 365
British India S.N.Co. 42, 81, 90, 106, 174
Brocklebank Line 228
Castle Line 105–6, 119, 121, 123, 125
China Mutual S.N.G. 28, 43, 49, 56, 107–9, 120, 124, 126–9, 140, 161, 167, 298, 313, 333; acquisition of, 23, 108, 138–9
China Navigation Co. 43–4, 59, 63–8, 70, 79–81, 121, 140, 142, 163, 280, 297, 336, 340–1
Cunard 10, 142, 212, 217, 365
DADG 53, 185
Danish East Asiatic Co. 190
De La Rama S.C. 294
Dolphin Line 360
Dutch Mails 56, 131, 185, 207, 357
East India Ocean S.S.Co. 43, 50, 79–80, 106, 108
Eastern Shipping Co. 80–1
Elder Dempster 56, 174–5, 209, 226–31, 241, 244, 332
Ellerman Line 32, 135, 365
Furness Withy 230, 231, 365
German East Africa Line 297
Glen Line 7, 28, 31, 56, 70, 105–6, 119–29, 174–5, 212, 224, 231–3,

261–2, 277, 284, 299, 300, 304, 321–4, 327, 333, 358; fleet 234; acquired by Holts 226–35
Greenshields, Cowie and Co. 162
Hamburg-Amerika 52, 143, 185, 224, 368
Henderson Line 336, 352
Ho Hong S.Co. 69, 81, 278
Holt, Alfred and Co. *see* Ocean Steam Ship Co.
Indra Line 52, 158, 162
Kim Seng Co. 78
Kingsin Line 124
Knight Line 158, 162
Koe Guan 80
KPM 41–2, 52, 56, 131, 207
KUK 106
Lamport and Holt 91, 93, 107, 155, 174–5
Matson Navigation Co. 362
MBK 183
Messageries Maritimes 90, 106, 123, 125, 132
Mitsubishi Mail S.S.Co. 106
Mitsui 284, 301–2
Mogul Line 107, 109, 124, 125, 128
Nederland (SMN) 41, 52
New Zealand Shipping Co. 246
North German Lloyd (NDL) 43, 50, 52, 79–80, 109, 124, 132, 143, 185, 368
NSMO ('OCEAAN') 42, 56–8, 108, 120, 131–2, 162, 167–8, 243, 258, 262, 275, 282, 299, 313, 345–6, 357
NYK 106, 124, 128, 129, 130, 183, 224, 302, 339, 368
OSK 188, 302, 368
P and O 10, 32, 61, 73, 81, 90, 103, 106, 123–5, 128, 130, 348, 364, 367–8
Pacific Mail S.Co. 105
Pacific S.N.G. 91, 95
Rathbone Lines 150
Rickmers 124
Rotterdam Lloyd 41, 52, 56, 131, 207
Royal Mail Group 155, 175, 205, 217, 226–31
Royden, Thomas and Co. 44
Russell and Co. 64, 87–8, 121
Sarawak S.S.Co. 81
Sea-Land 362, 367
Shanghai S.N.Co. 64

Shipping Lines and Companies (*cont.*)
Shaw Savill and Albion 217
Shire Line 31, 65, 105–6, 119, 121–5,
129, 156, 174, 175, 226–7, 231
Silver Line 264, 321
Skinner, Thomas and Co. 106
Straits Steamship Co. 156–7, 162,
186, 258, 276–8, 282, 284, 326
Swedish East Asia Co. 294, 357
United Baltic 232
Vestey Group 217–18
West Australia S.N.Co. 44, 218
White Star 132, 187, 217
Williamson, Charles and Co.
123
Ships' Husbands 145, 208, 284
Sino–Japanese War, 1937 195
Staff Bulletin 317
Steamship Dept. 18, 203, 249
Steamshipping 91
Stewart and Thompson 127, 139
Stock Exchange *see* Ocean S.S.Co.,
public quotation
Stowage Plans 35
Straits Trading Co. 43, 78
Strikes
Australia 199
Coal miners 185
Dockyards 142, 147, 151, 187
General strike 199
Hong Kong 199
Seamen 142, 194
Tilbury 367
Submarine Cable 77, 83

Taikoo Dockyard Co. 51, 67, 68, 161,
245, 280
Taikoo Sugar Refinery 67, 109, 280
Taiping Rebellion 86
Tientsin Lighter Co. 67, 140
Trades and Services (*see also* Pilgrim
Trade)
America 24, 26–7, 51, 161–2, 184,
194, 200, 203, 215, 304
Australia 24, 26, 34–5, 44, 47, 52,
56, 132, 138, 156, 165, 184, 187,
194, 200, 203, 215, 216–18, 255,
295, 301, 307, 358
Bangkok Rice Trade 40–1
Barber Blue Sea 70, 358

Benocean 358
Blue Sea 71, 357–8
cruises 53
De la Rama 294, 357
'East Coast' 53, 128–9, 134, 233–4,
323
Far Eastern Trade (China) 26–34,
39, 46, 50, 52, 96, 131, 136–8, 157,
161, 183–4, 187, 194, 201, 203, 213
table, 215, 220, 295 table, 307, 324;
containerised 360, 367–9
Java 26–7, 31, 43, 46, 56, 111, 116,
126, 131, 137–8, 157, 161, 176,
183–4, 187, 192, 194, 199, 201, 209,
215–16, 305, 357
Java–New York 131, 357
Malaysia–Indonesia 357
North China 28, 30–1
Pacific 24, 27, 31, 52, 55, 132, 138,
161–2, 184, 200, 215–16, 224
Polish 208
'Round the World' 52, 53, 132,
161
Straits 31–5, 37, 50, 123, 126,
129–31, 161, 176, 219
West African 256, 336
West Australia 27, 39, 44–7, 71, 79,
157, 218, 235, 312, 324–5
'West Coast' 128–30, 135, 323
Yangtze River 64–5, 86–7, 100, 121,
176, 197, 280
Training and Apprentices 268–70
Transglobe Airways 340–1, 343
Treaty of Peking, 1860 87, 96
Treaty of Tientsin, 1858 87
Trio Consortium 368
Troilus Class 321

Unions 149–50
Unitarianism *see* Nonconformity
United Africa Co. 229
University of Liverpool 154

Victoria Point Agreement 80–1
Victory Ships 250, 261, 262,
321
Victualling 209, 306, 312
Vietnam War 254, 255, 362
Vittoria Dock, Birkenhead 29, 137,
271

INDEX OF PLACES

Adelaide 47
Aden 104
Amoy 28, 37, 65, 86
Amsterdam 26, 30, 42, 52, 56–7, 133, 187
Antwerp 30, 143, 339, 367
Australia 21, 27, 44–5, 87, 90, 133, 135, 141, 148, 183, 187, 222–3, 300, 303, 326, 362
Avonmouth 30

Bangkok 40–1, 44, 55, 81, 87, 89, 101, 280
Barrow 30
Batavia (Djakarta) 31, 42, 44, 87, 137, 296
Belawan 41, 44, 140, 163; Holt property 140, 163
Belfast 29
Borneo 40, 43, 50, 51, 71, 78, 81, 88, 156, 278
Brazil 96
Bremen 30, 52, 133, 185
Brisbane 47, 247
Burma 88, 278, 352

Canada 49
Canton 85–6, 89
Cape Town 48
Cardiff 30
Ceylon 28, 90, 111, 222
Chefoo 31
Chemulpo 31
Cheribon 31
China 6, 27, 30, 34, 39, 63, 70, 77, 83–4, 88–91, 96, 106–7, 111, 121, 126–31, 137–8, 142–3, 148, 176, 219, 220–1, 295, 304, 316; economic conditions 85–8, 195–7, 253
Colombo 28, 31, 35
Chingkiang 87

Dairen (Dalny) 31, 55, 137, 197
Deli (*see also* Belawan) 41, 44, 46, 131

Egypt 93, 223

Foochow 62, 86, 96–7, 102
France 27
Fremantle 44, 326

Fusan 31

Gdynia 53, 58, 208, 223
Glasgow 30, 35, 47, 56, 107, 126, 129, 135, 138, 143, 301, 352

Hamburg 52, 133, 185
Hankow 60, 68, 84, 87, 100, 157; Holt property 140, 162, 196
Havre 30
Hong Kong 4, 28, 32–6, 41, 51–2, 58, 61–2, 66–9, 77, 86–7, 90, 96–9, 106, 117, 123–4, 131, 136–7, 140, 158, 161, 183, 195, 197, 222, 240, 245–6, 295, 304; Holt's Wharf 140, 162–3, 197, 245, 340

India 85, 88–90, 111
Indochina 87, 89, 304
Indonesia (Dutch East Indies) 30, 37, 41, 56, 78, 88, 90, 136, 176, 215, 255, 275, 295, 296, 305, 322, 326, 357; Holts property (*see also* Belawan) 275

Japan 27–30 34, 50, 60, 62–5, 68–9, 83, 109, 122–3, 126, 129–30, 136, 143, 148, 161, 173, 176, 183, 219, 220, 222, 255, 275, 295, 300, 305, 327, 368; opened to trade 89; earthquake 169; war with China 219, 222
Java 27, 30, 41, 46, 52, 90, 137, 147, 183, 185, 313
Jeddah 30, 37–8, 41, 103, 295

Kobe 62, 66, 106, 137, 227, 275
Kuala Lumpur 278
Korea 30, 67, 300, 369

Liverpool 11, 14, 28–9, 35, 74, 89, 92, 94, 96–7, 98, 102, 107, 124, 126, 129, 135, 138, 142–3, 145–6, 155, 173, 205, 295, 333; Docks 4, 13, 18–19, 26, 31, 35, 147–8, 149, 200; Liberals 20, 23; in Second World War 247
London 30, 35, 53, 122, 124–5, 128, 175, 233, 256, 323, 333–4

Macassar 31

Malaya (Malaysia) 37, 77–80, 87, 136,
    220, 253–4, 295, 300, 305, 324, 369
Malta 246
Manchuria 30, 62, 64–5
Manila 31, 35, 52, 66, 131, 161, 227,
    245, 301
Marseilles 29
Maryport 30
Mauritius 6, 73, 84, 96, 97
Melbourne 47, 71
Menad 41

Nagasaki 30, 106, 227
Netherlands 27, 136
Newport 30
New York 52, 183, 294
Ningpo 86

Oporto 168

Padang 42–4
Panama Canal 27, 52, 161, 166, 301
Peking 86
Penang 28, 35–6, 39, 41–3, 80–1, 88,
    90, 97, 99, 123, 137, 169, 227, 278
Philippines 31, 32, 52, 55, 65, 81, 131,
    254, 294, 304
Poland 53, 223
Portugal 35
Port Said 15, 103
Port Swettenham 31, 34–5, 55

Rangoon 81, 88
Rotterdam 30, 52, 56, 58, 133, 185,
    223, 301, 323, 367

Saigon 31, 133
Samerang 31, 137
Sandakan 41
Sarawak 278
Shanghai 28, 31–2, 35–6, 60, 62, 64,
    66, 67, 77, 86–7, 96, 97, 99, 123,
    131, 137, 142, 149, 196, 197, 208,
    232–3, 304–5; opening to trade 86;
    international settlement 86, 196;

conditions in 1945 274; Holt's
    Wharf 140, 162–3, 196, 275, 304
Siam (*see also* Thailand) 27, 50, 65, 72,
    78–80, 87, 89, 216
Singapore 4, 17, 27–8, 31, 34, 37,
    39–44, 50, 71, 72, 73, 75–80, 87,
    89–90, 97, 99, 106, 117, 123, 136,
    137, 156–7, 183, 195, 216, 220, 222,
    233, 240, 254, 274, 305, 324, 327;
    founding of 88; conditions after
    1945 254; Holt property 163, 165,
    196, 275, 340'
Sourabaya 31, 137
South Africa 21, 48–9, 170, 194,
    222
Southampton 34, 368
Suez Canal 4, 6, 26–8, 37, 52, 62, 76,
    84, 96, 98, 103–4, 106, 161, 174,
    221, 226, 301
Sumatra 27, 41, 55, 81, 156, 278;
    Holt's property (*see also* Belawan),
    140, 163, 196
Swansea 30, 143
Swatow 37, 63, 65, 87, 280
Sydney 47, 132

Taiwan 87, 183, 253, 255, 295, 369
Taku Bar 31, 67, 274
Ternate 41
Thailand (*see also* Siam) 278, 281, 304,
    357, 369
Tientsin 31, 62, 63, 67, 87, 137, 197
Tsingtao 137, 197
Turkey 39

United States 49, 173, 204, 342

Vancouver 52
Vladivostock 30, 31, 59, 63, 65, 77,
    143

West Indies 94–5, 340

Yokohama 27, 30, 63, 66, 106, 137,
    188, 227, 274–5

INDEX OF PERSONS

Adams, A. P. 78, 79
Adrian, P. C. 57
Ainsworth, Thomas 93, 94
Alexander, Lindsay 22, 117, 149, 257,

284–5, 287, 289, 335, 337, 342–3,
    351–5, 357, 363, 370
Anderson, Sir Donald 342, 348, 364–5
Anderson, E. 51, 79–80

Arnold, A. G. 192
Barber, N. C. F. 373
Beggs, Capt. E. 36
Bennett, A. M. 351
Berkhoff, K. 275
Blaase, J. 58
Bogaardt, Theodore Cornelius 40, 43, 50, 71, 77, 78, 79, 313
Bragg, Capt. J. T. 104
Brouwer, Capt. H. 275
Brunel I. K. 84, 90, 94
Burkhuyzen, J. G. 79
Butterfield, R. S. 61
Buxton, Viscount 170

Campbell, Randolph 104
Cayzer, Sir Nicholas 366
Chrimes, H. B. 284, 286, 319, 335, 351
Churchill, Winston 209, 234, 247
Clunies-Ross, George 96
Cohen, Sir Robert Waley 229
Colenso, A. H. 292
Conrad, Joseph 103
Cool, W. N. 56
Cornwell, R. 369
Crake, R. T. 360
Crichton, Sir Andrew 291, 365
Cripps, Leonard 11, 13, 19, 24, 57, 117, 143, 179, 182, 192
Cripps, Stafford 207, 241, 274
Crompton Albert 11, 12, 13, 18, 23, 34, 41, 42, 56, 92, 105, 113, 150, 207–9, 220, 230, 242, 244, 257, 259, 273, 325
Cunard, Samuel 91

Davies, H. O. 294
Davies, Maurice Llewellyn 12, 17, 23, 42, 56, 113, 143, 153
Davis, J. 348
Dickie, W. H. 259, 262, 267, 268, 272, 282, 290, 314
Disley, H. R. 293, 351
Dixon, Capt. 150
Dodds, T. J. 150
Dodwell, G. B. 139
Duif, Capt. A. A. 58
Dunphie, Sir Charles 329

Elder, D. 354, 373
Elder, John 91
Ellerton, G. 350–1

Evans, D. 293
Evans, Capt. D. L. C. 240, 316

Falconer, W. H. 332, 351
Flett, H. 148, 203, 268, 325
Flynn, Capt. George 170
Foot, Michael 274

Gardner, E. J. C. 278
Garton, H. W. 293, 351
Glasier, Malcolm 241, 284, 294
Gleichman, T. G. 58
Gould, Capt. 301
Greenwood, J. 294, 349, 351

Hagen, W. 150
Hahn, Kurt 147
Harmer, Sir F. 365
Hawkins, C. R. 128
Heathcote, Brian 148, 282, 289, 294
Hills, Ernest 273
Hobhouse, John Richard 7, 9, 13–15, 19, 57, 59, 68, 81, 117, 143, 151, 179, 201, 207, 233, 241, 256–7, 273, 274, 275, 276, 290, 318, 324; career with Holts 282–3
Hobhouse, Richard 8, 284, 286, 335, 351
Holland, Charles 228
Holt, Alfred 9, 11, 12, 15, 18, 27, 59–61, 64, 66, 67, 69, 73–4, 83–4, 87, 89–91, 95, 97, 99, 103, 105–7, 111, 113, 117, 120, 122, 164, 181; early life 92–4, marine engineer 84, 91, 94; attitude to conference 124–5; resignation 113; death 143
Holt Anna (Philip's wife) 98
Holt, George I (Alfred's father) 19, 90, 92, 94, 112, 155
Holt, George II (Alfred's son) 12, 13, 17, 23, 113, 143, 153
Holt, George Palmer 7, 8, 13, 57, 203, 242, 275, 277, 283, 284, 286, 287, 289, 296–8, 318, 335, 350–1, 355
Holt, Julian 355
Holt, Lawrence Durning 5, 7, 9, 10, 13–16, 19, 25, 68, 142, 151, 175, 179, 183, 192, 208, 231, 242, 248, 257, 267–8, 282, 287, 289, 316; life and career 133
Holt, Philip Henry 9, 11–13, 15, 18, 31, 40, 59, 66, 67, 91–2, 94, 97–8, 105, 113, 117, 120, 122, 148, 150,

Holt, Philip Henry (*cont.*)
159; visits far east 1877 40; helps
found Ocean 96; attitude to
conference 124–5, death 143, 145;
Philip Holt Trust 159, 318
Holt, Richard Durning 5, 6, 9, 10, 13,
16–24, 32, 47, 51, 54, 60, 67, 68, 79,
98, 113, 117, 123, 127, 133–6, 139,
141, 142–4, 147–9, 151, 153, 162,
170, 175, 178, 179, 183–7, 197, 199,
201, 203, 205, 206–9, 215–6, 220–1,
225, 229, 235, 242, 248, 258, 273,
283, 289, 291, 299; early career 18,
19; Unitarianism 19, 20, 22;
Liberalism 19, 22, 23; MP 20, 22,
23, 166; Mersey Dock Board 21,
175, 200; Baronetcy 24; death 24,
241; First W. W. 158–9;
conscription 166; Shipping
Controller 166–7; tours Far East
185; depression policy 206; Elder
Dempster 230; Glen 232–3
Holt, William Durning 112
Hughes, W. M. 132
Hyde, B. R. 56
Hyde, F. E. 64, 65, 121, 142, 153

Inchcape, Lord 132, 181
Inverforth, Lord 232

Jackson, F. P. 203, 292
Jenkins, Capt. David 106, 227
Jones, Sir Alfred 228
Jones, Capt. D. R. 351
Jones, Charles Sydney 10, 15, 18, 19,
143, 151, 155, 179, 242

Karsten, H. O. 128
Kennan, T. 352
Kidd, Capt. Alexander 73, 95, 97, 99,
103, 114
Kylsant, Lord (Sir Owen Philipps) 24,
226

Lane, F. L. 276, 277, 350, 351
Lang, William 61, 62
Leathers, Lord 372
Lee, Norman 241
Lenox-Conynham, C. 351
Lewis, Sir Frederick 230
Lindsay, W. S. 28, 95
Lloyd, George D. 20, 157, 166,
175

Mackenzie, Sir Compton 203
Maclay, Sir Joseph 166, 167
MacTier, R. S. 242, 277, 284, 285, 287,
321, 324, 328, 332, 335, 349–51, 366
Main, Henry 164
Mansfield, Capt. George 71, 72
Mansfield, George John 40, 71, 72,
74, 77
Mansfield, Walter 72, 73, 74, 75, 76,
77
Marriner, Sheila 64, 65, 121
Matchett, J. 292
McDavid, Herbert 117, 128, 234, 242,
325
McGregor, Cameron 233
McGregor, James 227
McLean, M. 362
McLennan, A. 277
McLintock, Sir William 232
McNeill, W. H. 351
Meek, M. 332, 351, 366, 368
Melly, Family 19, 98
Meyer, J. B. 56
Middleton, Capt. Isaac 73, 74, 75, 93,
95, 97, 100
Middleton-Smith C. A. 58
Miller, J. T. 292
Moore, Capt. W. J. 197
Morris, Capt. R. O. 245
Muirhead, A. E. 351
Murphy, Capt. 150
Murray, Capt. G. A. 247
Myles, Capt. 170

Nelson, W. 292
Nicholson, Sir John 10, 13, 15, 21, 22,
25, 60, 112, 117, 128, 146, 203,
234–5, 241–2, 257–59, 283–4, 285,
287, 302, 305, 314, 317, 322, 325,
329, 335, 340–1, 342–3, 348, 349,
351, 353–4, 360, 363–4, 365–6, 369;
career with Ocean 288–91
Nienhuys, J. 40

Phillips, Capt. T. 296, 301
Pirrie, Lord 228
Pitts, Capt. Frank 44
Potter, Beatrice 19, 22, 147, 175
Potter, Family 144, 207
Potter, Lawrencina (Robert's wife)
16, 19, 22
Price, E. G. 284, 302
Price, H. E. 293

Pridgeon, George Albert 136, 149, 151
Propert, Capt. 188

Raffles, Sir Stamford 88, 277
Rahusen, David 57–8
Rathbone, family 98
Romenij, J. E. 79
Roskill, Capt. S. W. 146, 236
Rowe, William 97
Rowse, H. J. 200
Royden, Thomas 162
Runciman, Walter 229, 231
Russell, Capt. J. A. 246

Scott, C. C. 62
Scott, James Henry 62, 70
Scott, James Hinton 70
Scott, John 95
Sexton, James 142
Smythe, H. N. 352
Somerville, H. E. 51, 276
Sparks, Capt. 301
St Johnston, K. 284, 286, 287, 335,
    340, 351, 353, 355, 365, 367
Stapledon, William Clibbett 10, 13,
    15, 38, 58, 103, 145, 203, 208
Stewart, Donald 139
Storey, E. 293
Storrs, C. D. 248, 259, 282, 289, 318
Sukarno, President 254
Sun Yat-sen 142, 195
Swayne, R. O. C. 283–5, 326, 349,
    351, 353–5, 362–4, 367–8
Swire, J. A. 70, 281
Swire, J. K. 280
Swire, John Samuel 'Senior' 4, 13, 26,
    31, 58–67, 78, 87, 104, 117, 121–2,
    125–6, 128; founder of conferences
    120; and conference system 124;
    relations with O.S.S. Co. 121–2
Swire, Warren 21, 60, 61, 67, 69, 70,
    184, 197, 242, 280

Swire, William Hudson 60, 61, 97–8,
    101

Taets Van Amerongen, Baron 57
Tan Keong Saik 78
Taylor, C. C. 293
Taylor, R. J. F. 284, 287, 335,
    366–7
Thornton, R. H. 7, 12, 13, 15, 21, 22,
    60, 68, 117, 143, 148, 165, 179, 202,
    241, 243, 250, 257–8, 282, 287
Threlfall, W. 293, 324
Tieman, J. 58
Tod, Alan 231
Toone, A. 292
Toosey, P. J. D. 335, 351
Townley, S. 203, 292
Tregonning, K. G. 51, 80, 82, 224
Turner Russell, Capt. 13, 73, 75, 97,
    150, 180
Turner, Graham 10, 11
Tytler, A. B. 293

Umpleby, Capt. 304
Utley, J. 293

Waller, P. J. 20
Ward, Col. John 170
Watson, Capt. J. E. 274, 317
Wells, Eliza (R. D. Holt's wife) 20,
    22
Williamson, Tom 155
Woods, Edward 93
Wortley, H. B. 10, 13, 16, 142–3,
    151, 153, 164
Wright, K. 343
Wright, R. J. 72, 74
Wurtzburg, Charles 143, 224, 234,
    242, 277, 278, 325
Wylie, H. 248, 293

Zimmern, Alfred 127

INDEX OF SHIPS

Achilles I 1, 30, 42, 61, 62, 70, 73,
    76, 96, 100, 101, 103, 106, 116,
    181
Achilles III 183, 193
Achilles IV 272
Aden 61
Aeneus I 47, 52, 165, 222

Agamemnon I 1, 42, 61, 75, 77, 96, 97,
    100, 101, 106, 142
Agamemnon III 193, 235
Agapenor II 262, 272, 300, 317
Ajax I 1, 30, 42, 74, 96, 99, 100, 101,
    103, 106, 114
Ajax II 21, 165

Ajax III   235
Ajax IV   272
Alpha   93
Anchises III   47, 52, 165
Anchises IV   274, 316
Antenor I   103–4
Antenor III   192, 222, 262
Ascanius II   47, 52, 165
Ascanius III   193, 301
Asphalion I   189, 192, 212
Astyanax I   52
Australind   44
Autolycus I   67, 161
Autolycus III   272, 300
Ayuthia   81

Belle of Southesk   73
Bellerophon II   49, 153, 272
Benmohr   169
Breconshire   235

Calchas III   193, 269, 313, 314
Carway   336
Centaur II   191, 218, 245, 246
Centaur III   312, 324, 327, 328
Charon II   191, 218, 235, 326
Circe   44, 81
Cleator   94, 95
Conway, HMS   147
Cyclops II   114
Cyclops III   164

Dardanus II   152
Denbighshire   234
Diomed I   75, 101, 102, 103
Diomed II   152, 169
Diomed V   35, 270, 272
Dolius I   192
Dolius III   310
Dumbarton Youth   94, 95

Elpenor I   193
Elve   162, 168
Emden   157, 169
Encounter Bay   366
Enterprise   71
Eumaeus I   170
Eumaeus II   34
Euryades   162
Eurybates II   192, 205
Eurymedon II   193
Eurymedon III   262

Fairland   362
Flintshire   106, 277
Foochow   64

Ganymede   40
Glaucus II   47
Glenalmond   329
Glenartney   106, 227, 235, 240
Glenearn   193, 234
Glenfinlas   329
Glengyle   106, 227, 234
Gleniffer   301, 307
Glenlyon   193
Glenogle   106
Gorgon II   218, 235, 326
Great Eastern   91, 94
Gunung Djati   297, 298

Hector I   108, 313
Hector III   152, 165
Hecuba   41
Helenus I   53
Hyacinth, HMS   170

Idomeneus II   35, 192, 212, 217
Ixion I   116, 152, 313

Jason II   38
Jason III   310
Jeddah   103, 104
Jervis Bay   366

Kajang   51
Kamuning   51
Keemun   156
Kepong   51
Kewsick   245
Knight Companion   162
Knight of the Garter   162
Knight Templar   162
Knight of the Thistle   162, 313
Kooringa   362

Laertes II   168
Laertes III   67
Laestrygon   140
Lycaon I   142, 188, 191
Lycaon II   272

Machaon II   193
Maron I   262
Maron III   272
Medon I   192

Medon II (*Empire Splendour*) 260
Medusa 44, 81
Melampus II 317, 320
Memnon IV 235
Menelaus II 152, 165
Menelaus IV 193
Menestheus I 269, 313
Mentor I 53
Minderoo 218
Myrmidon IV 262

Navajo Victory 313
Nestor I 76, 102
Nestor II 47, 52, 116, 165
Nestor III 48, 303
Nestor V 342, 343, 344
Nile 101
Normanby 313

Orari 246
Orestes I 104, 108, 313
Orestes III 47, 152
Orestes IV 192, 193
Orontes 93
Oxfordshire 170

Patroclus I 107
Patroclus II 36
Patroclus III 186, 189, 195, 222
Patroculus V 58
Peisander II 329, 330
Pembrokeshire 329
Perseus III 301, 310
Phemius II 316 *and* Appendix IV
Philoctetes I 188
Ping Suey 313
Plantagenet 95, 164
Polydorus I 57, 192, 262
Priam I 102, 313
Priam IV 236
Priam V 193, 329, 330
Prometheus II 67
Prometheus IV 329, 330
Protesilaus I 238
Protesilaus II 329, 330
Pyrrhus I 116, 152
Pyrrhus II 53
Pyrrhus III 314, 315

Queen Elizabeth 165

Radnorshire 329
Rhesus I 212
Rhexenor I 67
Rhexenor II 260, 262
Saladin I 94
Saladin II 44, 46
Sarpedon I 104, 108, 313
Sarpedon III 47, 152
Sarpedon IV 186, 189
Stentor I 94
Stentor III 192, 193, 212, 222
Stentor IV 260
Stirling Castle 106
Sultan 44
Swatow 64
Sydney, HMAS 169

Talisman 95
Talthybius I 183
Tantalus I 57, 106, 152
Tantalus II 192, 193, 245
Tantalus V 339
Teiresias I 53
Telamon I 116
Telemachus (HMS *Activity*) 235
Telemachus III 236, 258
Teucer I 313
Theseus II 301
Titan I 29, 36
Titan IV 339, 341, 344
Titanic 168, 170
Tokyo Bay 368
Troilus I 169
Troilus III 238, 246
Troilus V 339, 341, 344
Tyndareus 49, 161, 168, 170, 193, 296, 297

Ulysses I 313
Ulysses II 109, 313
Ulysses III 116, 152
Ulysses IV 48, 52, 165, 223, 245, 246

Veghtstroom 162, 168